Inverse Problems: In Memory of Professor Zbigniew Oziewicz

CONTEMPORARY MATHEMATICS

824

Inverse Problems: In Memory of Professor Zbigniew Oziewicz

AMS Special Session
Fall Western Sectional Meeting
October 23–24, 2021
Virtual

Hanna E. Makaruk
Robert M. Owczarek
Editors

AMERICAN MATHEMATICAL SOCIETY

2020 *Mathematics Subject Classification.* Primary 57K10, 57K14, 57K16, 57K31, 18F99, 55U15, 68Q01, 68Q04, 53C05.

Library of Congress Cataloging-in-Publication Data

Names: American Mathematical Society. Fall Western Sectional Meeting (2021 : Online), j author. | Makaruk, Hanna E., editor. | Owczarek, Robert M., 1963- editor. | Oziewicz, Zbigniew, honoree.

Title: Inverse problems : in memory of Professor Zbigniew Oziewicz : Fall Western Sectional Meeting, October 23-24, 2021, virtual / Hanna E. Makaruk, Robert M. Owczarek, editors.

Description: Providence, Rhode Island : American Mathematical Society, [2025] | Series: Contemporary mathematics, 0271-4132 ; volume 824 | Includes bibliographical references.

Identifiers: LCCN 2025005093 | ISBN 9781470470111 (paperback) | ISBN 9781470480448 (ebook)

Subjects: LCSH: Oziewicz, Zbigniew–Congresses. | Inverse problems (Differential equations)–Congresses. | AMS: Category theory; homological algebra – Categories and geometry – None of the above, but in this section. | Algebraic topology – Applied homological algebra and category theory – Chain complexes. | Computer science – Theory of computing – General. | Differential geometry – Global differential geometry – Connections, general theory. | LCGFT: Conference papers and proceedings. https://id.loc.gov/authorities/genreForms/gf2014026068

Classification: LCC QA378.5 .A44 2025 | DDC 515/.357–dc23/eng/20250515

LC record available at https://lccn.loc.gov/2025005093

Contemporary Mathematics ISSN: 0271-4132 (print); 1098-3627 (online)

DOI: https://doi.org/10.1090/conm/824

∞ The paper used in this book is acid-free and falls within the guidelines
established to ensure permanence and durability.
Visit the AMS home page at `https://www.ams.org/`

10 9 8 7 6 5 4 3 2 1 30 29 28 27 26 25

Dedicated to the memory of
Professor Zbigniew Oziewicz, our friend.

Contents

Preface

This volume contains proceedings from the AMS Special Session on Inverse Problems: In Memory of Professor Zbigniew Oziewicz at the AMS Fall Western Sectional Meeting #1172, held virtually on October 23–24, 2021.

The series of our AMS Special Sessions started in 1999 at the Fourth International Joint Meeting of the AMS and the Sociedad Matematica Mexicana (SMM) in Denton, Texas. This series of special sessions has continued for almost 25 years, approximately annually, mainly at AMS Western Section Meetings. The sessions were interdisciplinary from the beginning, meant to present high-quality research at a cross-section of pure mathematics and mathematical methods of physics, computer science, and other sciences. This interdisciplinarity was at the heart of our recently passed friend, Zbigniew Oziewicz, who participated in most sessions. He worked as a researcher and an academic teacher in these areas, primarily mathematical physics, also including some pure mathematics, like differential geometry, Frobenius algebras (that are important in topology, among others), foundations of physics, philosophy of science, logic, and computer science. He was an excellent debater who never hesitated to ask questions during the sessions, pushed all participants out of their comfort zones, and instigated constructive discussions. We wanted to honor his memory with the dedication of the session and these proceedings to his memory.

Hanna E. Makaruk
Los Alamos National Laboratory

Robert M. Owczarek
University of New Mexico

Contemporary Mathematics
Volume **824**, 2025
https://doi.org/10.1090/conm/824/16458

Remembering Zbigniew Oziewicz's activities in Mexico

Larissa Sbitneva

ABSTRACT. In this composition we give a brief review of some activities of Professor Zbigniew Oziewicz in Mexico.

The presentation is devoted to two themes. The first one describes the exceptional features of Lecture Courses delivered by Professor Oziewicz at the Undergraduate level, the second one is devoted to the rival alternative to the Einstein's special relativity theory suggested by Dr. Oziewicz and to the models based on the non-associative law of the summation of Einstein's velocities, finally we expose almost unknown ideas of Sophus Lie related to the axioms of existence of a neutral element, which is the principal controversy that gave rise to the alternative theory proposed by Zbigniew Oziewicz.

The relation of non-associative structures to the geometry underlying some models of composition of velocities in Relativity is outlined.

1. Introduction

During the period of 1995-2020 in Mexico, Professor Zbigniew Oziewicz served as a leading professor at the scientific seminars of the Faculty of Higher Studies Cuautitlan (FESC), at the National Autonomous University of Mexico (UNAM), organizing International Conferences, teaching courses and directing research for thesis.

Our scientific interactions, related to non-associative algebraic structures associated to the Relativity, culminated during the period of September–December 2019 when I was participating at the scientific seminars conducted by Dr. Oziewicz.

In this work I prefer to make an emphasis on his pedagogical performance during teaching the general courses as well as on his influence and inspiration in the process of future researchers formation.

2. Lectures of Professor Oziewicz

In this section there are exposed several fragments of the lectures of the general course "Theory of Radiation" dedicated to the Fourth Semester students of the Undergraduate Studies, at the Faculty of Higher Studies Cuautitlan (FESC), at the National Autonomous University of Mexico (UNAM).

2020 *Mathematics Subject Classification.* Primary 83A05, 20N05, 53C35, 53C22, 22A30.

Zbigniew was tireless in explaining his own understanding of laws of the Universe as four-dimensional Space-Time, although, as he confessed sometimes in his scientific seminars, not so many specialists were ready to share his vision.

Here I will try to expose his influence and inspiration in the formation of future researchers. In the classroom during the breaks he was always surrounded by students showing their admiration and affection.

As we may appreciate through the several fragments, during the lecture course there were always different kind of deviations dedicated to explanation of historical context, variety of approaches or biographical details of the researchers involved, about the Mathematical foundations, Physical interpretations, as well as modern treatment.

I will omit the themes with technical details, although their presentations by Dr. Oziewicz are of great value, and concentrate on his philosophical vision.

As far as the lectures environment, as a rule nobody is permitted to assist the class-sessions unless it is a so-called Magistral lecture. The class-sessions are considered as an intellectual interaction between the professor and students (Liberty of the Cathedra).

Nevertheless I had a privilege to be invited to assist the course on Theory of Radiation for the students of the 4-th Semester of the Undergraduate Studies during the semester, being a participant at his scientific seminars.

In the case of Prof. Oziewicz's courses the ambient was more remarkable: Inside there is the realm (world) of free personal ideas, outside, the so-called institutional knowledge: It is always emphasized on the whiteboard

OTROS-(Others)——————NOSOTROS (We)

2.1. Themes on duality. The themes on Duality: Assignments of tasks and revisions. The initial period of the class-session is usually dedicated to revision of the previous Exams, giving comments to all the details on the answers of every student (Exams are performed every week).

Always encouraging the authors of original solutions, with exclamations. Criticizing those who just follow traditional presentations from books.

Always remembering the complete names of every of his more than 60 students.

During this period usually all students are arrived to the classroom. Thus, Professor Oziewicz starts with the next Exams questions: The formulation of new problems to think over and requiring their own points of view.

But sometimes there happened more questions on the previous themes. An example of October 10, 2019 session: As a response on the difficulties with the Exam 4.

Dr. Oziewicz realizes the detailed revision of the theme of wave-particle duality with the emphasis that it is fundamental for the physics.

In his opinion all the textbook on physics and chemistry should start with the duality from the very first chapters (it is important to revise different books in library).

He invites to think over the proposals of two physicists: Louis de Broglie and Arthur Compton and explain the difference in their results.

Arthur Compton gave the experimental proof in 1923 (Nobel Prize 1927).

Louis de Broglie (1923) proclaimed: if Light is a photon, electron is a wave (Nobel Prize 1929), due to the following reasoning: Possesses energy implies possesses mass, consequently must possess the frequency, and expresses this statement by means of the formula, with the definition of the wavelength. So that electron possesses the longitude of the wave due to de Broglie.

What is the difference between the two formulas?

It is necessary to keep in mind that by that time the theory of sound waves was very well developed.

Dr. Oziewicz draws attention that in Internet they say that there is no difference! (at one Forum). But at the University we are obliged to understand and to be able to think and to distinguish.

There follow some advises: One should practice to write the letters to their parents, friends, etc. Intelligence is to describe the one's own opinion, in contrast to read from the computer screen.

He invites the students to read Mechanics of Lagrange which has wave nature and Mechanics of Newton which is corpuscular.

The Exam 4 consists of several difficult questions about the Dualities indicated with letters, A, B, C, D, etc.

The first part is Duality wave–particle: It is required to give the personal opinion with respect to two results:

Arthur Compton realized an experiment and deduced a formula. Louis de Broglie writes the formula where c is a speed of light which is constant in vacuum, that is, independent of the election of the body which is considered.

But it is relative in different matters, there is a table of speed of light in different media (gas, liquid, air).

Professor draws the attention that there is an article in Internet where there is explained that these two results are the same. But it is not so, emphasizes the professor.

The second part concerns the duality vector–covector, due to G. Ricci (1884). 4-Vectors do not exist without covectors.

The Universe has dimension 4, the Universe of the dimension 3 is illusion, it is not true. The wave-particle duality only is seen in dimension 4, Minkowski (1908). The pair vector–covector only has three possibilities to obtain the magnitude: negative, zero, positive (only in the Euclid's geometry).

Here appeared the terms time-like, light-like, space-like.

These considerations lead to distinguish the material world and the spiritual one (i.e., without mass).

2.2. Light-cone. There comes the construction of Light-cone, which is the theme for the Exam 6.

At this moment there are given suggestions to obtain the corresponding knowledge: To see in Internet about tensors for beginners, as well as to revise different books in the library. But it is not sufficient to read, it is necessary to write down their own notes. The ability to elaborate notes is very important.

As usual he is joking: one is not taking notes in movies, or theater. Notes on class should be sentences which have sense, accompanied with a title and data.

There follows proposal for the Final evaluation: revision an example of November 27, 2019 session.

Comments: To achieve 6 (pass) it is enough to perform 5 exams perfectly. First, there are given some observations concerning the Exam 4: Duality.

Proposal for the Final evaluation: Revision an example of November 27, 2019 session. Final evaluation: Exams 9 and 10 (which will be formulated only at the next session.

2.3. Concept of the process. There is an invitation to think over the concept of the Process.

One should keep in mind that every process conserves something (but there is a change in something that changes!). There is no process that changes everything (isothermal process conserves temperature, for example). "Everything is the same, nothing is changed in the Universe". This is expressed by partial derivatives.

These reflections are accompanied by historical consideration of the ancient philosophers: Heraclitus vs Parmenides. But both thinkers were mistaken!!!

Conservation and changes are always together.

There comes an introduction to the most important concept of the course: Grassmann algebra (1844). Indication to explore Internet.

The explication of the Grassmann algebra (which was the theme of Exam 8, is accompanied with the formulation of the Exam 9:

The processes of "creation" and "evaluation" (annihilation).

Evaluation is a derivation of Grassmann algebra. Evaluation is a derivation: This is the most important result of the course.

There is mentioned the Exam 8 to give an example and to apply the new results. Finally the Exam 9 is formulated:

Differential d (which appeared in the 1-st ley of Thermodynamic) with the emphasis that d is applied only to covectors.

The following development of this theme leads to the Lie super-algebras.

Continuation of November 27, 2019 session: December 5 2019. The Exam 10 is dedicated to the metric tensor as an Ambient. The detailed explanation can be seen in the article: "Science can never be objective" (see [**1**]).

The Four dimensional Universe. Class-session October 3, 2019. The 4-dimensional Universe Scheme. From the very beginning the attention is drawn to the development of space-time notion. Since the Aristotle's treatise one can read "Space and Time". Only in 1908 Hermann Minkowski gave the definition of 4-dimensional Space–Time.

Many interesting references due to Hegel (Universe is only the matter in movement), Lenin (Universe is matter), Newton (suspected the corpuscular nature of light), then Einstein (the Light energy is the same as frequency). Energy is the property of a particle (corpuscular nature). (This is a theme of Exam 4, above).

2.4. Dichotomies. There are two dichotomies. The universe where we live (during this course on Optics) is 4-dimensional (as well as in Physics, Mechanics, Chemistry).

All the concepts are relative. There is a dichotomy: Absolute–Relative

Absolute: Everything is independent on the rest state chosen. If it is independent on the state of rest (resting state) there is no need to choose a body at rest.

Relative: Everything depends on the election (choice) of a rest state. Kinetic energy depends on resting state, as well as Velocity. Resting state separate the absolute and relative processes.

Material world is everything which possesses a mass (the first law of the Thermodynamics, those which possess the internal energy $E = mc^2$).

Spiritual World is non-material. The concepts which do not refer to the internal energy are not those which possess internal energy.

There are two dichotomies: Material world is a possessor of internal energy.

Spiritual World is non-material. there are no possessors of internal energy.

Absolute: Material body, Atom, Molecule (are independent on the resting state).

Relative: Vacuum.

Absolute is almost vacuum but Einstein supposed in his theory that it possesses a mass.

Velocity: has no internal Energy, but possesses magnitude, positive scalar, Force.

Velocity of molecules: The Energy of molecules only enter. Do not possess the internal energy!

Speed of light in vacuum is zero, does not possess a mass,. But it is independent on resting state (body): one may assume the body to be Mars, Earth, Sol, no matter.

The Concept "Event" does not possess a mass, does not depend on resting state. There is presented the concept of Light Cone: Light is material, possesses internal energy, internal energy possesses a mass, internal energy of Light in vacuum is zero.

Any energy is Kinetic, is maximal, is positive.

Speed of Light in vacuum is c, in water, it is $\frac{3}{4}c$, the velocity decreases in different media.

The Total Energy is the sum of [Internal Energy, here is the mass] and [Kinetic Energy, velocity, speed].

Light Cone: 0 is a mass! One says that light in vacuum has no mass. No, it possesses mass, only it is zero.

Mass is a material magnitude, may be positive, negative, zero. With respect to the light cone, negative or zero is material, positive is non-material, "spiritual".

EXAM 11: Grassmannian algebra approach.

Previously Exam 8 also was dedicated to the Grassmann's abstract Gramma.

Here Professor starts with a marvelous story, very impressive...

Justus Grassmann, Professor of Mathematics and Physics and the Director of a German High School (Stettin), was a friend of Alexander von Humboldt. In 1820 there was introduced the program with 1 hour classes of mathematics daily. Before there was only 1 hour per week!

So, the Justus Grassmann posed a question to his son: Hermann Grassmann: A particle, when moves along a curve, produces a tangent vector. What can you say about the movement of a pencil along a surface? What may be suggested as its velocity?

The question is: What can be named as a velocity of this object? Something which would play the role of the tangent vector to the trajectory of the particle which is a curve.

The answer of a young Grassmann was: BIVECTOR. The notation was the "WEDGE" $\wedge$ of Grassmann, $X \wedge Y$. And in the case of motion of a piece of plane there should be a volume expressed by tri-vector, etc.

Thus we have a scheme: a point (of dimension 0) in a movement produces a curve (of dimension 1).

But a pencil (of dimension 1) in its movement produces a surfaces (dimension 2) tangent to the surface on which it is moving.

If one moves a door (2 dimensional object) there is produced a volume. Consequently if something of dimension k is moving then the velocity which is produced should be of dimension $k + 1$.

Thus Hermann Grassmann introduces the axiom $X \wedge X = 0$, which is interpreted as "no one letter is repeated", which furthermore leads to the axioms of creation.

Dr. Oziewicz establishes the relations of the laws of the Grassmann's gramma with the principle of exclusion of Pauli (1926) (Noble Prize, 1946).

The evaluation is seen as the derivation in Grassmannian's Algebra. Thus the Exam 8 requires the derivation.

Making emphasis: These are very important in the process of the ionization of an atom (i.e., elimination of an electron, annihilation).

In Mathematics they are called: Creation and annihilation operators.

2.5. Pedagogue. Thus we can see that the way of teaching of Zbigniew is to draw the attention to the more important structures applied in physics but the students should achieve the corresponding knowledge by their own efforts.

These ideas are concentrated in his article: Science must never be objective published in Studium Vilnense A (10) (2013), ISSN 1648-7907 XVIt International Conference on the Science and Quality of Life, Vilnius, Lithuania, where he cited Peztalozzi 1797. (See in [**1**].)

A pedagogue must allow that the students study for themselves and express their personal vision on Science: Science must be individual understanding. Science does not exist outside of personal opinions of scientists.

3. Underlying axioms in relativity and in the S. Lie's theory

3.1. Proposals for future investigation. Our scientific interactions were related to non-associative algebraic structures associated to the models of composition of velocities in Relativity.

In our previous works there were suggested the geometric interpretation for the algebraic model of Einstein's composition of velocities.

But in the last few years there originated some controversy, which Zbigniew was expressing in various conferences and publications concerning to a rival alternative to the Einstein's special relativity theory [**1**].

The main controversy consists in the fact that the non-associative constructions presume the existence of the neutral element, which is zero in the model of A. Ungar and is the unit element in a loop describing the underlying geometry [**2**].

These ideas are presented explicitly in his article: Science must never be objective, published in Studium Vilnense A (10) (2013), ISSN 1648-7907, XVIth International Conference on the Science and Quality of Life, Vilnius, Lithuania [**1**].

The proposal of Zbigniew was to write an article dedicated to all these controversies. Therefore here we will give a brief introduction on how a set with binary operation, called quasigroup, is related to symmetric spaces and their generalizations as a underlying geometry for the Einstein's summation law and their relation with the loop theory.

3.2. Relativity theory is about relativity of velocity: Alternative ideas by Z. Oziewicz. The theory of relativity is about the relativity of velocity. Thus, if v denotes the vector of the velocity of the flow of the river, relative to riverside, and w, the vector of velocity of the barge relative to the river, then the composition of these velocities, $v \oplus w$, will be the velocity of the barge relative to the riverside. The relativity of velocity implies the relativity of the position-spaces [Copernicus 1543; Galileo 1632, Descartes 1650; Berkeley 1710] (see in [**1**]), and, for the finite light-speed, also the relativity of simultaneity [Einstein 1905; Minkowski 1908] (see in[**1**]). Zbigniew considers experimental clock-synchronizations as not essential.

Relativity theory is about relativity of velocities, the composition of these velocities, $v \oplus w$, will be the velocity of the barge relative to the riverside.

The position-space of the water of the river is distinct from the position-space of the places of the riverside. These are two different spaces.

The velocity of the flow of the river relative to the river is zero-velocity vector, and this zero is a one-sided unit for the composition of velocity-vectors, which is crucial. The zero velocity-vector should be neither unique nor priviledged among many zero's-velocity vectors (see [**1**]).

Galileo is still misinterpreted and misunderstood by the false believing that the velocity v is a vector on three-dimensional position-space (position-space of the riverside or of the river?).

Copernicus in 1543 asked: who is in the motion? a river in the motion relative to riverside, or a riverside in a motion relative to a river? In fact relative velocity must be a vector-process on four-dimensional space-time of events. The three- versus four- dimensionality of the Universe is a subject to an ongoing sharp debate [Petkov 2005, 2007; Ivezic 2010] (see in[**1**]).

Hermann Grassmann invented bi-vector and wedge-Grassmann product, $\wedge$, in 1844,

$$v \wedge v = 0 \Rightarrow u \wedge v = -v \wedge u.$$

Isometries are generated by the Lie algebra of Clifford bi-vectors. This Lie algebra is isomorphic to the Lie algebra of tri-potents, $p^3 \sim p$, satisfying some relations of the Temperley-Lieb algebra.

The composition of not necessarily collinear (not parallel) velocities was derived by [MØller 1952] (see in[**1**]). MØller deduced composition from the Lorentz isometry-group.

Unfortunately MØller considered that relative velocity is a vector on three-dimensional position-manifold.

The MØller's finite-light-speed composition has the following 'relativistic' form,

$$u \oplus v = \frac{(v+u)}{(1+v\cdot u)} + \frac{\gamma}{(\gamma+1)}\frac{v\cdot(u\wedge v)}{(1+v\cdot u)}, \quad \gamma^2(1-\frac{v^2}{c^2}) = 1. \tag{3.1}$$

The maximal speed is that of zero-mass radiation, $|v| < c$.

We learn from Romuald Brazis that contrary to general belief the radiation-speed in vacuum needs not to be this maximum radiation speed. Brazis observed that a dispersive medium of nano-particles can amplify radiation, leading to speed higher than the speed in vacuum, [Brazis 2012 and 2013] (see in[**1**]).

If relative velocities are collinear, the Grassmann product vanishes, $u \wedge v = 0$, and the above finite-light-speed composition (3.1) takes the form

$$u \oplus v|_{(u \wedge v = 0)} = \frac{(v+u)}{(1+v \cdot u)}. \tag{3.2}$$

We have two options. We can add w to $u \oplus v$, or we can add $w \oplus u$ to v.

In 1988, 33 years after Einstein passed away, it was observed that these two options for above compositions give different results, the isometric composition (3.1) is non-associative

$$w \oplus (u \oplus v) \neq (w \oplus u) \oplus v) \tag{3.3}$$

What would composition look like if we add a sailor walking on the barge with the velocity w relative to the barge? What would be the velocity of this walking sailor relative to the riverside?

Thus, the walking sailor possesses two different velocities relative to riverside.

Yet, is there an alternative theory that predicts the associative composition of velocities? In fact, it is rare to meet anyone who wonders from what postulates the MØller's non-associative composition (3.1) follows.

What is postulated? What are the axioms, what are hypotheses? Zero velocities play the role of two-sided neutral [**2**],

$$u \oplus 0 = u, \quad 0 \oplus v = v. \tag{3.4}$$

When v is the velocity of the river relative to the riverside, we denote the unary inverse velocity, the velocity of the riverside relative to the river, by v^{-1}, and such unary inverse operation is a unipotent involution and of the same magnitude [**1**],

$$\begin{aligned} v^{-1} \oplus v = 0, &\quad v \oplus v^{-1} = 0, \\ (v^{-1})^{-1} = v, &\quad [v^{-1}]^2 = v^2. \end{aligned} \tag{3.5}$$

The fundamental postulate hypothesis for non-associative composition (3.1) is that the inverse of velocity must be reciprocal [**1**], i.e., the inverse is given by

$$(v^{-1})^{-1} = v. \tag{3.6}$$

However, each velocity we observe is always in the present (and not in the past or future), so one can imagine the velocity vector as tangent to the simultaneity of the observer.

Therefore the velocity v of the river, as observed from the riverside, and the inverse velocity v^{-1} of the riverside, as observed from the river, must be tangent to simultaneities of these two observers in mutual relative motion.

In 1905 Einstein established simultaneity (and clock synchronization) as the key issue of relativity theory: simultaneity of the riverside and simultaneity of the flowing river are distinct,

$$dt' \wedge dt \neq 0. \tag{3.7}$$

These distinct simultaneities contradict the hypothesis of reciprocity of inverse of relative velocities,

$$v^{-1} \wedge v = 0.$$

Therefore, contrary to the general belief, significant contradictions exist in the foundations of the hypotheses and axioms (3.6)–(3.7) of the relativity theory as it has now been accepted for over a century based on the hypothesis of the Lorentz isometry-group generated by tripotents:

(a) Mathematical incompatibility among composition of velocities and hypothesis of reciprocity of inverse of velocity,

$$(v \oplus u)^{-1} \neq u^{-1} \oplus v^{-1},$$

discovered by Mocanu in 1985 [Mocanu 1992] (see in[**1**]).

(b) Contradiction between non-associative quasi-group (Bol's loop) composition (3.1) (deduced from reciprocal inverse hypothesis) and experience of the physical reality. In fact, the velocity of one material body relative to another material body is unique.

(c) Reciprocal inverse of velocity, $v^{-1} \wedge v = 0$, contradicts the relativity of simultaneity, $dt' \wedge dt \neq 0$, i.e., the relativity of the proper time is not compatible with reciprocal inverse of velocity.

The alternative to the reciprocal inverse and to the tri-potents is non-reciprocal inverse of velocities (implied by split-idempotents $p^2 \sim p$), implicit in Minkowski [1908], and proposed explicitly in [Swierk 1988, Oziewicz 2005, ' 2007] (see in[**1**]),

$$R \cdot v = 0 \Rightarrow v^{-1} \equiv -\gamma v - (\gamma - \frac{1}{\gamma})R \tag{3.8}$$

$$\Rightarrow \quad v^{-1} \cdot v = -\gamma v^2 \iff [v + v^{-1}]^2 = -2(\gamma - 1)v^2, \tag{3.9}$$

$$[v^{-1} \wedge v]^2 = (\gamma^2 - 1)v^4. \tag{3.10}$$

The reciprocal inverse holds in the Galilean–Newtonian limit of infinite light-speed only, $\gamma|_{c=\infty} \equiv 1$.

The composition of velocities with non-reciprocal inverse is associative, MØller's quasi-group composition (3.1) is modified to the groupoid associative composition.

The above mathematical and physical contradictions (a), (b) and (c) disappear, but a groupoid category that disturbs existing habit of the Lorentz group (generated by Lie algebra of tri-potents) then enters the game.

For a groupoid generated by split-idempotents to replace the Lorentz group is not accepted by present consensus of Elites and textbooks.

Zbigniew Oziewicz quoted Lord Francis Bacon (1561–1626) in [**1**]:

"Books must follow sciences, and not sciences books."

Lord Francis Bacon was a contemporary of Galileo Galilei (1564–1642).

3.3. Sophus Lie's considerations with respect to a neutral element. In the last few years there originated some controversy, which Zbigniew was expressing in various conferences and publications concerning a rival alternative to the Einstein's special relativity theory.

In our previous works there was suggested the geometric interpretation for the algebraic model of Einstein's summation of velocities.

The main controversy of Z. Oziewicz consists in the fact that the non-associative constructions presume the existence of the neutral element, which is zero in the model of A. Ungar [**2**] and is a unit element in a loop describing the underlying geometry of Bol loops with Bruck identity [**3**],[**4**].

This is the mathematical Model, considered by A. Ungar[**2**], A. Nesterov [**5**], [**6**], [**7**], [**8**], J. Kocik [**9**], [**10**], and in our research group [**3**], [**4**], [**11**], which is a non-associative algebraic structure: Smooth left Bol Loop with the left Bruck identity.

We will describe how a set with binary operation, called quasigroup, is related to symmetric spaces of E. Cartan as an underlying geometry for the Einstein's summation law and their relation with the loop theory.

Zero velocitiy plays the role of two-sided unit, which implies the consequences for the inverse elements.

We will emphasize the role of a neutral element in the Einstein's addition of velocities in our model and in the smooth binary systems.

If in a set with a smooth binary operation there exists the neutral element then due to the theorem of Mal'cev, 1955, the left and right division do exist [**12**].

The main controversy consists in the fact that the non-associative algebraic constructions suggested in different models of the composition of relativistic velocities presume the existence of the neutral element, which is zero in the model of A. Ungar and is the unit element in a loop describing the underlying geometry.

AXIOMS: problems.

The treatise of Lie and Engel was translated and published by Editor J. Merker: J. Merker, editor/translator of Theory of Transformation Groups by Lie and Engel (Vol. I, 1888). Modern Presentation and English Translation [**13**].
`https://www.imo.universite-paris-saclay.fr/~merker/I-engel-lie.pdf`

The modernized English translation grew out of old simultaneous interest in the mathematics itself and in the metaphysical thoughts governing the continued development.

Then it gradually appeared to the editor that the mathematical thought of S. Lie is universal and transhistorical, hence deserves per se to be translated.

The archetypal objects of Lie's Theory of Continuous Transformation Groups are point transformation equations:

$$x_i \to f_i(x_1, x_2, x_3, \dots x_n; a_1, a_2, a_3, \dots a_r),$$

parametrized by a finite number r of real or complex parameters $(a_1, a_2, a_3, \dots a_r)$,

$$x'_i = f_i(x_1, x_2, x_3, \dots x_n; a_1, a_2, a_3, \dots a_r) =: f_i(x, a).$$

Namely, each map $x'_i =: f_i(x, a) =: f_a(x)$ is assumed to constitute a diffeomorphism from some domain in the source space into some domain in a target space of the same dimension equipped with coordinates.

The main idea is not to apply the axioms for neither identity nor inverse element.

A finite or infinite family of transformations between the x and the x'_i is called a group of transformations or a transformation group when any two transformations of the family executed one after the other give a transformation, which again belongs to the family.

Observation: It follows, in particular, that for the fractional-linear substiutions the denominators must appear with the same coefficients.

A finite continuous transformation group will be represented by a system of n equations:

$$x_i \to x'_i = f_i(x_1, x_2, x_3, \ldots x_n; a_1, a_2, a_3, \ldots a_r),$$

$(i = 1 \ldots n)$, where f_i denote analytic functions of the variables $x_1, x_2, x_3, \ldots x_n$ and of the arbitrary parameters $a_1, a_2, a_3, \ldots a_r$.

Since we have to deal with a group, two transformations:

$$\begin{aligned} x'_i &= f_i(x_1, x_2, x_3, \ldots x_n; a_1, a_2, a_3, \ldots a_r), \\ x''_i &= f'_i(x'_1, x'_2, x'_3, \ldots x'_n; a'_1, a'_2, a'_3, \ldots a'_r), \end{aligned}$$

when executed one after the other, must produce a transformation which belongs to the group, hence it has the form:

$$\begin{aligned} x''_i &= f'_i(f_1(x,a), f_2(x,a), f_3(x,a), \ldots f_n(x,a); b_1, b_2, b_3, \ldots b_r) \\ &= f''_i(x_1, x_2, x_3, \ldots x_n; c_1, c_2, c_3, \ldots c_r). \end{aligned}$$

Here, the c_k are naturally independent of the x and so, are functions of only the a_k and the b_k.

The known groups of this sort are the following: The group of fractional linear substituion in one variable. and a group with the parameters a_{ik} is not less known: The group of rotation in a space.

Both groups are isomorphic.

3.4. Sophus Lie's axioms. For S. Lie, the basic, single group axiom should just to require that such a family be closed under the composition, namely, that one always has

$$\begin{aligned} x'_i &=: f_i(x, a) =: f_a(x), \\ f_b(f_a(x)) &= f_c(x), \end{aligned}$$

for some c depending on a and on b.

The well known example is the Euclidean transformations on the plane, where the law of dependence $c = c(a, b)$ comes from trigonometric formulas.

More precisions on this definition is given in Chapter 3 of the treatise [**13**], but at present, we discuss whether one can really economize the other two group axioms: existence of an identity element and existence of inverses.

Every analytic diffeomorphism of an n-times extended space, i.e., $\mathbb{R}^n$, permutes all the points in a certain differentiable, invertible way.

Although they act on a set of infinite cardinality, diffeomorphisms can thus be thought to be sorts of analogs of the substitutions on a finite set.

In fact, in the years 1873–80, Lie's Fixed Idea was to build, in the geometric realm of n-dimensional continua, a counterpart of the Galois Theory of substitutions of roots of algebraic equations.

LEMMA 3.1. *If H is any subset of some abstract group G with* $\mathbf{card}(H) < \infty$, *which is closed under the group multiplication $h_1 \circ h_2 \in H$, whenever $h_1, h_2 \in H$, then H contains the identity element e of G, and every $h \in H$ has an inverse in H, so that H itself is a true subgroup of G.*

PROOF. Indeed, picking $h \in H$ arbitrary, the infinite sequence

$$h, h^2, h^3, \dots h^k \dots$$

of elements of the finite set H must become eventually periodic: $h^a = h^{(a+n)}$ for some $a > 1$ and for some $n > 1$, whence $e = h^n$, so $e \in H$ and $h^{(n-1)}$ is the inverse of h.

For more than thirteen years, S. Lie was convinced that a purely similar property should also hold with $G = \mathrm{Diff}_n$ being the (infinite continuous pseudo) group of analytic diffeomorphisms and with H subset of Diff_n, being any continuous family closed under composition. □

We quote a characteristic excerpt, pp. 444–445 from the treatise of Lie and Engel translated and published by Editor J. Merker (see [**13**]):

"As is known, one shows in the theory of substitutions that the permutations of a group can be ordered into pairwise inverse couples of permutations. Now, since the distinction between a permutation group and a transformation group only lies in the fact that the former contains a finite and the latter an infinite number of operations, it is natural to presume that the transformations of a transformation group can also be ordered into pairs of inverse transformations.

In previous works, I came to the conclusion that this should actually be the case. But because in the course of my investigations in question, certain implicit hypotheses have been made about the nature of the appearing functions, then I think that it is necessary to explicitly add the requirement that the transformations of the group can be ordered into pairs of inverse transformations. In any case, I conjecture that this is a necessary consequence of my original definition of the concept of transformation group. However, it has been impossible for me to prove this in general."

Engel devised a counterexample in the first year he worked with Lie (1884).

Observation: In Vol. I of the Theory of Transformation Groups, this example appears only in Chapter 9, on pp. 163-165 (see [**13**]), and it is written in small characters. In fact, Lie still believed that a deep analogy with substitution groups should come out as a theorem.

Hence the structure of the first nine chapters insists on setting aside, whenever possible, the two axioms of existence of identity element and of existence of inverses.

S. Lie considered the so-called Differential equations assumption substituting the Axiom of Inverse. He emphasize that the fundamental differential equation

$$\frac{\partial x_i'}{\partial a_k} = \sum_{j=1}^{r} \psi_{kj}(a)\xi_{ji}(x'),$$

which can be deduced from the composition axiom only, should take the place as being the main continuous group assumption.

A technically central theorem states that if some transformation equations $x' = f(x;a)$ with essential parameters a satisfy such kind of differential equations $\frac{\partial x'_i}{\partial a_k} = \sum_{j=1}^{r} \psi_{kj}(a)\xi_{ji}(x')$, for x and a running in appropriate domains $\mathcal{X} \subset \mathbb{K}^n$ and $\mathcal{A} \subset \mathbb{K}^r$, then every transformation $x' = f(x;a)$, whose parameters a lie in a small neighbourhood of some fixed $a^0 \in \mathcal{A}$, can be obtained by firstly performing the initial transformation $\bar{x} = f(x, a^0)$ and then, secondly, by performing a certain transformation:

$$x'_i = \exp(t\lambda_1 X_1 + \cdots + t\lambda_r X_r)(\bar{x}_i), \quad (i = 1 \dots n),$$

of the one-term group (see Chapter 4 in [**13**]) generated by some suitable linear combination

$$\lambda_1 X_1 + \cdots + \lambda_r X_r$$

of the n infinitesimal transformations

$$X_k := \sum_{i=1}^{n} \xi_{ki}(x) \frac{\partial}{\partial x_i}.$$

This theorem is crucial for establishing the so-called Second Fundamental Theorem: *To any Lie algebra of local analytic vector fields is associated a local Lie transformation group containing the identity element.* (See Chapter 9 in [**13**].)

Thus the Lie's answer to the Engel's counterexample shows that starting from transformation equations $x' = f(x;a)$ that are only assumed to be closed under the composition, one can always obtain the identity element and all the inverses of transformations near the identity by appropriately changing coordinates in the parameter space (Theorem 26 on p. 178 in [**13**]).

To do justice to this great treatise of S. Lie and F. Engel, the editor J. Merker translates in Chapter 9 on page 161 in [**13**] how Master Lie managed to produce the Theorem 26 on p. 178 in [**13**], which he considered to provide the sought analogy with finite group theory, after taking Engel's counterexample into account.

Afterwards, it is proved the equally important proposition that r independent infinitesimal transformations which satisfy the concerned relations do always determine an r-term group with the identity transformation (See Theorem 21: 1st direct in [**13**]).

This theorem can now be immediately applied to all r-term groups, whether or not they contain the identity transformation.

3.5. Quasigroups and loops in geometry. One can find different definitions of a group, for example in [Guggenheimer, Differential Geomety Ch. 5 Introduction to transformation groups] one can read:

"In algebra, a group is a set G for which a composition (usually called multiplication) is defined that associates with each pair a, b of elements of G a unique element $c = a \cdot b \in G$.

This map $a, b \to a \cdot b$ must satisfy

$$x \cdot (y \cdot z) = (x \cdot y) \cdot z, \quad \text{for all} \quad x \in G.$$

For all $x, y \in G$ there exist unique z_1 and z_2, such that $x \cdot z_1 = y$ and $z_2 \cdot x = y$. Also there exists an element e such that

$$x \cdot e = e \cdot x = x \quad \text{for all} \quad x \in G.$$

From these axioms it may be shown that there exists in G a unique unit (neutral element).

Thus, there exists a unit element e and a unique element $x^{(-1)}$,

$$(x \cdot x^{(-1)} = x^{(-1)} \cdot x = e, \quad \text{for all} \quad x \in G.$$

The group is commutative or abelian if $x \cdot y = y \cdot x$

In this case the composition is usually written as addition $x + y = x \cdot y$."

There is discussed the difference between this definition and the traditional one in algebra.

We can emphasize that the equivalence of the two definitions is demonstrated in [Pontryagin, Continuous groups].

Observation: Associativity has underlying geometric meaning as Desargues configurations. Pappus configuration, for the commutativity.

The geometry of symmetric spaces of E. Cartan associated with left loops corresponds to the structure of the left-distributive quasigroups.

A set Q with a binary operation $(x, y) \mapsto x \cdot y$ is called a quasigroup if for any given $a, b \in Q$ the equations $a \cdot y = b$ and $x \cdot a = b$ have precisely one solution $y = a\backslash b$ (left division) and $x = b/a$ (right division).

Thus, any quasigroup $\langle Q, \cdot \rangle$ possesses the left and right divisions.

The operations can be expressed equivalently by means of bijections: The left translations and right translations, so that the left and right divisions may be considered as further two operations on $(Q, \cdot)$.

If in a quasigroup there is an element ε such that $\varepsilon \cdot x = x \cdot \varepsilon = x$ then Q is called a loop and the element ε is the unit (or neutral) element of Q.

In the loop theory there is a relation between quasigroups and loops established by means of isotopism of operations.

For any quasigroup $(Q, \cdot)$ and arbitrary elements $a, b \in Q$ the multiplication

$$x * y = x/a \cdot b\backslash y$$

determines a loop $(Q, *)$, on the set Q, which is an isotopic loop, having $b \cdot a$ as the unit element.

O. Loos formulated an algebraic analog of symmetric spaces of E. Cartan in algebraic language equivalent to terms of local quasigroups [**14**].

Furthermore, applying the construction of a unit element, we obtain the caracteristic identities for a geodesic loop of a symmetric space [**15**].

These Identities are the so called Left Bol Identity and the Left Bruck Identity, which are valid for the model of A. Ungar. This was demonstrated in [**11**], and thus permits considerations of the underlying geometry for this model.

A left Bol loop is a left loop with the Left Bol Identity

$$L_{a \cdot (b \cdot a)} = L_a \circ L_b \circ L_a.$$

The left Bruck identity

$$(x \cdot y) \cdot (x \cdot y) = x \cdot (y \cdot (y \cdot x))$$

expresses the left gyrocommutative property of the operation of addition of velocities in the A.Ungar's model.

This relation expresses the equality of the magnitudes of the vectors $\vec{u} \oplus \vec{v}$ and $\vec{v} \oplus \vec{u}$, $\|\vec{u} \oplus \vec{v}\| = \|\vec{v} \oplus \vec{u}\|$, and is due to Silberstein's 1914 book (See [**2**], p2, pxxxviii):

In the construction of A. Ungar a neutral element, zero-velocity, is stated as an axiom.

In loop theory, a loop can be associated to any guasigroup be means of the isotopism theory constructing a corresponding unit element from the binary operation of the quasigroup.

The necessity of introducing the axioms concerned the existence of neutral element, or unit, and inverses was questioned since the works of S. Lie on Transformations groups, where he elaborated an idea (during 13 years of investigations) that the existence of these entities can be obtained as the consequence of the only principal axiom of the closure of the operation determined by the compositions of transformations of a given family taking into account the so-called Differential equations assumption.

We hope that the controversies expressed by Z. Oziewicz may be resolved within the recent considerations of locality and distant observer described in the recent article "Principle of relative locality" by G. Amelino-Camelia, L. Freidel, J.Kowalski-Glikman, L. Smolin, *arXiv:1101.0931v2 [hep-th] 31 Jan 2011*, where the book on Smooth Quasigroups and Loops by L. Sabinin [**3**] is cited.

We would like to close our composition with the favorite quotations of Zbigniew Oziewicz:

"Axioms of the physics. Mathematicians must consider all theories logically possible."

(David Hilbert, Congress of Mathematicians, Paris 1900, The sixth problem.)

"The axiomatic basis of theoretical physics must not be obtained from experiments, but must be freely invented by the human mind."

(Albert Einstein, Herbert Spence Lecture at Oxford 1933.)

References

[1] Z. Oziewicz. *Science must never be objective* Studium Vilnense A I0 (2013) 3-8, ISSN 1648-7907 XVIt International onference on the Science and Quality of Life, Vilnius, Lithuania

[2] A. A. Ungar, *Beyond the Einstein addition law and its gyroscopic Thomas precession*, Fundamental Theories of Physics, vol. 117, Kluwer Academic Publishers Group, Dordrecht, 2001. The theory of gyrogroups and gyrovector spaces, DOI 10.1007/978-94-010-9122-0. MR1978122

[3] L. V. Sabinin, *Smooth quasigroups and loops*, Mathematics and its Applications, vol. 492, Kluwer Academic Publishers, Dordrecht, 1999, DOI 10.1007/978-94-011-4491-9. MR1727714

[4] P. O. Miheev and L. V. Sabinin, *Quasigroups and differential geometry*, Quasigroups and loops: theory and applications, Sigma Ser. Pure Math., vol. 8, Heldermann, Berlin, 1990, pp. 357–430. MR1125818

[5] A. Nesterov and L. V. Sabinin, *Smooth loops and Thomas precession*, Hadronic J. **20** (1997), no. 3, 219–237. MR1676951

[6] A. I. Nesterov and L. V. Sabinin, *Smooth loops, generalized coherent states, and geometric phases*, Internat. J. Theoret. Phys. **36** (1997), no. 9, 1981–1989, DOI 10.1007/BF02435954. MR1476347

[7] A. I. Nesterov and L. V. Sabinin, *Non-associative geometry and discrete structure of spacetime*, Comment. Math. Univ. Carolin. **41** (2000), no. 2, 347–357. Loops'99 (Prague). MR1780876

[8] A. I. Nesterov, *Gravity within the framework of nonassociative geometry*, Non-associative algebra and its applications, Lect. Notes Pure Appl. Math., vol. 246, Chapman & Hall/CRC, Boca Raton, FL, 2006, pp. 299–311, DOI 10.1201/9781420003451.ch24. MR2203715

[9] J. Kocik, *The loop of relativistic velocities as a deformation of the menhir loop*, arXiv:1910.06556v1 [math-ph], 2019.

[10] J. Kocik, *Cromlech, menhirs and celestial sphere: an unusual representation of the Lorentz group*, arXiv:1604.05698 [math-ph], 2019.

[11] L. V. Sabinin, L. L. Sabinina, and L. V. Sbitneva, *On the notion of gyrogroup*, Aequationes Math. **56** (1998), no. 1-2, 11–17, DOI 10.1007/s000100050039. MR1628291

[12] A. I. Mal′cev, *Analytic loops* (Russian), Mat. Sb. (N.S.) **36(78)** (1955), 569–576. MR69190

[13] S. Lie, *Theory of transformation groups. I*, Springer, Heidelberg, 2015. General properties of continuous transformation groups; A contemporary approach and translation; With the collaboration of Friedrich Engel; Edited and translated from the German and with a foreword by Joël Merker. MR3331126

[14] O. Loos, *Symmetric spaces. II: Compact spaces and classification*, W. A. Benjamin, Inc., New York-Amsterdam, 1969. MR239006

[15] M. Kikkawa, *On local loops in affine manifolds*, J. Sci. Hiroshima Univ. Ser. A-I Math. **28** (1964), 199–207. MR187173

Centro De Investigación En Ciencias, Universidad Autónoma del Estado de Morelos, 62209 Cuernavaca, Morelos State, Mexico

Email address: larissa@uaem.mx

Contemporary Mathematics
Volume **824**, 2025
https://doi.org/10.1090/conm/824/16459

Accurate geometrical inferences from proton radiography data using the CpRad forward model

Elise Tang, Sky Sjue, Fesseha Mariam, and Zhaowen Tang

ABSTRACT. Proton radiography (pRad) is an experimental platform at Los Alamos National Lab that uses protons with 800 MeV kinetic energy to image dense materials. Images can be made for both static objects and dynamic processes. Accurate simulations of pRad aid experimental design and enable inferences about dynamic geometries. This model is also useful for experimental planning. The CpRad model incorporates important aspects of the experimental proton radiography system at LANL to accurately model chromatic blur in the images. In this paper we demonstrate on multiple static images that CpRad simulations are more reliable at reconstructing geometries and finding edges than gradient-based techniques.

1. Introduction

Los Alamos National Lab is home to a proton radiography facility that uses 800 MeV protons for imaging [**3**]. The protons are produced by the linear accelerator at the Los Alamos Neutron Science Center (LANSCE). Proton radiography (pRad) images both static objects and dynamic processes. The protons interact with the object materials via Coulomb interactions and nuclear interactions [**4**]. The scattering from these interactions is managed by a magnetic lens system that refocuses the protons at the image location.

Nuclear interactions lead to large angle scattering, but the cross section is small. Due to the lower proton energy used at the LANL proton radiography (pRad) facility, the effect is assumed to be simple beam attenuation.

The proton-nucleus interaction results in Multiple Coulomb Scattering (MCS) due to protons interacting multiple times before scattering out of the bulk material, leading to significant blurring of the image [**4**] [**5**]. This is addressed by the magnetic lens which refocuses the protons at a further location. This magnetic lens also allows for collimation of the beam, to generate contrast. It sorts the protons into radial positions proportional to their polar scattering angle at the Fourier plane. A collimator placed here then cuts out the protons with large angle scattering, reducing the blur and generating contrast for different areal densities [**4**]. The lateral change in position for a proton due to chromatic blur from MCS is

$$\delta_x = T_{126}\phi_x\delta, \tag{1.1}$$

2020 *Mathematics Subject Classification.* Primary 68U10.

where T_{126} is the energy-dependent transport matrix element corresponding to angle [**1**], ϕ_x is the new angle from MCS scattering, and δ is the energy loss.

The protons also interact with the electron clouds of the atoms in the material, which leads to energy loss. This gives an energy straggling distribution. Since some protons are now off the tune energy, they will not be transported ideally by the magnetic lens, which is one source of chromatic blur[**5**].

2. CpRad

Proton radiography has been simulated in various ways prior to our model [**6**][**7**]. CpRad is a ray tracing simulation of proton radiography. It is written in C and has the advantage of being able to run on GPU, MPI, or with threading. An object is input as a density file, typically in TIFF format [**8**]. An input file is used to set values for the necessary parameters for the simulation, including number of pixels, magnetic lens parameters, matching angles, initial and tune energies of the protons, collimator cut angle, magnetic lens parameters, detector blur, and various geometrical parameters. Additionally, values for each material in an object are input, such as the radiation length, nuclear attenuation length, Z/A, and the stopping power. Objects are simulated with the option of 1-dimensional spherical symmetry, 2-dimensional vertical or beam axis symmetry, and 3-dimensions. Invoking the symmetry of the object constrains the model by reducing the degrees of freedom. CpRad then simulates the final radiograph by ray tracing through the density field and calculating the MCS and detector blurs as well as the energy loss. In the absence of chromatic blur and detector blur, the transmission through the object is given as the product of the nuclear interactions and the MCS transmission:

$$T = T_\lambda T_{MCS} = \exp(-\frac{\rho l}{\lambda})\frac{1-\exp(-\frac{\theta^2_{MCS}}{2\theta^2})}{1-\exp(-\frac{\theta^2_{MCS}}{2\theta_0^2})}, \tag{2.1}$$

where ρl is the areal density, λ is the nuclear interaction length, θ_0 is the beam emittance angle, $\theta^2 = \theta_0^2 + \theta^2_{MCS}$. The scattering angle of the MCS interaction is approximated well by:

$$\theta^2_{MCS} = \left(\frac{13.6}{\beta cp}\right)^2 \frac{\rho l}{X_0}\left(1 + 0.038\ln\left(\frac{\rho l}{X_0\beta^2}\right)^2\right). \tag{2.2}$$

To get an accurate simulation it's important to begin with reasonable values for the input parameters. Some of the values are accurately known from the experimental setup at the pRad beamline. Other necessary values are fitted roughly before being used in the CpRad simulation to model images. These images are then compared to the actual data image and then the values are refined through a fitting process. CpRad can also be used to forward model images from hydrocode simulation for experimental planning purposes.

2.1. Imaged objects. Two objects were used for the edge finding tests. One is a tantalum plate, 3 mm thick with nine 0.4 mm slits. This analysis focuses on the middle set of three slits. Each slit is 0.1 cm apart. Some images were taken with a 4 cm thick aluminum plate placed behind the tantalum plate. This object provides a very simple geometry for testing edge finding. With the bare tantalum plate there is high contrast between the slits and the bulk material. The addition of the aluminum plate behind the tantalum is a better approximation to a real experiment where the thick object generates more blur and the contrast is reduced.

The other object is a set of nested cylinders, starting on the outside with teflon, then Al, Cu, and air. The radii are 3 cm, 2 cm, 1 cm, and 0.5 cm and it is 8 cm tall. This geometry represents more of the complexity of a real dynamic experiment, with overlapping materials and continuously varying path lengths.

2.2. Methodology. We will compare results of edge finding from a simple edge finding algorithm, a Canny algorithm [**2**], and optimization of density models simulated with CpRad. The simple edge finding algorithm is implemented in Python. First the magnitude of the gradient with gaussian filter is found, then the image is made binary with a threshold, then it is skeletonized to create the edge.

3. Results

3.1. Tantalum slits. Proton radiography data was taken on the tantalum plate with slits, both without and with a 4 cm aluminum mitigation plate behind it. Data on and off tune energy is here used for edge finding tests.

3.1.1. *No Al.* Imaging the bare Ta plate on tune energy is the most straightforward situation. An edge finder gives a width of slit size of 0.5 mm, slightly wider than the true slit size. The extra width ($\sim 100\mu$m) given by the edge finder can be explained by detector blur. Images were also taken of the bare Ta plate but with off-focus protons. This leads to more chromatic blur. CpRad was used to fit for the slit width, with 4 points allowed to vary in position and thickness along each of the 6 edges. The points on each edge are connected by linear interpolation. A geometric model of the object densities is used as the initial input to CpRad, along with educated guesses for the values of other variables, such as beam spread and tune energy. A comparison of the results from the CpRad simulation and the edge finder are shown in Table 1. The density model simulated with CpRad clearly finds closer values for the slit sizes.

TABLE 1. Slit sizes found for bare plate at an off-focus magnetic lens current of 1525 Amps, instead of the on-focus current of 1575 Amps. Expected slit size is 0.04 cm.

CpRad	Δ	**Edge Finder**	Δ
0.0405	0.0005	0.198	0.158 cm
0.0406	0.0006	0.158	0.118 cm
0.0463	0.0063	0.184	0.144 cm

3.1.2. *With Al.* The Ta plate with 4 cm Al plate behind it reproduces a characteristic of dynamic experiments where the features of interest are obscured by a material overburden. Data was taken with a magnetic lens current of 1522 Amps, compared to the ideal of 1548 Amps. This tests chromatic blur modeling capabilities. The data image is shown in Figure 1a. A naïve Canny fit finds too many edges due to the bimodal transmission pattern from off-tune protons, shown in Figure 1b. Using the simple edge finder still results in a slit width of 3.6 mm, 9 times wider than the actual slit width, as seen in Figure 1c. Another method is to invert the transmission equation and solve for for the density. Figures 2a and 2b show an overfit density inversion with no regularization. When the simple edge finder is used it correctly returns a result of 0.4 mm. Direct inversion is not possible for multiple materials and more complex geometries.

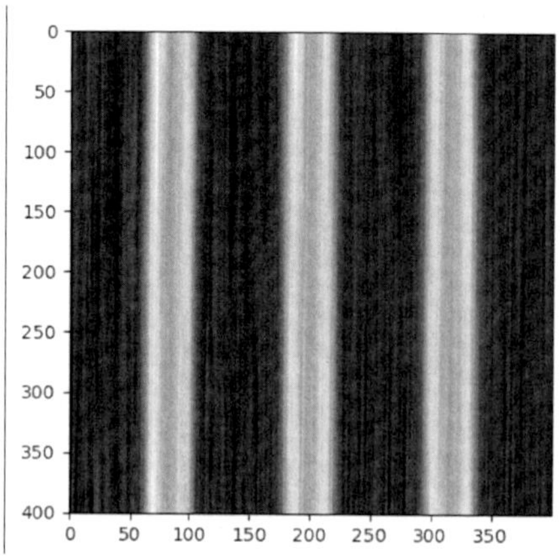

(A) The off-focus protons lead to an image with an interesting bimodal pattern.

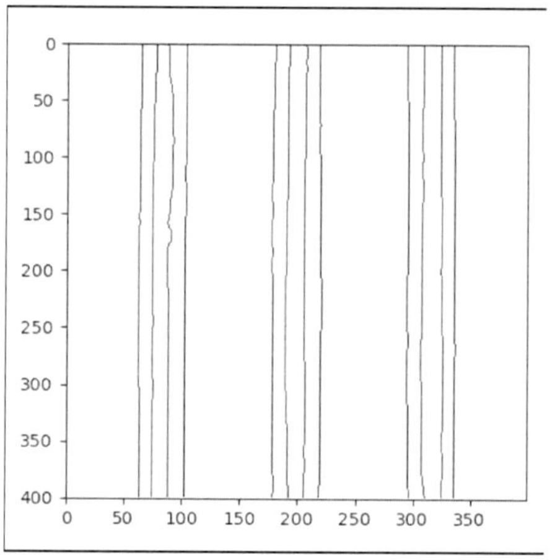

(B) Edge detection for Ta slits with Al plate, off-focus.

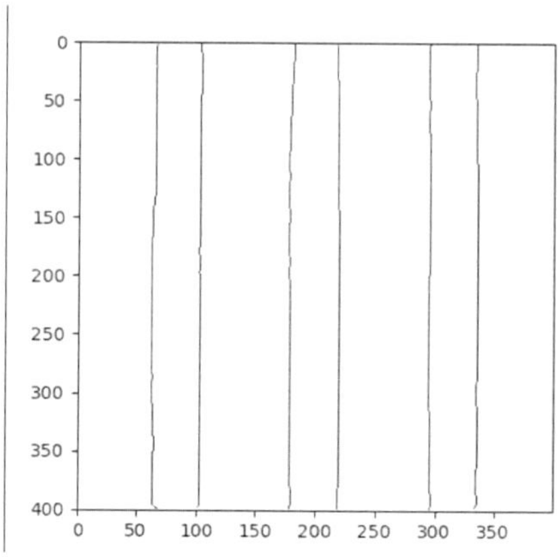

(C) Thresholded to find one pair of edges for each slit.

FIGURE 1. Data and edge finding for Ta with Al plate at off-focus current.

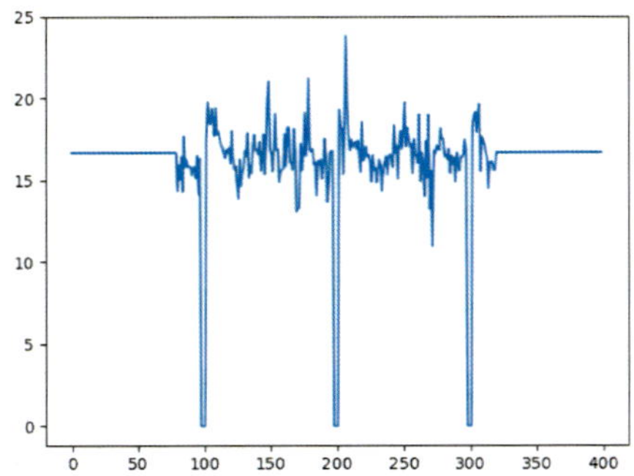

(A) Unconstrained density model resulting in a best fit to the transmission image.

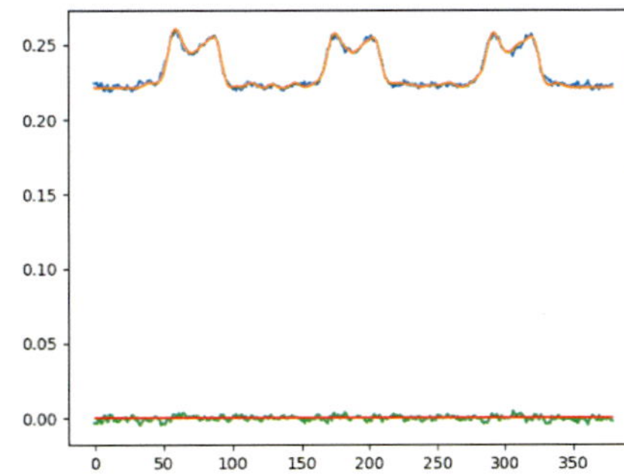

(B) Unconstrained density model compared to data, along with residuals.

FIGURE 2. Inverting to get density.

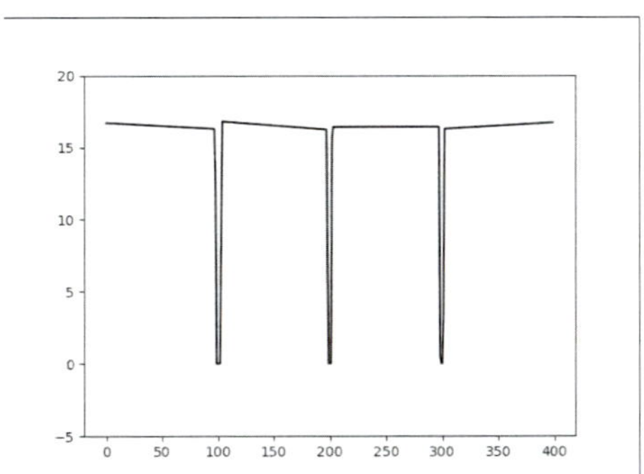

(A) Density of slitted Ta plate found by CpRad when Al plate is placed behind.

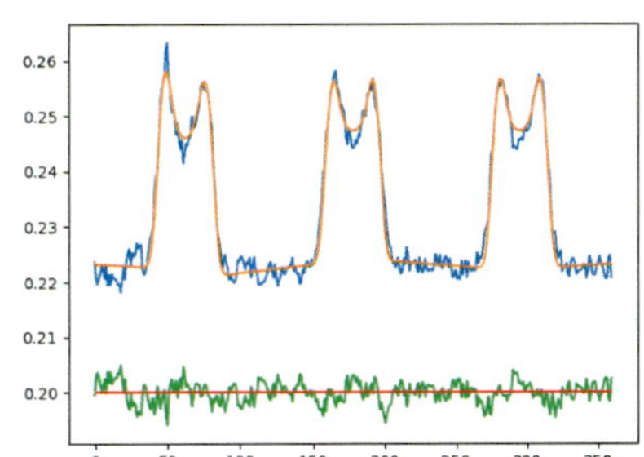

(B) Transmission from CpRad simulation compared to data, along with residuals. Constant 0.2 added to residual values.

FIGURE 3. CpRad simulation of Ta slit plate with Al plate.

Better results are obtained with a 48-parameter density model simulated with CpRad, with 4 location and 4 thickness parameters for each of the 6 lines. This is shown in Figures 3a and 3b. The results for the 3 slit widths are 0.0461 cm, 0.0373 cm, 0.0374 cm. This is still not a perfect solution, however it is less noisy than an inversion density fit and can be used for complex geometries and well as simple ones.

3.2. CTO. Images were also taken on a cylindrical test object (CTO) composed of shells of teflon, aluminum, and copper, with an inner core of air. In this case with multiple materials the protons can only be focused to one interface. We looked at the CTO focused at two different interfaces. When the simple edge finder is applied, it finds different values for the edge locations depending on the focus current, due to the changing chromatic blur. Figure 4 shows the proton radiography images of the nested cylinders with different focus currents. In Figure 4a the outer teflon edge is in focus, whereas in Figure 4b the aluminum/copper interface

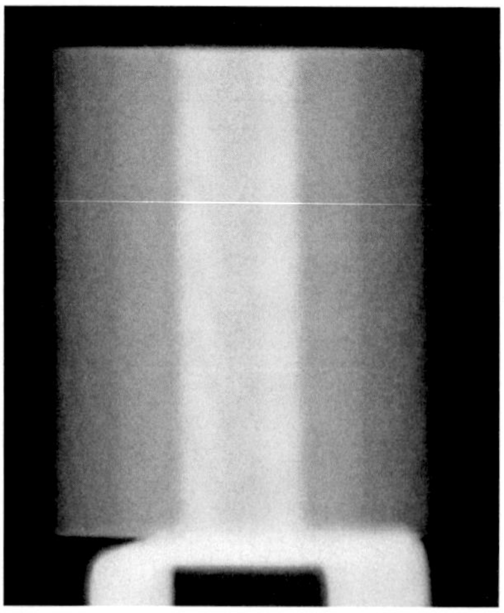

(A) Proton radiography image of nested cylinder with magnetic lens current of 1542 Amps, close to teflon edge energy loss.

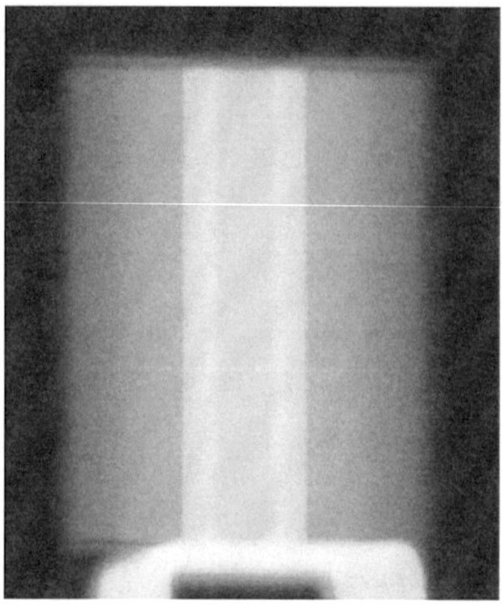

(B) Proton radiography image of nested cylinder with magnetic lens current of 1522 Amps, close to copper edge energy loss.

FIGURE 4. Two images of the nested cylinders at different focus currents.

is in focus. Proton radiography can only have one focus current at a time, which means that complex objects will have spatially varying chromatic blur.

When the simple edge finder is used to find the edges of the nested cylinders, different results are found for the static radii depending on the focus current, as shown in Figure 5a. Quantitative interpretation of experimental results requires a method that robustly handles chromatic effects.

Figure 5b shows a series of pictures of the edges of the nested cylinders. The top image is the actual pRad image with the different shells. The next picture down is the edge finder results for the two tune energies, as in Figure 5a. The last two images are results from density models fit to data after simulation with CpRad. Table 2 shows the results from the edge finder and the density models fit via CpRad, compared to the machined value. CpRad, which accounts for the chromatic blur, performs much better than the edge finder.

TABLE 2. Diameters of nested cylinders.

Material	**CpRad**	**Edge Finder**	**Actual**
Cu 1	0.954	0.783	1.0 cm
Cu 2	0.99	0.72	1.0 cm
Al 1	2.052	1.89	2.0 cm
Al 2	2.07	1.899	2.0 cm
Teflon 1	4.05	2.871	4.0 cm
Teflon 2	4.05	2.862	4.0 cm
Outer 1	6.102	5.652	6.0 cm
Outer 2	6.03	5.724	6.0 cm

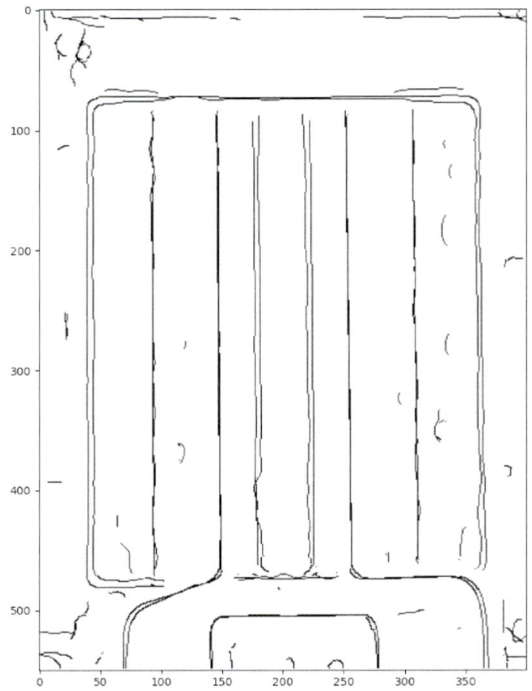

(A) Edges of nested cylinders found via edge detection.

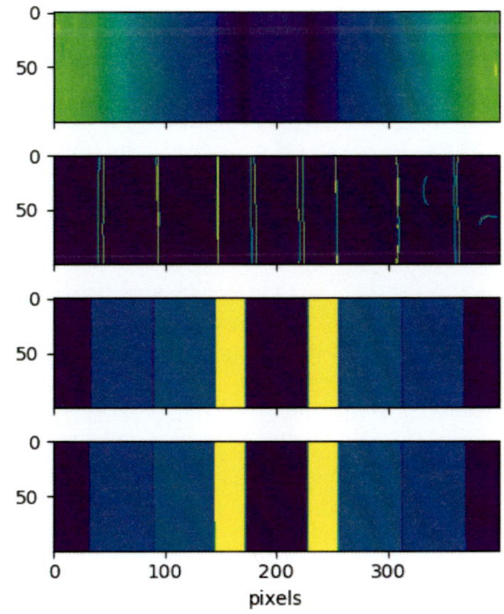

(B) Top image is pRad data, then the simple edge finder results for both tune energies, then the CpRad simulation results for each tune energy.

FIGURE 5. Nested cylinders

4. Conclusions

High fidelity simulations give more accurate results than using a edge finders to find geometric edges in proton radiography images. This chromatic blur is also important for extracting density from pRad images. The CpRad model enables us to extract more accurate densities and geometries from proton radiography data.

References

[1] K. L. Brown, D. C. Carey, C. Iselin, and F. Rothacker, *Transport a computer program for designing charged particle beam transport systems*, SLAC-Ql, Rev. 3, May 1983.

[2] J. Canny, *A computational approach to edge detection*, IEEE Transactions on pattern analysis and machine intelligence (1986), no. 6, 679–698.

[3] F. E. Merrill, *Flash proton radiography*, Reviews of Accelerator Science and Technology (2015), 165–80.

[4] C. L. Morris, N. S. P. King, K. Kwiatkowski, F. G. Mariam, F. E. Merrill, and A. Saunders, *Charged particle radiography*, Reports on progress in physics. Physical Society (Great Britain) **76** (2013).

[5] C. T. Mottershead and J. D. Zumbro, *Magnetic optics for proton radiography*, Proceedings of the 1997 Particle Accelerator Conference (Cat. No.97CH36167), vol. 2, 1997, pp. 1397–1399 vol.2.

[6] Alexis E. Schach von Wittenau, Maurice Aufderheide, and Gary Henderson, *Predicting image blur in proton radiography: Comparisons between measurements and monte carlo simulations*, Nuclear Instruments and Methods in Physics Research Section A: Accelerators, Spectrometers, Detectors and Associated Equipment **652** (2011), no. 1, 901–904, Symposium on Radiation Measurements and Applications (SORMA) XII 2010.

[7] S. K. L. Sjue, F.E. Merill F. G. Mariam, C. L. Morris, and A. Saunders, *High order magnetic optics for high dynamic range proton radiography at a kinetic energy of 800 mev*, Review of Scientific Instruments **87** (2016).

[8] Sky Sjue, Elise Tang, and Erin Connor, *Cprad user's manual*, 2021.

Physics Division, Los Alamos National Laboratory, Los Alamos, New Mexico, 87545
Email address: etang@lanl.gov

Physics Division, Los Alamos National Laboratory, Los Alamos, New Mexico, 87545

Physics Division, Los Alamos National Laboratory, Los Alamos, New Mexico, 87545

Physics Division, Los Alamos National Laboratory, Los Alamos, New Mexico, 87545

Contemporary Mathematics
Volume **824**, 2025
https://doi.org/10.1090/conm/824/16453

Cauchy polynomials: Traces, symmetric functions, and a raising operator

Jerzy Kocik

ABSTRACT. The polynomial relationship expressing elementary symmetric functions in terms of power sums is formulated via a "raising operator" and Fock space construction. A simple graphical proof of this relation is proposed. The new operator extends the Heisenberg algebra. Reciprocal relations are also expressed in terms of another raising operator. This study is motivated by the natural appearance of these polynomials in the theory of invariants for Lax equations and in classical and topological field theories, but it extends to various combinatorial aspects.

Introduction

The original motivation for these notes lies in mathematical physics and was to better understand the origin of the relations between the different invariants of endomorphism: generalized determinants, traces and their powers. One may equivalently see them as the relations between the different families of symmetric functions.

As a result, the relations, understood as polynomials, obtain a new life. They turn out to be a generalization of Hermite polynomials. The construction via a raising operator proposed here leads also to a generalization of Heisenberg algebra.

NOTE. The raising operator introduced here should not be confused with the raising operator appearing in a related context, like in [**5**].

1. Cauchy polynomials

The formulas (1.1) might look familiar. They represent the **relations** between the determinants and the power traces of matrices (endomorphisms). Here, however, we want to strip them of the context and ignore the meaning of terms, and view them as **polynomials** in the infinite number of variables $(x_1, x_2, \dots)$ forming an interesting family on its own. In the lack of a better name, we shall call them

2020 *Mathematics Subject Classification.* Primary 05E05, 15A15; Secondary 15A75.
Key words and phrases. Symmetric functions, endomorphism invariants, traces and determinants, Fock space, raising operator.

"**Cauchy polynomials**" for the reason given at the end of this section.

$$
(1.1)\qquad
\begin{aligned}
q_0(x) &= 1\\
q_1(x) &= x_1\\
q_2(x) &= x_1^2 - x_2\\
q_3(x) &= x_1^3 - 3x_1x_2 + 2x_3\\
q_4(x) &= x_1^4 - 6x_1^2x_2 + 8x_1x_3 + 3x_2^2 - 6x_4\\
q_5(x) &= x_1^5 - 10x_1^3x_2 + 20x_1^2x_3 + 15x_1x_2^2 - 30x_1x_4 - 20x_2x_3 + 24x_5\\
q_6(x) &= x_1^6 - 15x_1^4x_2 + 40x_1^3x_3 + 45x_1^2x_2^2 - 90x_1^2x_4 - 120x_1x_2x_3\\
&\quad +144x_1x_5 - 15x_2^3 + 90x_2x_4 + 40x_3^2 - 120x_6
\end{aligned}
$$

In particular, we define them via a raising operator, described in the next section. In the following sections, we also review the connections with combinatorics, symmetric functions, and more.

By the conjugated Cauchy polynomials we will understand the same polynomials with all coefficients of positive sign.

1.1. Alternative definition of the Cauchy polynomials. It might be a right place to report that the above polynomials may be defined via determinants [**9**]. Define matrices by the formula

$$
M_{ij} \;=\; \begin{cases} x_{i-j+1} & \text{if } i \geq j\,,\\ i & \text{if } j = i+1\,,\\ 0 & \text{otherwise,}\end{cases}
$$

and the condition $i, j \leq n$ for matrix M_n. Here are the first five such matrices:

$$
M_1 = \begin{bmatrix} x_1 \end{bmatrix}, \quad M_2 = \begin{bmatrix} x_1 & 1\\ x_2 & x_1 \end{bmatrix}, \quad M_3 = \begin{bmatrix} x_1 & 1 & 0\\ x_2 & x_1 & 2\\ x_3 & x_2 & x_1 \end{bmatrix},
$$

$$
M_4 = \begin{bmatrix} x_1 & 1 & 0 & 0\\ x_2 & x_1 & 2 & 0\\ x_3 & x_2 & x_1 & 3\\ x_4 & x_3 & x_2 & x_1 \end{bmatrix}, \quad M_5 = \begin{bmatrix} x_1 & 1 & 0 & 0 & 0\\ x_2 & x_1 & 2 & 0 & 0\\ x_3 & x_2 & x_1 & 3 & 0\\ x_4 & x_3 & x_2 & x_1 & 4\\ x_5 & x_4 & x_3 & x_2 & x_1 \end{bmatrix}.
$$

Direct inspection reveals the following:

$$
\det M_1 = x_1, \quad \det M_2 = x_1^2 - x_2, \quad \det M_3 = x_1^3 - 2x_1x_2 + 3x_2, \quad \text{etc.}
$$

The following proposition may be viewed as an alternative definition of the Cauchy polynomials.

THEOREM 1.1. *The Cauchy polynomials coincide with the determinants of the above matrices*

$$q_i(x) = \det M_i.$$

A small adjustment to the matrices produces the adjoint Cauchy polynomials. If

$$M_1^+ = \begin{bmatrix} x_1 \end{bmatrix}, \quad M_2^+ = \begin{bmatrix} x_1 & -1 \\ x_2 & x_1 \end{bmatrix}, \quad M_3^+ = \begin{bmatrix} x_1 & -1 & 0 \\ x_2 & x_1 & -2 \\ x_3 & x_2 & x_1 \end{bmatrix},$$

$$M_4^+ = \begin{bmatrix} x_1 & -1 & 0 & 0 \\ x_2 & x_1 & -2 & 0 \\ x_3 & x_2 & x_1 & -3 \\ x_4 & x_3 & x_2 & x_1 \end{bmatrix}, \quad M_5^+ = \begin{bmatrix} x_1 & -1 & 0 & 0 & 0 \\ x_2 & x_1 & -2 & 0 & 0 \\ x_3 & x_2 & x_1 & -3 & 0 \\ x_4 & x_3 & x_2 & x_1 & -4 \\ x_5 & x_4 & x_3 & x_2 & x_1 \end{bmatrix},$$

then

$$q_i^+(x) = \det M_i^+ .$$

In both cases, the reversing the order of the sequence $(1, 2, 3, \dots)$ in the matrices to the opposite does not affect the determinant. To exemplify:

$$\begin{vmatrix} x_1 & 1 & 0 & 0 \\ x_2 & x_1 & 2 & 0 \\ x_3 & x_2 & x_1 & 3 \\ x_4 & x_3 & x_2 & x_1 \end{vmatrix} = \begin{vmatrix} x_1 & 3 & 0 & 0 \\ x_2 & x_1 & 2 & 0 \\ x_3 & x_2 & x_1 & 1 \\ x_4 & x_3 & x_2 & x_1 \end{vmatrix} = q_4(x).$$

1.2. Hermite polynomials generalized. Before we continue, let us observe that, quite remarkably, the Hermite polynomials may be obtained by a simple reduction of our polynomials. Namely, under substitution $x_1 = x$, $x_2 = 1$, and $x_i = 0$ for all $i > 2$, they become the so-called "probabilist's Hermite polynomials" $H(x)$:

Probabilist's Hermite polynomials

$$x_1 = x,\ x_2 = 1,\ \text{and}\ \forall i > 2\ \ x_i = 0 \quad \Rightarrow \quad \begin{aligned} H_0(x) &= 1 \\ H_1(x) &= x \\ H_2(x) &= x^2 - 1 \\ H_3(x) &= x^3 - 3x \\ H_4(x) &= x^4 - 6x^2 + 3 \\ H_5(x) &= x^5 - 10x^3 + 15x \\ H_6(x) &= x^6 - 15x^4 + 45x^2 - 15 \\ &\text{etc.} \dots \end{aligned} \tag{1.2}$$

As it is known, there are two types: the above "probabilist's Hermite polynomials" $H(x)$ and "physicist's Hermite polynomials". A slightly different substitution leads to the latter, denoted here as $\widetilde{H}$:

Physicist's Hermite polynomials

$$x_1 = 2x,\ x_2 = 2,\ \text{and}\ \forall i > 2\ \ x_i = 0 \quad \Rightarrow \quad \begin{aligned} \widetilde{H}_0(x) &= 1 \\ \widetilde{H}_1(x) &= 2x \\ \widetilde{H}_2(x) &= 4x^2 - 2 \\ \widetilde{H}_3(x) &= 8x^3 - 6x \\ \widetilde{H}_4(x) &= 16x^4 - 24x^2 + 12 \\ \widetilde{H}_5(x) &= 32x^5 - 160x^3 + 120x \\ \widetilde{H}_6(x) &= 64x^6 - 480x^4 + 7205x^2 - 120 \\ &\text{etc.} \dots \end{aligned}$$

Usually, the two types of Hermite polynomials are considered as mutual re-scaling:

$$\widetilde{H}(x) = 2^{n/2}\, H(\sqrt{2}x)$$

while we get both directly from the Cauchy polynomials.

Consequently, Hermite polynomials may be also expressed as determinants:

$$H_1 = \begin{vmatrix} x \end{vmatrix}, \quad H_2 = \begin{vmatrix} x & 1 \\ 1 & x \end{vmatrix}, \quad H_3 = \begin{vmatrix} x & 1 & 0 \\ 1 & x & 2 \\ 0 & 1 & x \end{vmatrix},$$

$$H_4 = \begin{vmatrix} x & 1 & 0 & 0 \\ 1 & x & 2 & 0 \\ 0 & 1 & x & 3 \\ 0 & 0 & 1 & x \end{vmatrix}, \quad H_5 = \begin{vmatrix} x & 1 & 0 & 0 & 0 \\ 1 & x & 2 & 0 & 0 \\ 0 & 1 & x & 3 & 0 \\ 0 & 0 & 1 & x & 4 \\ 0 & 0 & 0 & 1 & x \end{vmatrix}$$

and, similarly,

$$\tilde{H}_1 = \begin{vmatrix} 2x \end{vmatrix}, \quad \tilde{H}_2 = \begin{vmatrix} 2x & 1 \\ 2 & 2x \end{vmatrix}, \quad \tilde{H}_3 = \begin{vmatrix} 2x & 1 & 0 \\ 2 & 2x & 2 \\ 0 & 2 & 2x \end{vmatrix}, \quad \tilde{H}_4 = \begin{vmatrix} 2x & 1 & 0 & 0 \\ 2 & 2x & 2 & 0 \\ 0 & 2 & 2x & 3 \\ 0 & 0 & 2 & 2x \end{vmatrix}.$$

Other restrictions will produce other families of polynomials, see Appendix C for examples.

1.3. A warm-up problem. Consider a café for n guests with round tables of different sizes and different number of seats, like the one shown in Figure 1. The question is, in how many ways one can seat n guests so that at least one person has a different neighbor on one of his sides. Here is the solution for the case shown in the figure: There are $n! = 16!$ permutations of the guests. We need to factor out the cases that are not considered different. Thus, swapping the 3-chair tables will

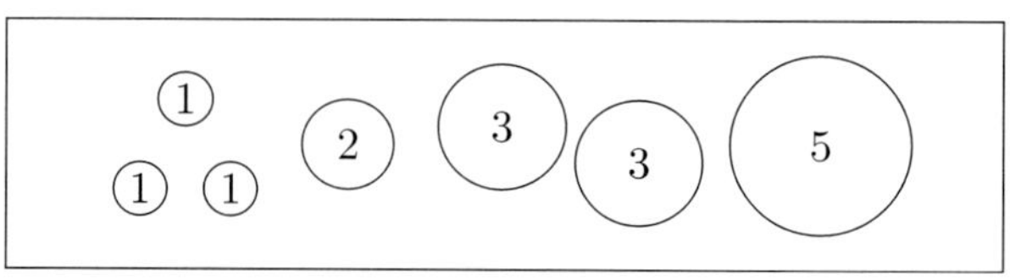

FIGURE 1. Tables in a café, $|3, 1, 2, 0, 1\rangle$.

not affect the neighboring pattern, as well as permuting the 1-chair tables. Also, rotating the chairs around each table preserves the pattern. Hence the answer is

$$\frac{16!}{(3!\,2!)\cdot(2\cdot 3\cdot 3\cdot 5)}.$$

In the case of a café with table configuration $[\alpha_1, \alpha_2, \alpha_3, \dots]$, where α_i denotes the number of tables with i chairs, we have

$$\frac{n!}{(\alpha_1!\,\alpha_2!\alpha_3!\cdot\dots)\cdot(1^{\alpha_1}2^{\alpha_2}3^{\alpha_3}\cdot\dots)} \tag{1.3}$$

distinct configurations of n guests, corresponding to the total number of chairs $n = \sum_i \alpha_i \cdot i$. (Clearly, some α's may be zeros.) For the above numerical case we have $[3, 1, 2, 0, 1]$ and therefore the numerical example is

$$\frac{16!}{(3!\,1!\,2!\,0!\,1!) \cdot (1^3 \cdot 2^1 \cdot 3^2 \cdot 4^0 \cdot 5^1)}.$$

The formula (1.3) appears originally in the theory of symmetric functions as a "Cauchy formula". These numbers (with appropriate signs) appear also as the coefficients of the polynomials we discuss, hence the proposed name "Cauchy polynomials."

1.4. Final note. These findings need to be considered in the wider context of the Bell polynomials [**9**]. They provide a framework for these and many other families of polynomials by various specifications.

2. Fock construction for Cauchy polynomials (main result)

In this section, we provide a definition of the Cauchy polynomials (1.1) via a certain raising operator.

Let $\mathcal{P}[x]$ be a linear space of polynomials over variables $\{x_1, x_2, x_3, \ldots\}$. We shall use the multi-index notation $(n) = (n_1, n_2, \ldots)$. The monomials

$$x^{(n)} = x_1^{n_1}\, x_2^{n_2}\, x_3^{n_3} \ldots$$

form a basis of $\mathcal{P}[x]$, and a general element of $\mathcal{P}[x]$ is

$$q(x) = \sum_{(n)} c_{(n)}\, x^{(n)}$$

with some coefficients $c_{(n)}$.

DEFINITION 2.1. Derivation operator $\delta \in \operatorname{End} \mathcal{P}[x]$ is defined by its action on a single variable and by the Leibniz rule

$$\text{(2.1)} \qquad \begin{array}{ll} \text{(i)} & \delta\, x_k = k\, x_{k+1}\,, \\ \text{(ii)} & \delta\,(ab) = (\delta a) \cdot b + a \cdot \delta b\,. \end{array}$$

THEOREM 2.2. *For a simple power x_k^n and for a general monomial, one has respectively the following formulae*

$$\text{(2.2)} \qquad \begin{array}{ll} \text{(i)} & \delta\, x_k^n = nk\, x_k^{n-1}\, x_{k+1}\,, \\ \text{(ii)} & \delta\, x^{(n)} = \displaystyle\sum_i i \left(\frac{d}{dx_i}\, x^{(n)} \right) \cdot x_{i+1}\,. \end{array}$$

(The sum runs over these i's for which n_i is not zero.)

We shall now define a **raising** operator $\Delta \in \operatorname{End} \mathcal{P}[x]$ by

$$\text{(2.3)} \qquad \Delta = \eta - \delta\,,$$

where $\eta \equiv x_1 \cdot$ denotes the operator of multiplication by variable x_1. The operator Δ let us form a sequence of polynomials (coherent states) generated by a consecutive application of the raising operator Δ, namely

$$\text{(2.4)} \qquad \begin{array}{ll} \text{(i)} & q_1\,(x) = x_1\,, \\ \text{(ii)} & q_{k+1}(x) = \Delta q_k = (\eta - \delta) q_k\,. \end{array}$$

The following family of polynomials is generated:

$$
\begin{array}{rcl}
q_1(x) & = & x_1 \\
q_2(x) & = & x_1^2 - x_2 \\
q_3(x) & = & x_1^3 - 3x_1x_2 + 2x_3 \\
q_4(x) & = & x_1^4 - 6x_1^2x_2 + 8x_1x_3 + 3x_2^2 - 6x_4 \\
q_5(x) & = & x_1^5 - 10x_1^3x_2 + 20x_1^2x_3 + 15x_1x_2^2 - 30x_1x_4 - 20x_2x_3 + 24x_5 \\
q_6(x) & = & x_1^6 - 15x_1^4x_2 + 40x_1^3x_3 + 45x_1^2x_2^2 - 90x_1^2x_4 - 120x_1x_2x_3 \\
 & & +144x_1x_5 - 15x_2^3 + 90x_2x_4 + 40x_3^2 - 120x_6 \\
 & \dots &
\end{array} \tag{2.5}
$$

One may actually start with $q_0(x) = 1$, which may be viewed as a "vacuum state", and $q_k = \Delta^k q_1$ as the k-th "excited state" obtained via the raising operator Δ. One may easily recognize in the above the polynomials (3.4):

Bracket notation. Let us rewrite the definitions in the Dirac-like notation. The monomials will be represented as

$$x^{(n)} \;\mapsto\; |n_1, n_2, \ldots\rangle\,,$$

so that

$$q(x) \;\mapsto\; \sum_{(n)} c_{(n)} |n\rangle\,.$$

The entries beyond the last nonzero entry are simply ignored. For instance

$$q(x) = 2x_1^3x_2^5x_4 + 7x_2x_3 \qquad \text{becomes} \qquad \mathbf{q} = 2\,|3\;5\;0\;1\rangle + 7\,|0\;1\;1\rangle.$$

Now, the raising operator is defined as

$$\Delta = \eta - \delta,$$

where

$$
\begin{aligned}
\eta\,|n_1, n_2, n_3, \ldots\rangle &= |n_1 + 1,\, n_2, n_3, \ldots\,\rangle \\
\delta\,|n_1, n_2, n_3, \ldots\rangle &= \sum_i (i\,n_i)\,|n_1,\, n_2, \ldots\,, n_{i-1},\, n_i - 1,\, n_{i+1} + 1,\, n_{i+2},\, \ldots\,\rangle,
\end{aligned} \tag{2.6}
$$

where the sum extends over the terms for which $n_i > 0$. (Or, equivalently, one can simply set $|n_1, n_2, \ldots\rangle = 0$ whenever $n_i < 0$ for any i).

The system of the polynomials becomes in the Dirac-like notation the following family of superposition of states:

$$
\begin{aligned}
\mathbf{q}_1 &= |1\rangle \\
\mathbf{q}_2 &= |2\rangle - |0\,1\rangle \\
\mathbf{q}_3 &= |3\rangle - 3\,|1\,1\rangle + 2\,|0\,0\,1\rangle \\
\mathbf{q}_4 &= |4\rangle - 6\,|2\,1\rangle + 8\,|1\,0\,1\rangle + 3\,|0\,2\rangle - 6\,|0\,0\,0\,1\rangle \\
\mathbf{q}_5 &= |5\rangle - 10\,|3\,1\rangle + 20\,|2\,0\,1\rangle + 15\,|1\,2\rangle - 30\,|1\,0\,0\,1\rangle - 20\,|0\,1\,1\rangle \\
&\quad + 24\,|0\,0\,0\,0\,1\rangle \\
\mathbf{q}_6 &= |6\rangle - 15\,|4\,1\rangle + 40\,|3\,0\,1\rangle + 45\,|2\,2\rangle - 90\,|2\,0\,0\,1\rangle - 120\,|1\,1\,1\rangle \\
&\quad + 144\,|1\,0\,0\,0\,1\rangle - 15\,|0\,3\rangle + 90\,|0\,1\,0\,1\rangle + 40\,|0\,0\,2\rangle - 120\,|0\,0\,0\,0\,0\,1\rangle \\
&\dots
\end{aligned} \tag{2.7}
$$

An inner product may defined so that $\langle n|m\rangle = 1$ if $n = m$, and vanishes if $n \neq m$. Consequently, $c_{(n)} = \langle\, n \mid p(x)\,\rangle$. (See Figure 2.)

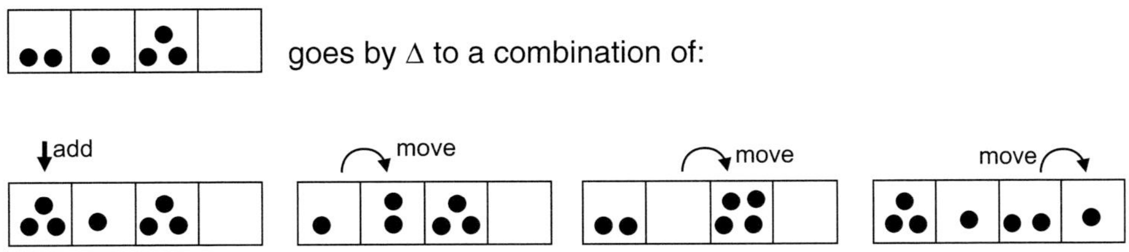

FIGURE 2. The action of the raising operator visualized via a moves of stones on a board illustrating the flavor of Turing machine of Δ. The coefficients are ignored.

We will view the above construction as the definition of abstract "Cauchy polynomials", and next prove that they represent the relations between the traces and determinants, or equivalently, between different types of symmetric polynomials.

REMARK 2.3. The family of polynomials p_i together with the raising operator Δ may be interpreted and studied as a so-called **Appell system** (see [**4**]).

3. Initial motivations

Here we recount the original motivation for this study. Recall that an endomorphism $A \in \text{End}\, L$ of a linear space L may be characterized by its invariants with respect to some symmetry group, most generally $GL(n)$, where $n = \dim L$. The basic invariants are the determinant and the trace, $\det A$ and $\text{Tr}\, A$:

$$\text{Tr}\ gAg^{-1} = \text{Tr}\ A\,, \qquad \det gAg^{-1} = \det A$$

for $g \in GL(n)$. These generalize into two families of invariants. The elements of one, called **power-traces**, are defined

$$I_k = \text{Tr}\, A^k, \quad k = 1, 2, \dots . \tag{3.1}$$

The other family generalizes the determinant. Its elements, denoted by J_i, are defined as coefficients of the characteristic polynomial for the endomorphism A, namely

$$\det(A - \lambda) = \sum_{i=0}^{n} (-\lambda)^{n-i}\, J_i\,, \tag{3.2}$$

where $n = \dim L$. We shall call J_k a **prodeterminant** of order k. In particular

$$\begin{aligned} J_1(A) \;&=\; \text{Tr}\ A \;\; = I_1(A)\,, \\ J_n(A) \;&=\; \det A\,. \end{aligned}$$

REMARK 3.1. Prodeterminant (generalized determinant) can also be defined also as

$$J_k = \frac{(-1)^{n-k}}{(n-k)!}\ \frac{d^{n-k}}{d\lambda^{n-k}}\ \det(A-\lambda)\big|_{\lambda=0}$$

or, in a more explicit form, as

$$J_k = \sum\ A[a_1, a_2, \dots, a_k]\,, \tag{3.3}$$

where $A[\ldots]$ denotes determinant of the $k \times k$ minor (submatrix) determined by the set of columns and rows indexed by the same subset $[a_1, \ldots, a_k]$ of the set $[1, \ldots, n]$. The summation in (3.3) extends over all $\binom{n}{k}$ possible selections of such subsets.

EXAMPLE 3.2. Let A be a 3×3 matrix

$$A = \begin{bmatrix} a & b & c \\ d & e & f \\ g & h & i \end{bmatrix}.$$

Then one has the following prodeterminants

$$\begin{aligned} J_1 &= a + e + i \\ J_2 &= \det \left[\begin{smallmatrix} a & b \\ d & e \end{smallmatrix}\right] + \det \left[\begin{smallmatrix} a & c \\ g & i \end{smallmatrix}\right] + \det \left[\begin{smallmatrix} e & f \\ h & i \end{smallmatrix}\right] \\ J_3 &= \det A \end{aligned}$$

while the power-traces are

$$\begin{aligned} I_1 &= a + e + i \\ I_2 &= a^2 + e^2 + i^2 + 2db + 2gc + 2fh \\ I_3 &= a^3 + e^3 + i^3 + 3abd + 3cdg + 3bde \\ &\quad +3efg + 3bfh + 3fgi + 3ach + 3chi\,. \end{aligned}$$

Particular applications in systems with symmetries often favor one of the two types of invariants, like Lax equations use rather I while the theory of Chern classes and some particle models—J (see Appendix A). But, in fact, the two are dependent:

$$\begin{aligned} J_1 &= I_1 \\ 2J_2 &= I_1^2 - I_2 \\ 6J_3 &= I_1^3 - 3I_1I_2 + 2I_3 \\ 12J_4 &= I_1^4 - 6I_1^2I_2 + 8I_1I_3 + 3I_2^2 - 6I_4 \\ &\cdots \end{aligned} \tag{3.4}$$

Increasing the dimension n brings new relations, but does not affect the relations (3.4) already established. Hence we may think of a single infinite family of such relations. Now, the idea is to ignore the inner meaning of the functions J and I, and consider the infinite family of abstract polynomials in infinite (countable and ordered) set of variables, as seen in (1.1).

THEOREM 3.3. *The system determined by raising operator Δ via Fock space construction* (2.4) *coincides with the determinant-trace relations,*

$$J_k(A) = \frac{1}{k!}\, p_k(I_1, I_2, \ldots, I_k)\,,$$

where $A \in \operatorname{End} L$ is an endomorphism of a linear space, $I_i = \operatorname{Tr} A^i$, and $J_k(A)$ denotes k-th prodeterminant.

A graphical proof of this theorem is provided in the next section.

These invariants appear naturally in many areas of physics, including the theory of integrable systems on Lie algebras (Lax equation), generalized determinants in classical field theory (see Appendix A).

As is well known, the same relations hold for symmetric polynomials, since they are realized via the eigenvalues of the endomorphism. In particular, prodeterminants can be expressed as polynomials in power traces, with the Cauchy enumeration formula giving the coefficients.

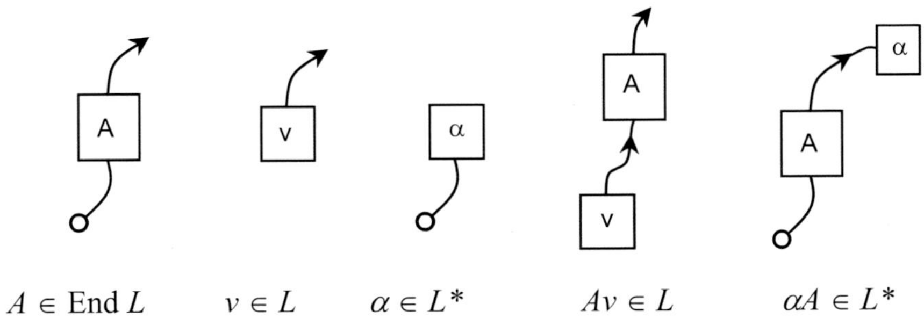

FIGURE 3. Contractions of simple tensors

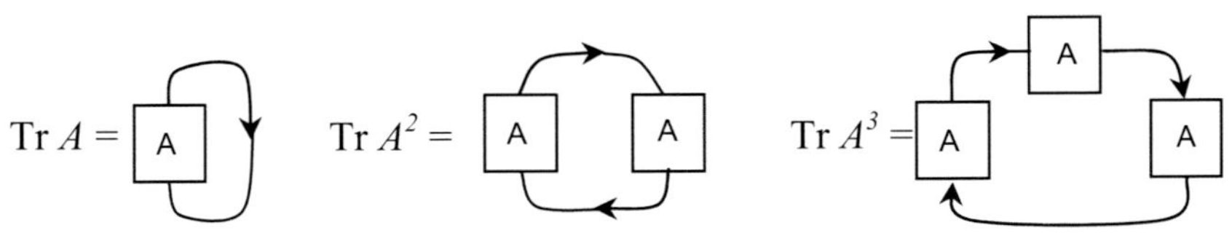

FIGURE 4. Power traces of an endomorphism A

4. Proof via graphic representation

In this section, we show the relations (3.4) for determinants and traces using the graphical language for tensor algebra. The graphical language for tensor contractions is increasingly popular [**3**, **6**, **11**], and—besides the conceptual value—may lead to nice simplifications of proofs, like the one we present.

Here is a rapid introduction: A vector is represented by a square with a single arrow out ("contravariant slot"), a linear form (covector), by a single arrow in ("covariant slot"). An endomorphism A will be presented by a square with two arrows, one going out and one going in. The arrows may be viewed as representing indices, upper and lower, respectively, if A is represented by a matrix. In general, the arrows represent "slots" of A viewed as a tensor, contravariant and covariant, respectively. Figure 3 shows graphical representation of basic linear operations (contractions).

The trace and power-traces can be viewed as seen in Figure 4.

Recall that prodeterminants (3.2) can equivalently be defined as "averages" over traces:

$$J_k = \frac{1}{k!} \sum_{\sigma \in S_k} \operatorname{sgn}(\sigma)\, A^{i_1}_{\sigma(i_1)}\, A^{i_2}_{\sigma(i_2)} \cdots A^{i_k}_{\sigma(i_k)}\,, \tag{4.1}$$

where an additional sum over repeated indices of terms is understood (Einstein's summation convention). Employing (4.1), the prodeterminants J_k are obtained as follows. Consider a tensor product of k copies of endomorphism A, namely the (k,k)-type tensor $A \otimes A \ldots \otimes A$ (left side of Figure 5). One can take a trace of this operator by closing the k *out*-arrows with the k *in*-arrows . There are $k!$ such possible pairings of the k arrows with the k slots.

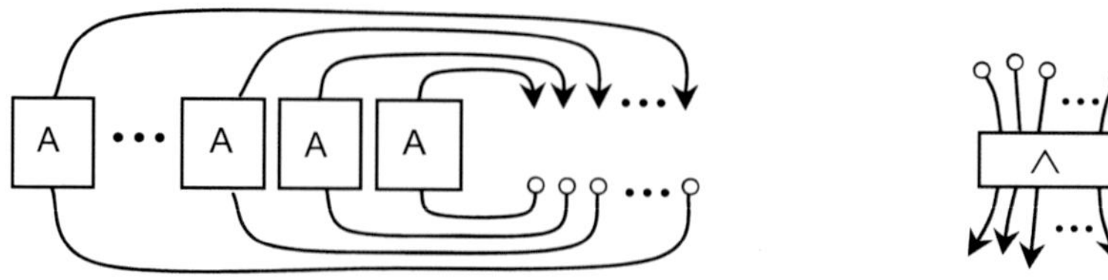

FIGURE 5. Tensor product $A^{\otimes n}$ and alternating tensor $\wedge$

Scalar J_k is obtained by taking a sum over all possibilities (permutations), each term assuming the sign corresponding to the parity of the permutation. In other words, we contract $A \otimes A \otimes \ldots \otimes A$ with the (k, k)-type alternating tensor "$\wedge$", totally antisymmetric in both sectors (represented in the right side of Figure 5). Figure 6 illustrates the case of J_2.

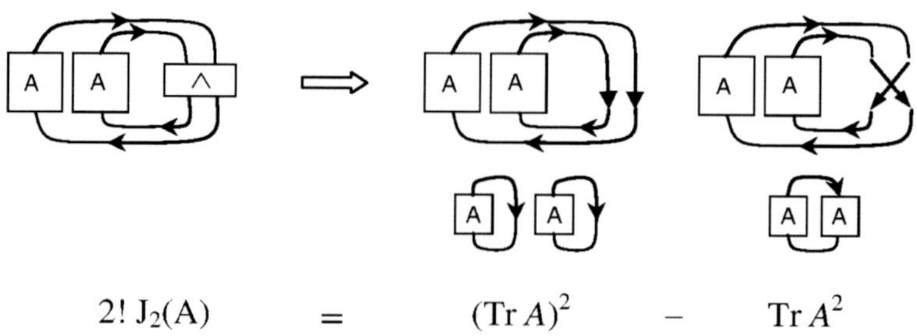

FIGURE 6. Unfolding the second Cauchy polynomial.

The formula relating prodeterminants to power-traces unfolds itself evidently! In a similar combinatorial play with strings one obtains the next invariant J_3, see Figure 7. The general form of Cauchy polynomials as a result of the action of the raising operator (2.3) emerges by induction and combinatorial meaning of the above tensor contractions. Indeed, increasing the number of the "tensor boxes" on the left side on any of the last three figures will add under contraction two types of new graphical loops: either the new tensor makes a single loop with itself, or it will get in the path of one of the existing loops increasing their length by one. The former case corresponds to the operator $\eta \equiv x_1\cdot$ in (2.3), the latter to δ. The next section illustrates the issue.

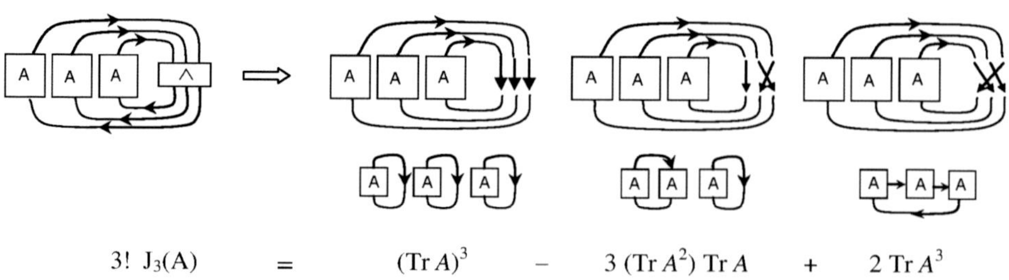

FIGURE 7. Unfolding the third Cauchy polynomial.

5. Combinatorial aspect of the raising operator

Here we reveal the meaning and the source of the raising operator, illustrated with several related examples of combinatorial nature.

5.1. Dynamical systems and the café problem. A finite dynamical system may be represented as a directed graph (Figure 8 left). A natural question arises how to generate and list (and enumerate) such systems in in some orderly fashion.

The café problem (see Section 1) is illustrated in Figure 8, right, by three tables each with a different number of chairs. Like in the previous case, one wants to generate all such configurations, and be able to evaluate the number of different ways one may set, e.g., $n = 6$ guests (different = at least one person has a different left of right neighbor).

FIGURE 8. Equivalence of dynamical systems and table settings

These two problems are evidently equivalent. and have the same answer, represented by the Cauchy polynomials (1.3). But more interesting, the numbers for both problems may be obtained via raising operator (2.3), in both cases getting a clear combinatorial illustration.

Below we continue with the café case. To add a new guest, we need to bring a new chair and either set a new 1-chair table or add it to an existent table of k chairs in one of the $(k-1)$ spaces between them. See Figure 9.

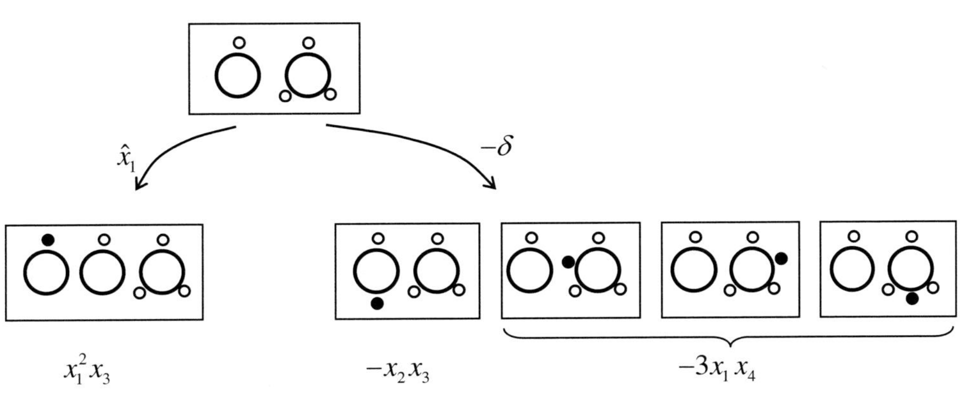

FIGURE 9. The combinatorial interpretation of the raising operator.

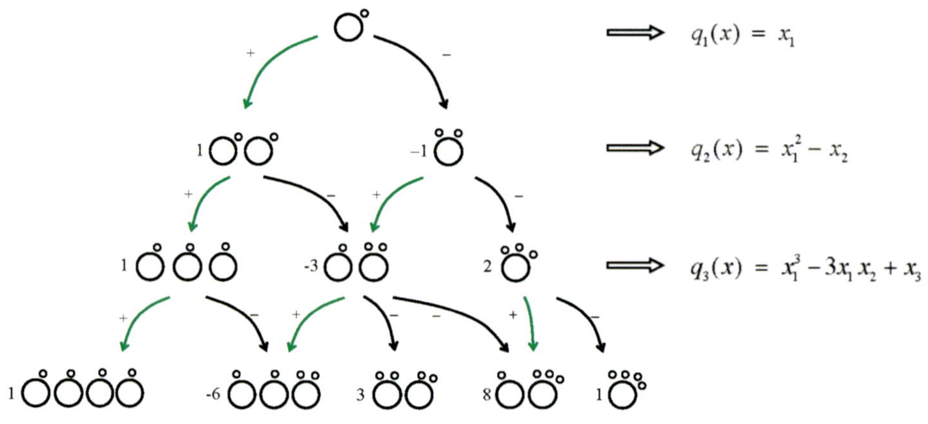

FIGURE 10. The tree of café states. Here the raising operator emerges.

Denote the first operation by η, and the second by δ. Then for the simple enumeration we would use the

$$\Delta^+ = \eta + \delta.$$

We may also add a twist to the construction and assign a negative sign to increasing the existent table size

$$\Delta = \eta - \delta.$$

Now, starting the empty room (or one 1-chair table), we get the following tree by consecutive application of the raising operator (see Figure 10).

The configurations that emerge may be represented by the Dirac like notation or by polynomials. Any particular arrangement of tables can be changed into a monomial by replacing an n-chair table by variable x_n. The levels in the tree (possibilities for a fixed number of guests) may be viewed as a "superposition" of states.

The absolute values of the coefficients for both operators Δ and Δ^+ will be the same, but the second operator is our raising operator for Cauchy polynomials.

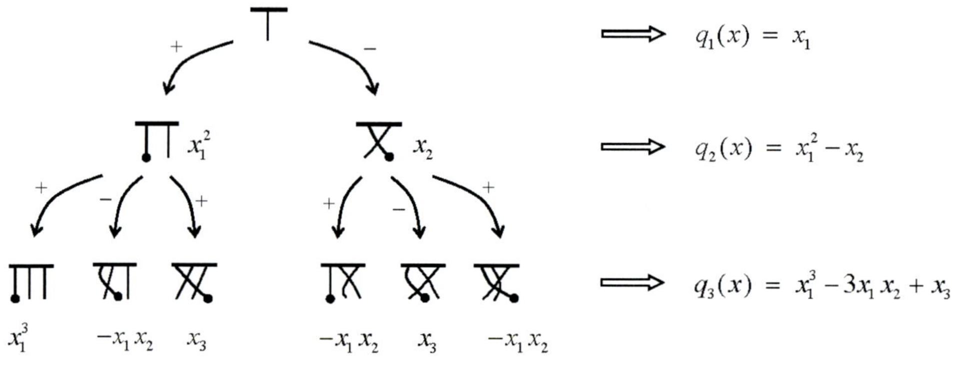

FIGURE 11. Permutation tree. Different process, same result.

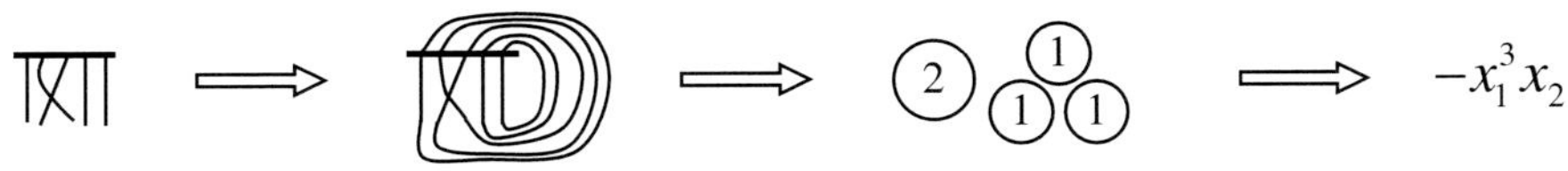

FIGURE 12. From a permutation to a monomial

5.2. Signed permutations. Permutations may be graphically represented as commutative braids. To list all permutations of different order, one may form a tree where new new levels are obtained from the previous by adding a new strand on the left side, and weave it to consecutive positions. In Figure 11, the new strand in a row is marked with a bullet at the end for an easier identification.

By closing the strands (top with corresponding bottom, Figure 12), the braids become systems of circles of different lengths, and as such, quite like the tables, may be interpreted as monomials.

Every level of the tree produces a different polynomial, namely Cauchy polynomial. The signs are implied by the parity of the permutations. Only the first three polynomials are depicted. For the fourth, only one item is resolved. Note that contributions to a particular monomial appear at different places.

5.3. Conjugated raising operator. By the **adjoint raising operator** we shall mean

$$\Delta^+ = \eta + \delta \,,$$

and the corresponding sequence of polynomials

$$\begin{array}{ll} \text{(i)} & r_1(x) = x_1 \,, \\ \text{(ii)} & r_{i+1}(x) = (\hat{x}_1 + \delta) r_i \end{array}$$

will be called adjoint (or sign-free) Cauchy polynomials. Interestingly enough, the absolute values of the coefficients of these two types of polynomials coincide, i.e.,

$$\langle\, \mathbf{n} \mid r_i(x) \,\rangle = |\, \langle\, \mathbf{n} \mid q_i(x) \,\rangle \,| \;.$$

Thus Δ^+ enumerates directly the unsigned café problem, the number of dynamical systems, as well as other related questions, like the number of conjugacy classes of symmetric group S_n.

6. Lie algebra

For each i, the operators $\hat{x}_i$ of multiplication by x_i, the partial derivatives $\partial_i \equiv \partial/\partial x_i$ with respect to x_i and identity, all acting in the space of polynomials $\mathcal{P}[x]$, form Heisenberg Lie algebra

$$\mathcal{H}_i \;=\; \text{gen}\{\partial_i, \hat{x}_i\} \,,$$

the algebra of the harmonic oscillator. By including the derivation δ defined in (2.1), this algebra may be extended to a Lie algebra

$$\bar{\mathcal{H}} = \text{gen}\,\{\partial_i,\, \hat{x}_j,\, \delta\} \tag{6.1}$$

with the following commutation relations for the generators:

$$
\begin{aligned}
[\, \hat{x}_i,\, \hat{x}_j \,] &= 0 \\
[\, \partial_i,\, \hat{x}_j \,] &= \delta_{ij} \\
[\, \delta,\, \hat{x}_j \,] &= j \cdot \hat{x}_{j+1} \\
[\, \partial_j,\, \delta \,] &= (j-1) \cdot \partial_{j-1}
\end{aligned}
$$

It is easy to calculate the Lie bracket of the two raising operators:

$$[\, \Delta,\, \Delta^+ \,] = 2\hat{x}_2\delta\,.$$

Consider a subspace $\mathcal{J}[x] \subset \mathcal{P}[x]$ spanned by the Cauchy polynomials $\{p_1, p_2, \ldots\}$. By definition, Δ is a raising operator in the subspace $\mathcal{J}[x]$. The derivative with respect to the first variable acts as a lowering operator:

$$\partial_1 q_n = n \cdot q_{n-1}\,.$$

Derivative with respect to the k-th variable lowers the index by k:

$$\partial_k q_n = \frac{(-1)^{k+1} n!}{k \cdot (n-k)!} \cdot q_{n-k}\,.$$

From this follows that

$$[\, \partial_1,\, \Delta \,]\, q_n = q_n\,,$$

and, in general:

$$[\, \partial_k,\, \Delta \,]\, q_n \;=\; \frac{(-1)^{k+1}\, n!}{(n-k+1)!}\, q_{n-k+1}\,.$$

More on this will be developed elsewhere.

7. Reciprocal Cauchy polynomials

Interpreting the Cauchy polynomials (1.1) as equations, and solving them for x's as expressed by variable q_i, we get:

$$
\begin{aligned}
0!\, x_1 &= q_1 \\
1!\, x_2 &= q_1^2 - q_2 \\
2!\, x_3 &= 2q_1^3 - 3q_1q_2 + q_3 \\
3!\, x_4 &= 6q_1^4 - 12q_1^2q_2 + 4q_1q_3 + 3q_2^2 - q_4 \\
4!\, x_5 &= 24q_1^5 - 60q_1^3q_2 + 20q_1^2q_3 + 30q_1q_2^2 - 5q_1q_4 - 10q_2q_3 + q_5 \\
5!\, x_6 &= 120q_1^6 - 360q_1^4q_2 + 120q_1^3q_3 + 270q_1^2q_2^2 - 30q_1^2q_4 - 120q_1q_2q_3 \\
&\quad -30q_2^3 + 6q_1q_5 + 15q_2q_4 + 10q_3^2 - q_6
\end{aligned}
$$

It seems appropriate to define the "reciprocal Cauchy polynomials" by the right-hand side polynomials and dropping the factorial in front. Here is the result in the Dirac-like notation:

$$
\begin{aligned}
\mathbf{x}_1 &= |1\rangle \\
\mathbf{x}_2 &= |2\rangle - |0\ 1\rangle \\
\mathbf{x}_3 &= 2\,|3\rangle - 3\,|1\ 1\rangle + |0\ 0\ 1\rangle \\
\mathbf{x}_4 &= 6\,|4\rangle - 12\,|2\ 1\rangle + 4\,|1\ 0\ 1\rangle + 3\,|0\ 2\rangle - |0\ 0\ 0\ 1\rangle \\
\mathbf{x}_5 &= 24\,|5\rangle - 60\,|3\ 1\rangle + 20\,|2\ 0\ 1\rangle + 30\,|1\ 2\rangle - 5\,|1\ 0\ 0\ 1\rangle - 10\,|0\ 1\ 1\rangle \\
&\quad + |0\ 0\ 0\ 0\ 1\rangle \\
\mathbf{x}_6 &= 120\,|6\rangle - 360\,|4\ 1\rangle + 120\,|3\ 0\ 1\rangle + 270\,|2\ 2\rangle - 30\,|2\ 0\ 0\ 1\rangle - 120\,|1\ \ 1\ 1\rangle \\
&\quad - 30\,|0\ 3\rangle + 6\,|1\ 0\ 0\ 0\ 1\rangle + 15\,|0\ 1\ 0\ 1\rangle + 10\,|0\ 0\ 2\rangle - |0\ 0\ 0\ 0\ 0\ 1\rangle
\end{aligned}
$$

One wonders if a similar "raising operator" may be formed to obtain the above polynomials. The answer is yes. Define new versions of the δ and η operators as follows:

$$
\begin{aligned}
\eta^*\ |n_1, n_2, n_3, \ldots\rangle &= \left(\sum_i n_i\right)\ |n_1{+}1, n_2, n_3, \ldots\rangle \\
\delta^*\ |n_1, n_2, n_3, \ldots\rangle &= \sum_i n_i \cdot |n_1,\ n_2, \ldots, n_{i-1},\ n_i{-}1,\ n_{i+1}{+}1,\ n_{i+2}, \ldots\rangle \\
&= n_1\ |n_1{-}1, n_2{+}1, n_3, \ldots\rangle \\
&\quad + n_2\ |n_1, n_2{-}1, n_3{+}1, \ldots\rangle + \ldots
\end{aligned}
\tag{7.1}
$$

Now, define

$$
\Delta^* = \eta^* - \delta^*\,. \tag{7.2}
$$

THEOREM 7.1. *The reciprocal Cauchy polynomials are generated by the above raising operator* (7.2)*:*

$$
\mathbf{x}_{k+1} = \Delta^*\,\mathbf{x}_k\,.
$$

Or, starting with the "vacuum state",

$$
\mathbf{x}_{k+1} = (\eta^* - \delta^*)^k\ |1\rangle\,.
$$

The proof is based on direct but meticulous calculations.

Note the difference: the original operator η is defined simply as a multiplication operator. In the above case, it is more involved. In the polynomial representation, the operators are

$$
\begin{aligned}
\eta^* &= p_1\,(p_1\partial_1 + p_2\partial_2 + p_3\partial_3 + \ldots), \\
\delta^* &= p_2\partial_1 + p_3\partial_2 + p_4\partial_3 + \ldots,
\end{aligned}
\tag{7.3}
$$

where $\partial_i \equiv \frac{\partial}{\partial p_i}$.

Clearly, we define the conjugated version of the reciprocal raising operator and the corresponding polynomials in the analogous way:

$$(\Delta^*)^+ = \eta^* + \delta^*\,, \qquad \mathbf{y}_{k+1} = \Delta^*\,\mathbf{y}_k = (\Delta^*)^k\,|1\rangle, \tag{7.4}$$

where $|1\rangle$ corresponds to $y_1(p) = p_1$.

8. Coda

We introduced two raising operators

$$\Delta = \eta - \delta \qquad \text{end} \qquad \Delta^+ = \eta + \delta\,.$$

The latter creates via Fock space construction polynomials called here the conjugated Cauchy polynomials, that enumerate the combinatorial problems like "café problem", enumeration of finite dynamical systems, permutation of a given degree, conjugation classes of a symmetry group, etc. The former does it with an extra twist where the objects carry a sign. In particular, the raising operators relate invariants of the endomorphisms as well as the related symmetric polynomials.

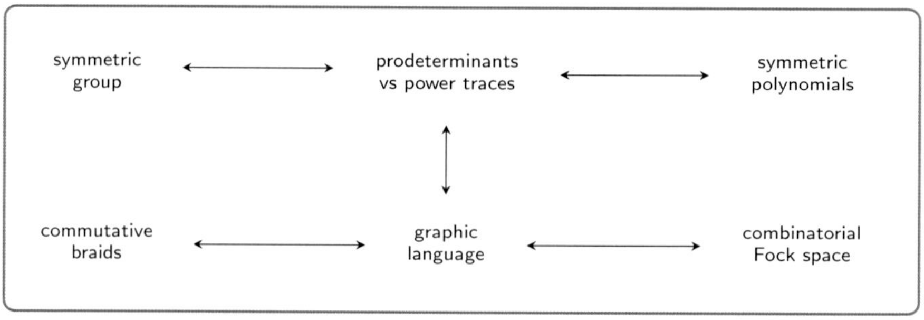

The Cauchy polynomials are interesting on their own. They are generalized Hermite polynomials and come with a related Lie algebra and are a source of other families of polynomials. One has to see this as a special case of many interesting polynomials that result as specifications of the Bell polynomials [**9**]. An example is to set $x_i = x$ for all i.

$$\begin{aligned} q_0(x) &= 1 \\ q_1(x) &= x \\ q_2(x) &= x^2 - x \\ q_3(x) &= x^3 - 3x^2 + 2x \\ q_4(x) &= x^4 - 6x^3 + 11x^2 - 6x \\ q_5(x) &= x^5 - 10x^4 + 350x^3 - 50x^2 + 24x\,. \end{aligned} \tag{8.1}$$

We have also defined the "dual" raising operators

$$\Delta^* = \eta^* - \delta^* \qquad \text{end} \qquad (\Delta^*)^+ = \eta^* + \delta^*$$

which give rise to reciprocal Cauchy polynomials that may also be of interest.

Appendix A. Applications of invariants

A.1. Geometric definition of prodeterminant. Given linear space L of dimension d. Consider Grassmann space (tensor space)

$$\wedge^k L = L \wedge L \wedge \ldots \wedge L\,.$$

Any endomorphism $A \in \text{End } L$ has a natural extension to an endomorphism $A^{\wedge k} \in \text{End } \wedge^k L$ which for a simple multi-vector $v_1 \wedge v_2 \wedge \cdots \wedge v_k$ is defined

$$A^*(v_1 \wedge v_2 \wedge \cdots \wedge v_k) = A(v_1) \wedge A(v_2) \wedge \cdots \wedge A(v_k)\,.$$

Note that if $\{e_i\}$ is a basis of L, and $\{\epsilon_i\}$ is the dual basis of L^*, then $A = \epsilon^i A_i^j e_j$, and

$$A^{\wedge k} = A \wedge A \wedge \cdots \wedge A = A_{i_1}^{p_1} A_{i_2}^{p_2} \ldots A_{i_k}^{p_k}\, e_{p_1} \wedge e_{p_2} \wedge \ldots \wedge e_{p_k} \otimes \epsilon^{i_1} \wedge \epsilon^{i_2} \wedge \ldots \wedge \epsilon^{i_k}\,.$$

Since multi-vectors $e_{p_1} \wedge e_{p_2} \wedge \ldots \wedge e_{p_k}$ form a basis in the Grassmann space, the formula for prodeterminant can be easily written as a trace of the induced endomorphism

$$J_k(A) = \text{Tr } A^{\wedge k}\,.$$

In particular, $J_1(A) = \text{Tr } A$ and $J_n(A) = \text{Tr } A^{\wedge n}$.

A.2. Matrix manifolds. Consider orbits of the adjoint action of the general linear group $GL(n)$ acting on the space of endomorphisms End (L) over some field of some space L of dimension $\dim L = n$:

$$g:\; M \longrightarrow gMg^{-1}\,.$$

The $n \times n$ dimensional space End $L \cong \mathbb{R}^{n\times n}$ is foliated by the orbits of this action, which we call **matrix manifolds** (see [**8**], where the case of complex spaces is considered). Thus the prodeterminants J_k, invariant with respect to this action, are natural objects to consider in this context. In particular, a set of n values determines an orbit (that is, the set of orbits is parametrized by the values of J's). If a matrix M is an element of some orbit $\mathcal{O}_M$, then

$$\text{Ker } J_1(M) \cap \text{Ker } J_2(M) \cap \ldots \cap \text{Ker } J_n(M) = T_M\mathcal{O}\,. \tag{A.1}$$

For more (the role played by the rank of M and for applications in field theory), see [**8**].

A.3. Lax equation. Consider a matrix representation of Lie algebra L and a dynamical system

$$\dot{M} = [M, B]\,.$$

The power-traces of M provide natural invariants (called *Casimir invariants*) of the Lax dynamical system. Indeed

$$\begin{aligned} \tfrac{d}{dt} \text{ Tr } M^n &= n \text{ Tr } \dot{M}M^{n-1} = n \,\text{Tr}\,[M,B]M^{n-1} \\ &= n(\text{Tr } MBM^{n-1} - \text{Tr } BMM^{n-1}) = 0\,. \end{aligned}$$

In particular, one can consider the adjoint representation of L and then M and B are directly elements of L. Let L be a Lie algebra. In [**7**] we consider the space of Lie algebra L as a manifold and define a (1,1)-type tensor field (a field of endomorphisms) by

$$A_x(\tilde{v}) = (\text{ad}_{\,x}\, v)^\sim = [x, v]^\sim\,.$$

If x^i are (linear) coordinates on L then

$$A = x^i \, c_{ij}^k \, \frac{\partial}{\partial x^k} \otimes dx^j \, .$$

At every point $a \in L$, tensor A can be viewed as an endomorphism of the tangent space, $A : T_a L \to T_a L$. One can define a distribution $\Im A \subset TL$. It is easy to show that this distribution is integrable; the integral manifolds $\mathcal{O}$ coincide with the orbits of the adjoint action of L, and

$$\Im A = T\mathcal{O} \, .$$

The power-traces of the adjoint representation provide a set of scalar functions $I_i : L \to \mathbb{R}$. One of them, I_2, is the Killing form (known in this context as Cartan quadratic function)

$$I_2(a) = K(a,a) = c_{ip}^q c_{qj}^p a^i a^j \, .$$

Constant value of the Killing function determine a pseudosphere (hyperbolic sphere—in the case of semi-simple algebras). Orbits of the adjoint action of the corresponding Lie group lie in these spheres. They lie inside the surfaces determined by all the higher-order power-traces. Thus, if for a set of n numbers r we define a submanifold of L

$$S(r_1, \ldots, r_n) = \{ \, a \in L \mid I_1(a) = r_1, \ldots, I_n(a) = r_n \, \},$$

then at any point $a \in L$

$$\mathcal{O}_a \subset S(I_1(a), \ldots, I_n(a)) \, .$$

Given vector field B on L, define a new vector field of a dynamical system

$$X_B = A \llcorner B \, .$$

Thus we have a corollary: every Lax dynamical system on L preserves each I_i as the first integral of motion. Indeed

$$X_B f = df \llcorner X_B = df \llcorner A \llcorner B = 0 \, .$$

Appendix B. Symmetric functions

Connection of the invariants with the symmetric functions follows directly from the fact that if an endomorphism A has a diagonal matrix form in some basis, with eigenvalues on the diagonal

$$A = \begin{bmatrix} \lambda_1 & 0 & \ldots & 0 \\ 0 & \lambda_2 & \ldots & 0 \\ \vdots & \vdots & \ddots & \\ 0 & 0 & & \lambda_n \end{bmatrix},$$

then clearly the power traces (3.1) become

$$I_k = \sum_i^n \lambda_i^k \, ,$$

while the prodeterminants (3.2), c.f. (3.3), become

$$J_k = \sum \lambda_{a_1} \cdot \lambda_{a_2} \cdot \ldots \cdot \lambda_{a_k} \, ,$$

where the last sum runs over all $\binom{n}{k}$ selections of k among the set of the n eigenvalues λ of the endomorphism. These expressions can easily be recognized as the symmetric functions evaluated on the set of eigenvalues of A. Since the invariants I and J corresponds to the two types of basic symmetric functions, relations (3.4) may be understood as a transformation formula for the change of basis of the space of symmetric functions.

Let us recall some basic facts. The space of symmetric functions of degree k is denoted by $\mathcal{S}_k$. The three most frequently used families of basic symmetric functions in variables $x_1, x_2, \ldots, x_n$ are as follows (using notation of [**9**]):

(a) ELEMENTARY SYMMETRIC FUNCTIONS:

$$e_k = \sum x_{i_1} x_{i_2} \ldots x_{i_k}, \quad 1 \le i_1 < i_2 < \ldots < i_k \le n\,.$$

(b) POWER SUMS:

$$p_k = x_1^k + x_2^k + \ldots + x_n^k\,. \tag{B.1}$$

(c) WRONSKI (OR COMPLETE HOMOGENEOUS) FUNCTIONS:

$$h_k = \sum x_{i_1} x_{i_2} \ldots x_{i_k}, \quad 1 \le i_1 \le i_2 \le \ldots \le i_k \le n\,.$$

The number n of variables in these definitions may be unrestricted—one can always set $x_i = 0$ for $i > n$. Here is an example of the basic symmetric functions of degree 1, 2 and 3 in three variables:

$$\begin{aligned}
e_1 = p_1 = h_1 &= x_1 + x_2 + x_3 \\
e_2 &= x_1x_2 + x_2x_3 + x_1x_3 \\
p_2 &= x_1^2 + x_2^2 + x_3^2 \\
h_2 &= x_1^2 + x_2^2 + x_3^2 + x_1x_2 + x_2x_3 + x_1x_3 \\
e_3 &= x_1x_2x_3 \\
p_3 &= x_1^3 + x_2^3 + x_3^3 \\
h_3 &= x_1^3 + x_2^3 + x_3^3 + x_1x_2x_3 + x_1^2x_2 + x_1x_2^2 \\
&\quad + x_1^2x_3 + x_1x_3^2 + x_2^2x_3 + x_2x_3^2
\end{aligned}$$

DEFINITION B.1. By $\lambda \vdash k$ we denote a fact that the set of positive integers $\lambda = (\lambda_1, \lambda_2, \ldots)$ is a **partition** of natural number k, that is:

$$\begin{aligned}
\lambda_1 + \lambda_2 + \ldots + \lambda_l &= k, \\
\lambda_1 \ge \lambda_2 \ge \ldots &\ge \lambda_l\,.
\end{aligned}$$

A partition may alternatively be described by a **partition symbol** $\alpha = [\alpha_1, \alpha_2, \ldots, \alpha_k]$ in which α_i indicates how many times the number i occurs among the elements of set λ. Thus

$$1 \cdot \alpha_1 + 2 \cdot \alpha_2 + \ldots + l \cdot \alpha_l = \alpha\,.$$

By $\alpha \models k$ we shall denote that α is a partition symbol of k.

EXAMPLE B.2. Number 10 admits partition $1 + 1 + 1 + 3 + 4 = 10$, thus we write $(1, 1, 1, 3, 4) \vdash 10$. For the partition symbol we write $[3, 0, 1, 1] \models 10$.

The **fundamental theorem of symmetric functions** states that a basis of the space of homogeneous symmetric functions of degree k may be composed from products of members of any of these families. That is, if $\{b_i\}$ is any of the three families (B.1), then the set of products

$$\{\, b_1^{\alpha_1} \cdot b_2^{\alpha_2} \cdot \cdots \cdot b_k^{\alpha_k} \mid \alpha \models k \,\}$$

forms a basis of the space $\mathcal{S}_k$ of symmetric functions of degree k. The relation between the types of symmetric functions is well known (see, e.g., [**9**]). In particular:

THEOREM B.3. *Any elementary symmetric function can be expressed as a linear combination of products of the power sums:*

$$e_k = \sum_{\alpha \models k} (-1)^{\alpha_2+\alpha_4+\dots} \cdot \frac{C(\alpha)}{k!} \cdot p_1^{\alpha_1} \cdot p_2^{\alpha_2} \cdot \dots \cdot p_k^{\alpha_k} ,$$

where the sum runs over all partitions of k and where $C(\alpha)$, known as Cauchy formula, is given by

$$C(\alpha) = \frac{k!}{\alpha_1!\alpha_2! \cdots \alpha_k! \,\cdot\, 1^{\alpha_1} 2^{\alpha_2} \cdots k^{\alpha_k}} .$$

Interestingly, the relation between Wronski functions h and power sums p utilizes the same Cauchy formula, but without alternating sign:

$$h_k = \sum_{\alpha \models k} \frac{C(\alpha)}{n!} \, p_1^{\alpha_1} \cdot p_2^{\alpha_2} \cdot \dots \cdot p_k^{\alpha_k} .$$

Proofs of this statement involve usually a rather unpleasant juggling with sums, indices, and logarithms [**9**]. In the previous section, an alternative simpler derivation of the Cauchy coefficients is given.

In conclusion, the raising operator and its conjugated version find application to the families of the symmetric polynomials.

Acknowledgment

I am most grateful to Philip Feinsilver for generously sharing his expertise in related areas and for inspiring discussions.

References

[1] V. I. Arnol′d, *The Hamiltonian nature of the Euler equations in the dynamics of a rigid body and of an ideal fluid* (Russian), Uspehi Mat. Nauk **24** (1969), no. 3(147), 225–226. MR277163

[2] D. H. Collingwood and W. M. McGovern, *Nilpotent orbits in semisimple Lie algebras*, Van Nostrand Reinhold Mathematics Series, Van Nostrand Reinhold Co., New York, 1993. MR1251060

[3] P. Cvitanović, *Group theory for Feynman diagrams in non-Abelian gauge theories*, Phys. Rev. D (3) **14** (1976), no. 6, 1536–1553, DOI 10.1103/PhysRevD.14.1536. MR421376

[4] P. Feinsilver, J. Kocik, and R. Schott, *Representations of the Schrödinger algebra and Appell systems*, Fortschr. Phys. **52** (2004), no. 4, 343–359, DOI 10.1002/prop.200310124. MR2045066

[5] A. Garsia and J. B. Remmel, *Breakthroughs in the theory of Macdonald polynomials*, Proc. Natl. Acad. Sci. USA **102** (2005), no. 11, 3891–3894, DOI 10.1073/pnas.0409705102. MR2139721

[6] L. H. Kauffman, *Knots and physics*, Series on Knots and Everything, vol. 1, World Scientific Publishing Co., Inc., River Edge, NJ, 1991, DOI 10.1142/9789812796226. MR1141156

[7] J. Kocik, *Canonical endomorphism field on a Lie algebra*, J. Gen. Lie Theory Appl. **4** (2010), Art. ID G100302, 15, DOI 10.4303/jglta/G100302. MR2795573

[8] J. Kocik and J. Rzewuski, *Structure of matrix manifolds and a particle model*, J. Math. Phys. **37** (1996), no. 2, 1004–1028, DOI 10.1063/1.531423. MR1371054

[9] I. G. Macdonald, *Symmetric functions and Hall polynomials*, 2nd ed., Oxford Classic Texts in the Physical Sciences, The Clarendon Press, Oxford University Press, New York, 2015. With contribution by A. V. Zelevinsky and a foreword by Richard Stanley; Reprint of the 2008 paperback edition [MR1354144]. MR3443860

[10] C. O'Sullivan, *De Moivre and Bell polynomials*, Expo. Math. **40** (2022), no. 4, 870–893, DOI 10.1016/j.exmath.2022.07.006. MR4518634

[11] R. Penrose, *Applications of negative dimensional tensors*, Combinatorial Mathematics and its Applications (Proc. Conf., Oxford, 1969), Academic Press, London-New York, 1971, pp. 221–244. MR281657

Department of Mathematics, Southern Illinois University, Carbondale, IL 62901
Email address: `jkocik@siu.edu`

Contemporary Mathematics
Volume **824**, 2025
https://doi.org/10.1090/conm/824/16452

Topology of vortex reconnection

Louis H. Kauffman

ABSTRACT. This paper studies how knotted vortices undergo reconnection in their process of simplification. We consider the world line of such a knotted vortex and show how the relationships of knot theory and four dimensional topology can be used to give lower bounds on its reconnection number and how to explicitly determine the reconnection number for torus links and positive links.

1. Introduction

This paper discusses knotted vortices and how they tend to degenerate to unknots and unlinks by reconnection processes. We define a reconnection number of knots and links and show how some aspects four dimensional topology allow the computation of the reconnection number. This allows us the see that in some cases of experiments with computer models of vortex knots, the models do find the least number of reconnections to undo the knot. In other cases the physical process is less efficient than the best mathematical reconnection. Our main result is an exact determination of the reconnection number $R(K)$ for K any positive link. In particular, we show:

If K is an unsplittable knot or link with a positive diagram with $c(K)$ crossings and $s(K)$ Seifert circles, then the reconnection number of K is exactly twice the genus of the Seifert spanning surface of K when K is a knot, and for both knots and links we have the formula

$$R(K) = c(K) - s(K) + 1.$$

If $K = T(a, b)$ is a torus knot of type (a, b) then the reconnection number of K is equal to the product $(a - 1)(b - 1)$.

The meanings of the above terms are given in the body of the paper.

This mathematical result about the topology of vortex reconnection is proved herein by using deep results about four-ball surfaces bounding knots and links in three-space. These results are due to Rasmussen [**20**] via his use of Khovanov homology. Results related to those of Rasmussen were originally proved by Kronheimer and Mrowka using gauge theory and so have their roots in relationships of

2020 *Mathematics Subject Classification.* Primary 57K10.

Key words and phrases. Alexander polynomial, Alexander-Conway polynomial, Khovanov homology, knotted vortex, reconnection, reconnection number, knot signature.

topology and physics. Kronheimer and Mrowka proved a conjecture of John Milnor about the torus knots and singularities of algebraic varieties in a space of two complex variables. We return to physics in this paper by applying a result about genera of surfaces in the four-ball to properties of vortices in the four dimensions of spacetime. In this way we also return to the ideas that generated knot theory at the hands of Lord Kelvin and his nineteenth century theory of atoms as vortices in the luminiferous aether.

The interest in these results about reconnection numbers for knots is both mathematical and physical. It is a very nice mathematical question to ask how many oriented saddle point reconnections are needed to unknot a knot. This is nearly as basic a question as the query about the unknotting number of knot – how many crossing switches will unknot the knot. On the physical side, vortices in fluids and superfluid vortices tend to unknot via physical reconnection. Knowing the least possible numbers of reconnections needed to undo a knot makes possible better comparisons with the way such knots behave in experiments.

2. Vortex knot reconnection

We consider knotted vortices. The main result of the paper is an exact determination of the *reconnection number* $R(K)$ for K any *positive* link. The meanings of these terms will be made clear in the body of the paper. Reconnection is a process that happens spontaneously to a one-dimensional vortex, creating a recombination of the vortex lines and a change in topology. The reconnection number is the least number of reconnections needed to transform the knot to an unknotted loop. Figure 1 gives a diagrammatic illustration of reconnection. Figure 2 and Figure 4 are photographs of actual reconnection of vortices in water [**1**, **12**]. The Figure 2 illustrates the experimental work of Kleckner and Irvine [**12**] and shows a water vortex in the form of a trefoil knot undergoing a cascade of reconnections that result in unknotted and unlinked circles in parallel with the diagrammatic illustration in Figure 1. Figure 3 is a drawing of the vortex reconnection that occurs in the photograph in Figure 4. In these two figures we see that a reconnection can,

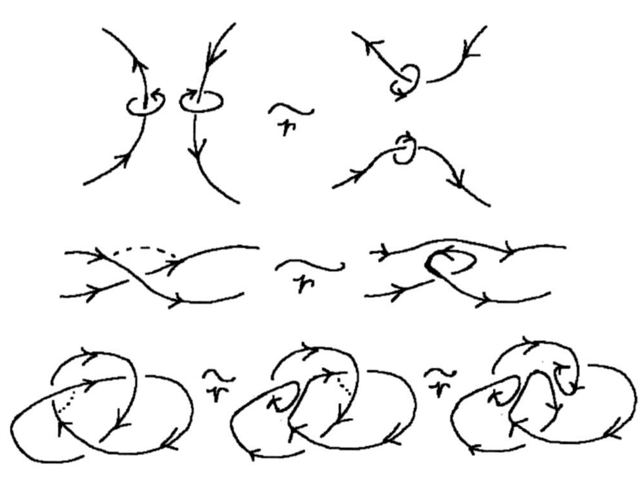

FIGURE 1. Diagrammatic vortex reconnection and topology change

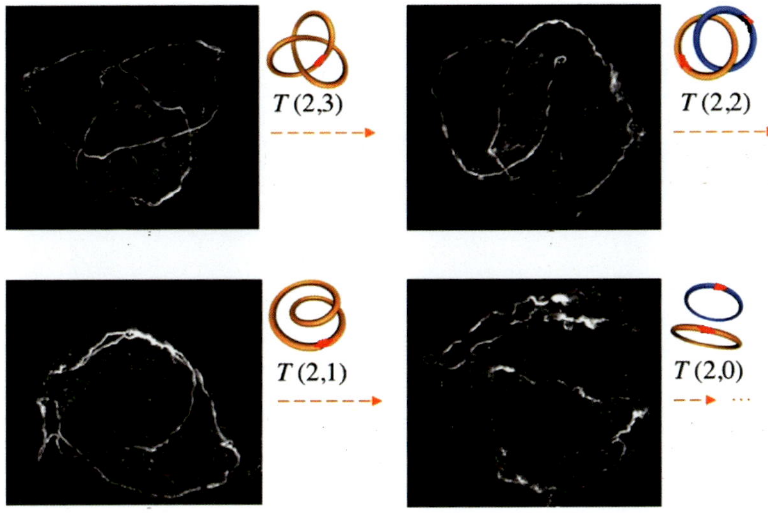

FIGURE 2. Trefoil cascade

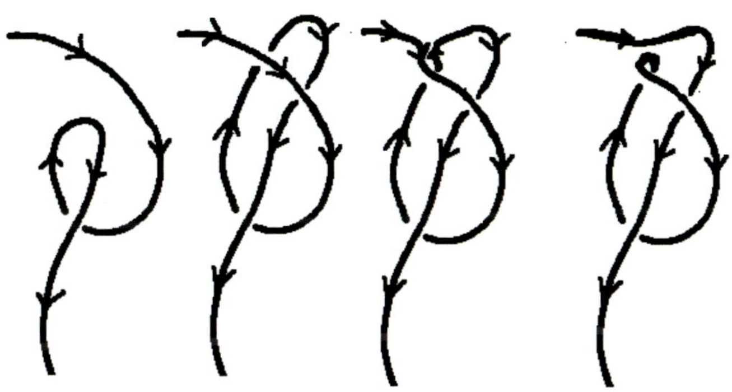

FIGURE 3. Vortex reconnection producing a link

under appropriate circumstances, cause the topology to become more complicated. In this case a loop linking the main vortex is produced by the reconnection. This occurs in a turbine experiment performed by Professor Alexander Alekseenko [**1**]. For other work on reconnection, see [**17–19**].

Lord Kelvin (Sir William Thompson) in the nineteenth century theorized that material atoms were knotted vortices in the luminiferous aether. His theory led him to request the making of tables of knots, and such tables were constructed by Peter Guthrie Tait, T. P. Kirkman, and C. N. Little. These early tables preceded the development of knot theory in terms of topological invariants, and were a strong motivation for their development. The Kelvin theory of knotted vortex atoms did not survive the criticisms of the aether. These criticisms were driven by the negative results of the Michelson-Morley experiment and the advent of Einstein's theory of special relativity. The possibility of knotting at the level of elementary particles has not been ruled out. Knotting of fluid vortices remains of great interest to physicists. For many years the possibility of knotted vortices in three dimensional fluid dynamics was a topic of theoretical interest, but no one had explicitly exhibited

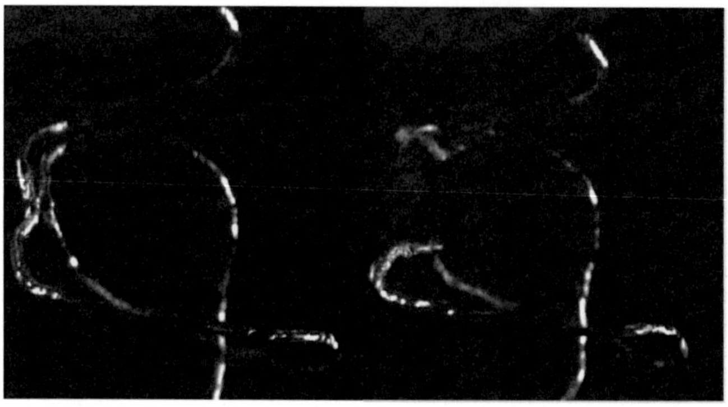

FIGURE 4. Vortex reconnection and topology change

knotted vortices in even such a ubiquitous fluid as water. This situation changed in 2012 with the work of William Irvine and Dustin Kleckner [**12**]. They produced repeatable experiments that yield knotted vortices in water. Their method is a generalization of earlier methods that produce ring vortices in smoke and indeed in water. Their method is to make a knotted template and to pull it quickly through still water. The edge of the template produces a knotted vortex that can be observed by high speed photography. The resulting visual evidence of knotted vortices in a palpable fluid is fascinating and very suggestive of many questions about the behaviour of such structures.

One can simulate knotted vortices and their dynamics using the Gross-Pitaevskii non-linear Schroedinger equation [**13**]. In both experiments with actual fluids and in computer experiements it is seen that the knotted vortices change by a process called *reconnection.* At the diagram level a reconnection is what we have called a re-smoothing in our discussions of knot and link invariants. Two parallel vortex filaments come close to one another. An exchange process occurs that results in a transition to a new topological connection corresponding to a re-smoothing as illustrated in Figure 1. The vortices must be locally spinning (around the one dimensional vortex line) in the same direction for reconnection to occur. This directionality is illustrated in the Figure 1. In this figure the reader will see a directionality along the vortex line and a spin of the vortex indicated by the right hand rule with respect to the vortex line. This same direction of spin is reinforced locally. Note that two parallel vortex lines have opposite directionality. In Figure 4 we illustrate photographs from a turbine experiment performed by Professor Alexander Alekseenko [**1**]. In the first photograph the vortex line is connected. In the seond photograph the reconnection has divided the vortex into two components with one component encircling the other. The topology has made a transition to the linking of two vortex lines.

There are consequences of reconnection for the related combinatorial topology. In particular, if we consider oriented knot and link diagrams under regular isotopy (Reidemeister moves two and three) and also oriented reconnection as described in the last paragraph, then one sees that the *writhe* of the diagram is invariant under these moves. Recall that for a diagram D, the writhe, $wr(D)$ is the sum of the crossing signs in the diagram. In Figure 6 the signs of crossings are indicated, with the diagram K_+ having a sign of $+1$ and the diagram K_- having a sign of

-1. In the Figure 1 we illustrate the result of reconnection near a diagrammatic crossing. The reader will have no difficulty seeing that the writhe is preserved under reconnection. At the bottom of the figure we illustrate two successive reconnections on a trefoil diagram. After the first reconnection, the diagram has transformed into a link of two circles. One more reconnection transforms this link to a connected and unknotted loop.

How many reconnections are needed to unknot a knot, or to unlink a link? This is a natural mathematical question and it is a key physical question about knotted vortices since their observed behaviour is to undergo reconnection and become a collection of unlinked and unknotted circles in three dimensional space. See [**13**] for experimental results about this question. In this section of the paper, we will construct lower bounds on this *reconnection number* by regarding the process of reconnection as the creation of a saddle point in a surface that represents the world line of the evolving vortex. Then the reconnection number can be compared with the genus of a world line surface for the vortex. A nontrivial application of four dimensional topology gives the lower bound.

We say that two oriented knots are *cobordant* if they form the ends of an embedded tube ($S^1 \times I$) in the cartesian product of three dimensional space with a unit interval. One can think of a cobordism as a process that transforms one knot into the other. The process consists in a sequence of births of unknotted circles, deaths of unknotted circles and passage through oriented saddle points. The saddle point passages correspond to local oriented arc reconnections as indicated by the figures. A process that creates a genus zero surface (a topological tube) is called a *concordance*. We can consider surfaces in the four-ball D^4 that bound a knot embedded in the boundary three-sphere S^3. If F is a surface in D^4 that bounds a knot K in S^3, then F can be described by a process of births and deaths of circles and passage through saddle points. For a given process, we can calculate the genus of the surface that is produced. The *four ball genus* $g_4(K)$ is the least genus among all surfaces that bound the knot K in the four-ball. From the point of view of processes, the four-ball genus is the least genus traced by any process that creates a surface bounding the knot. There are general results known about the four-ball genus. For example, it is known [**16**] that $|\sigma(K)|/2 \leq g_4(K)$ where $\sigma(K)$ denotes the classical signature of the knot. This is computed as the signature of $V + V^T$ where V is a Seifert matrix for the knot and will be discussed in Section 3. If K is a link with μ components, then the result generalizes to

$$|\sigma(K)| \leq 2g_4(K) + \mu - 1.$$

Also, if $u(K)$ is the unknotting number of a link K (the least number of crossing switches needed to convert K (from any diagram or in space) to the unlink, then

$$g_4(K) \leq u(K).$$

The relevance of the process view of cobordisms of knots is that the saddle passage is exactly a writhe-preserving oriented reconnection. In studying vortices one is interested in the least number of such reconnections that will transform a knot to a collection of unknotted circles. Let us define $R(K)$ to be this *reconnection number for the knot* K. That is $R(K)$ is the least number of reconnections that will transform K to a collection of unknotted circles. By thinking in terms of the four-ball genus, we see that $g_4(K) \leq R(K)/2$, since pairs of saddle moves contribute to the increment of one handle, whence an addition of one to the genus. Thus we find the result

THEOREM. *Let $\sigma(K)$ denote the signature of an oriented knot K and $R(K)$ the reconnection number of K. Then $|\sigma(K)| \leq R(K)$.*

The background and proof of this result will be discussed below. For example, if K is the trefoil knot, then $|\sigma(K)| = 2$, from which we conclude that it will take at least two reconnections to undo the trefoil, and indeed two reconnections suffice.

Cobordisms as defined below correspond to allowed reconnections of vortices. Writhe is preserved by these reconnections. The deaths and births of unknotted circles are relevant to the cobordism topology but not directly to the vortices. Bounds on the number of saddle points that will make a cobordism from a knot to a collection of unknots can be computed from the topology. We will detail more about the topology and how to use it to study reconnection in the next sections.

As an example, view Figure 8 and Figure 9. Here we see that two reconnections suffice to undo the trefoil knot. Each reconnection is interpreted as an oriented saddle cobordism. Writhe is preserved throughout. The surface produced by the two reconnections has genus equal to one. Thus any single reconnection for the trefoil will have to produce a non-trivial link.

The main mathematical result of this paper is the following Theorem, proved in Section 5.

RECONNECTION THEOREM. *If K is a positive unsplittable link diagram with with $c(K)$ crossings and $s(K)$ Seifert circles, then the reconnection number of K is given by the formula*

$$R(K) = c(K) - s(K) + 1.$$

With this result we give an exact formula for the reconnection numbers for positive links and knots and thus give a strong basis for further experiments with physical vortices and mathematically modeled vortices.

3. Spanning surfaces for knots and links

It is a well-known that every oriented classical (a one dimensional closed curve or curves embedded in three dimensional space) knot or link bounds an embedded orientable surface in three-space. A representative surface of this kind can be obtained by the algorithm due to Seifert (See [**6**–**8**]). We have illustrated Seifert's algorithm for a trefoil diagram in Figure 5. The algorithm proceeds as follows: At each oriented crossing in a given diagram K, smooth that crossing in the oriented manner (reconnecting the arcs locally so that the crossing disappears and the connections respect the orientation). The result of this operation is a collection of oriented simple closed curves in the plane, usually called the *Seifert circles.* To form the *Seifert surface* $F(K)$ for the diagram K, attach disjoint discs to each of the Seifert circles, and connect these discs to one another by local half-twisted bands at the sites of the smoothing of the diagram. This process is indicated in the Figure 5. In that figure we have not completed the illustration of the outer disc.

It is important to observe that we can calculate the genus of the resulting surface quite easily from the combinatorics of the link diagram K and knowledge of its number of components $\mu(K)$.

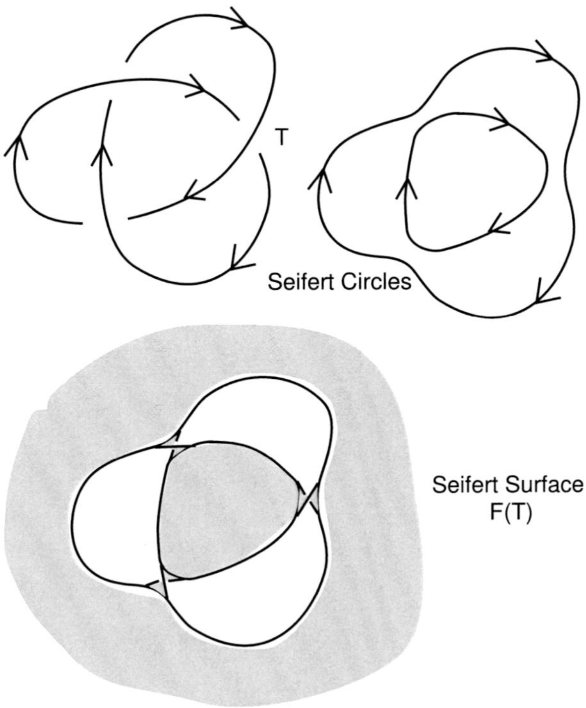

FIGURE 5. Classical Seifert surface

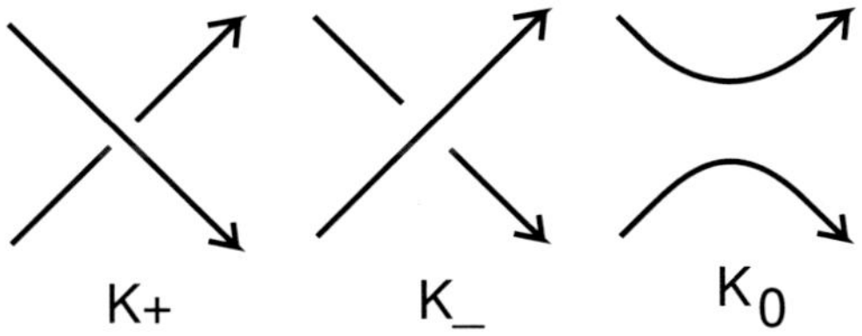

FIGURE 6. Crossing signs in a Skein triple

LEMMA. *Let K be a classical connected link diagram with $c(K)$ crossings and $s(K)$ Seifert circles. Then the genus of the Seifert Surface $F(K)$ is given by the formula*

$$g(F(K)) = (1/2)(c(K) - s(K) + 1 - (\mu(K) - 1)),$$

where $c(K)$ denotes the number of crossings in the diagram K, $s(K)$ denotes the number of Seifert circles for K, and $\mu(K)$ denotes the number of components of the link K.

PROOF. The surface $F(K)$, as described prior to the statement of the Lemma, is the homotopy type of a cell complex consisting of the projected graph of the knot diagram with 2-cells attached to each cycle in the graph that corresponds to a Seifert circle and 2-cells attached to each component of the link. Thus we have that the Euler characteristic of this surface is given by the formula

$$\chi(F(K)) = c - e + s + \mu,$$

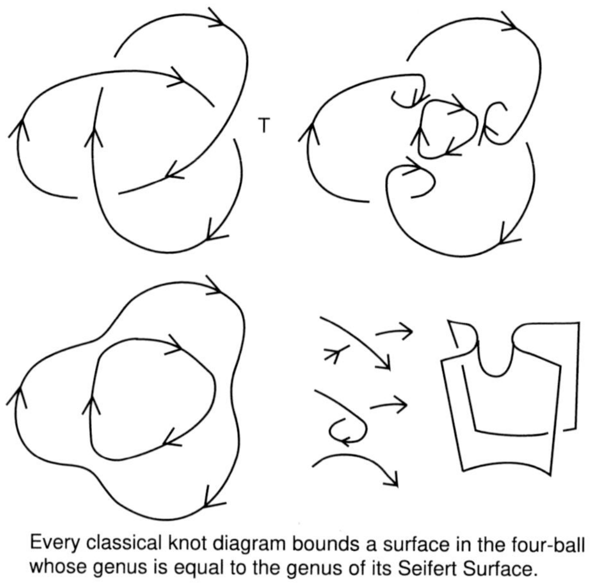

FIGURE 7. Classical cobordism curface

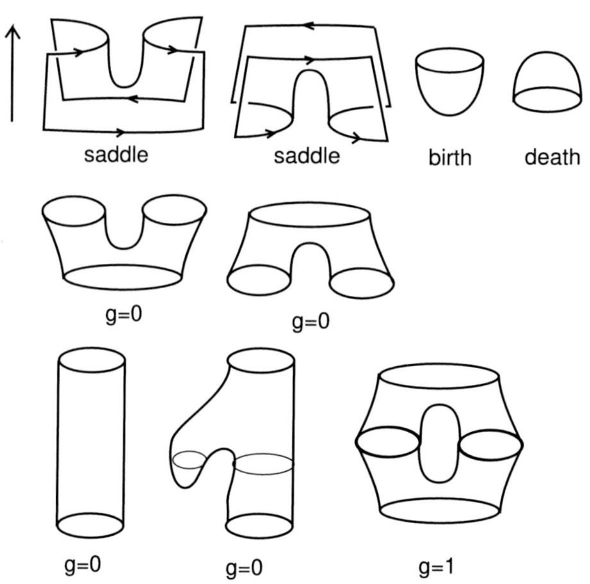

FIGURE 8. Saddles, births and deaths

where c, the number of crossings in the diagram, is the number of zero-cells, e is the number of one-cells (edges) in the projected diagram (from node to node), s is the number of Seifert circles and $\mu = \mu(K)$ as these are in correspondence with the 2-cells. We know that $4c = 2e$ since there are four edges locally incident to each crossing. Thus,

$$\chi(F(K)) = -c + s + \mu.$$

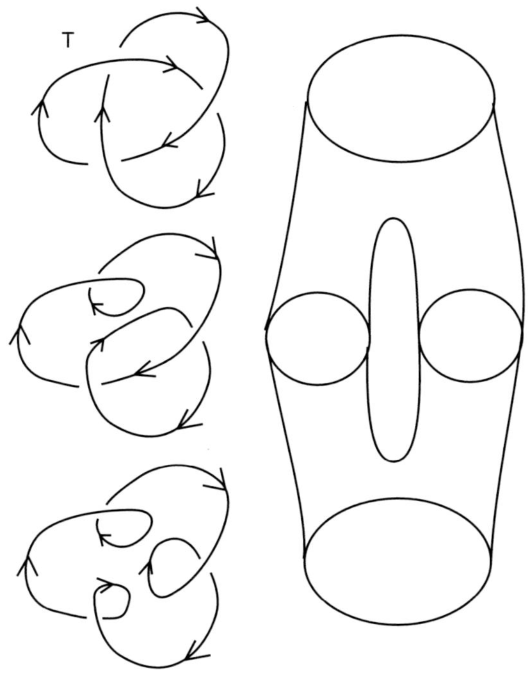

FIGURE 9. A genus one reconnection sequence

Furthermore, we have that $\chi(F(K)) = 2 - 2g(F(K))$, since this surface is orientable. From this it follows that $2 - 2g(F(K)) = -c + s + \mu$, and hence

$$g(F(K)) = (1/2)(c - s + 1 - (\mu - 1)) = (1/2)(c(K) - s(K) + 1 - (\mu(K) - 1)).$$

This completes the proof.

We now observe that *for any classical link K, there is a surface bounding that link in the four-ball that is homeomorphic to the Seifert surface.* One can construct this surface by pushing the Seifert surface into the four-ball keeping it fixed along the boundary. We will give here a different description of this surface as indicated in Figure 7. In that figure we *perform a reconnection at every crossing of the diagram.* The result is a collection of unknotted and unlinked curves. By our interpretation of surfaces in the four-ball obtained by saddle moves (reconnections) and isotopies, we can then bound each of these curves by discs (via deaths of circles) and obtain a surface $S(K)$ embedded in the four-ball with boundary K. As the reader can easily see, the curves produced by the saddle transformations are in one-to-one correspondence with the Seifert circles for K, and it easy to verify that $S(K)$ is homeomorphic with the Seifert surface $F(K)$ with boundary the link K. Thus we know that $g(S(K)) = (1/2)(c(K) - s(K) + 1 - (\mu(K) - 1))$. Note that the genus of the surface is one half the rank of its first homology group,

4. The Seifert pairing and the signature

Let F be a spanning surface in three-space for a knot or link K. We define a linking number measure of the embedding of F via the *Seifert paring* defined as an asymmetric bilinear form

$$\Theta : H_1(F) \times H_1(F) \longrightarrow Z,$$

given by the formula

$$\Theta(x, y) = Lk(x^*, y)$$

where Lk denotes linking number and x^* denotes the result of translating the chain representing x in the first homology group $H_1(F)$ along the positive normal to the surface F so that x^* is supported in the complement $S^3 - K$. The Seifert pairing gives invariant information about the topology of the embedding of the spanning surface for K, and it also gives information about the embedding of K. Below, we list properties of the Seifert pairing. See [**7**] for more information.

(1) Let $\Delta_K(t) = Det(\Theta - t\Theta^T)$ where $\Theta^T(x, y) = \Theta(y, x)$ is the *transpose* of Θ. Then $\Delta_K(t)$ is the *Alexander polynomial* of K.
(2) Let $\sigma(K) = Signature(\Theta + \Theta^T)$. This is the *signature* of the link K. It is the signature of the symmetric bilinear pairing

$$\Theta + \Theta^T : H_1(F) \times H_1(F) \longrightarrow Z.$$

A remarkable property of the signature of a link is that it forms a lower bound for the four-ball genus of that link [**5, 16**]. That is, we have the following inequality:

$$|\sigma(K)| \leq 2g_4(L) + \mu(K) - 1,$$

where $g_4(L)$ denotes the least genus among connected orientable surfaces in the four-ball D^4 that span the link K in the three-sphere S^3, seen as the boundary of the four-ball. Here $\mu(K)$ is the number of components of the link K. Thus, in the case of a knot we have $\mu(K) = 1$ and so

$$|\sigma(K)| \leq 2g_4(L).$$

5. Applications to vortex degeneration

If K is a knot or link representing a vortex that can undergo oriented writhe-preserving reconnection, then each act of reconnection can be interpreted as a saddle-point cobordism. Thus, if the knot undergoes N reconnections to produce a collection of unlinked circles, then we can interpret the cobordism as producing a surface in the 4-ball of genus no more than $N/2$.

The relevance of the process view of cobordisms of knots is that the saddle passage is exactly a writhe-preserving oriented reconnection. In studying vortices one is interested in the least number of such reconnections that will transform a knot or link to a collection of unknotted circles. We define $R(K)$ to be the least number of reconnections that will transform K to a collection of unknotted circles. When K is a link, with more than one component, then we are concerned with the reconnection number when K is unsplittable. A link K is said to be unsplittable if there is no isotopy in three dimensional space that can carry the link into disjoint non-empty parts that are in separate three-balls. Thus we are only interested in links that will require reconnection in order to become separated into a collection of unknotted circles. In speaking of an unsplittable link, we can include the case of knots, since no knot can be split apart into non-empty parts in disjoint three-balls.

Using the four-ball genus, we have:

THEOREM. *Let $g_4(K)$ denote the least genus of four-ball spanning surfaces of an unsplittable link K in the three-sphere. Let $R(K)$ be the reconnection number for K. Then*

$$2g_4(K) + \mu(K) - 1 \leq R(K).$$

For the signature of the link K, we conclude that

$$|\sigma(K)| \leq R(K).$$

PROOF. A generic connected surface that forms a cobordism between an unsplittable link K and an unknotted loop will require at least $\mu(K) - 1$ saddle points (reconnections). Pairs of saddle moves contribute to the increment of one handle, whence an addition of one to the genus. This implies that twice the four-ball genus plus $\mu(K) - 1$ is a lower bound on the reconnection number. Thus $2g_4(K) + \mu(K) - 1 \leq R(K)$. The result about the signature follows from the fact, quoted in the last section, that

$$|\sigma(K)| \leq 2g_4(L) + \mu(K) - 1,$$

where $\mu(K)$ is the number of components of the link K.

REMARK. The key point about this Theorem is that if we know the four-ball genus of a link K, then we have a strong bound on the reconnection number $R(K)$ of K. A case in point is the result of Rasmussen [**20**] (an application of Khovanov Homology. See also [**9**–**11**].) that tells us the following:

DEFINITION. An oriented knot K is said to be *positive* if all its crossings are of positive type. For reference, the crossings in the trefoil knot in Figure 7 are positive.

THEOREM (Rasmussen [**20**]). *Let K be a positive knot, then the four-ball genus of K is equal to the genus of the Seifert surface for K. Thus*

$$g_4(K) = \frac{1}{2}(c(K) - s(K) + 1).$$

From this we obtain the following results about reconnection:

THEOREM. *If K is a positive knot, then the reconnection number of K is bounded below by twice the genus of the Seifert spanning surface of K. That is,*

$$R(K) \geq 2g_4(K) = c(K) - s(K) + 1,$$

where $c(K)$ is the number of crossings of K and $s(K)$ is equal to the number of Seifert circles of K.

PROOF. This result follows from Rasmussen's result [**20**] that $g_4(K)$ is equal to the genus of a Seifert surface for K and our discussion prior to the Theorem.

THEOREM. *If K is a positive knot with a positive diagram with $c(K)$ crossings and $s(K)$ Seifert circles, then the reconnection number of K is exactly twice the genus of the Seifert spanning surface of K. That is, we have the formula*

$$R(K) = c(K) - s(K) + 1.$$

PROOF. Since we know that $R(K) \geq c(K) - s(K) + 1$, it suffices to show how to perform $c(K) - s(K) + 1$ reconnections and obtain an unknotted loop. From our description of the construction of the Seifert surface we showed that one reconnection at each crossing in K produces the set of Seifert circles. We can choose to reconnect at all but $s(K)-1$ crossings so that the result is a single unknotted loop. In order to understand the geometry of this reassembly it is helpful to phrase it more generally. The set of Seifert circles is a special case of an arbitrary collection of disjoint circles in the plane. Given two circles, let a reassembly between them be denoted by a straight line segment drawn from one to the other. Given s circles in the plane, we can ask how many reassembly lines need to be drawn to create a simple closed curve. The answer, of course, is $s-1$. In the general case, we are free to reassemble wherever we please, rather than at specific sites. It is easy to see that for the case of the Seifert circles for any link diagram the reassemblies can be chosen so that the resulting single loop is unknotted. This means that by performing $c(K)-(s(K)-1)$ reconnections on any connected link diagram, we can produce an unknotted curve. Such a reconnection is sufficient, for positive knots, to show that $R(K)=c(K)-s(K)+1$. This completes the proof.

Remark. The above result extends to positive links by using extensions of Rasmussen's work, as we shall see below. The work of Rasmussen for knots gives a combinatorial proof of Milnor's conjecture [**15**] that the four-ball genus of a torus knot of type (a,b) is $(a-1)(b-1)/2$. Milnor's conjecture was first proved by Kronheimer and Mrowka [**14**] by using gauge theory. The construction in the proof of the Theorem above shows that, *for any knot or link K, $R(K)\leq c(K)-s(K)+1$.* In the case of positive links, this upper bound is also a lower bound.

Generalization to links. We now generalize the above results to the case of unsplittable links.

Reconnection Theorem. *If K is a positive unsplittable link diagram with with $c(K)$ crossings and $s(K)$ Seifert circles, then the reconnection number of K is given by the formula*

$$R(K)=c(K)-s(K)+1.$$

Note that this formula generalizes our result for positive knots.

Proof. The result of Rasmussen for knots generalizes to links. This can be seen directly by examining the paper by Rasmussen, and it has been formalized by Cavallo [**4**]. See also [**2, 3**]. We state the result and refer the reader to Cavallo's paper for the details. This is the result: For a positive unsplittable link, $g_4(K)=(c(K)-s(K)+1-(\mu(K)-1))/2$. The reader will note that this is the genus of the Seifert spanning surface for K. Thus this statement is a direct generalization of the result of Rasmussen. Thus we have that

$$2g_4(K)+\mu(K)-1=c(K)-s(K)+1,$$

and we know from our previous Theorem that

$$2g_4(K)+\mu(K)-1\leq R(K).$$

Therefore $c(K)-s(K)+1\leq R(K)$, and by the same constructive method we have already explained, one can perform $c(K)-s(K)+1$ reconnections and transform K to an unknotted loop. Therefore $c(K)-s(K)+1=R(K)$. This completes the proof of the Theorem.

θ	a	b
a	-1	1
b	0	-1

FIGURE 10. Seifert pairing for surface bounding trefoil knot

EXAMPLE 1. If T is the trefoil knot, then $|\sigma(K)| = 2$, from which we conclude that it will take at least two reconnections to undo the trefoil, and indeed two reconnections suffice. The Seifert Pairing for the trefoil knot is illustrated in Figure 10. The matrix for the pairing is

$$\Theta = \begin{pmatrix} -1 & 1 \\ 0 & -1 \end{pmatrix}.$$

Thus

$$\Theta + \Theta^T = \begin{pmatrix} -2 & 1 \\ 1 & -2 \end{pmatrix},$$

from which it is easy to see that $|\sigma(K)| = 2$.

EXAMPLE 2. In this example, we first point out that the switching of a crossing in a knot or link can be accomplished with two saddle moves that create a single addition to the genus in a cobordism surface. See Figure 11 for a detailed illustration of the process that gives this result. In this figure we show how to accomplish this switch by using a curl nearby the crossing. After the process both the crossing and the curl have been switched and the total writhe remains the same as at the beginning. In a knot where there is no nearby curl we can, in principle, obtain one for this process by producing two opposite curls from nothing by using only the second and third Reidemeister moves, as shown in Figure 12. This example shows that if we are looking for the reconnection number of a knot or link we can examine how many crossings it will require to undo the knot. For example in the case of a trefoil knot, it requires one crossing switch to undo the trefoil and so we see that it can be undone by two reconnections as shown in Figure 13. Of course this method of undoing the trefoil by crossing switching is much more complex than our first method shown in Figure 1. Nevertheless, we can conclude that

THEOREM. *If a knot of link K has unknotting number $u(K)$, then $R(K) \leq 2u(K)$.*

Here the unknotting number $u(K)$ is the least number of crossing switches needed to unknot the knot or link K. The unknotting number is an invariant of K, but in general it is very difficult to compute this number.

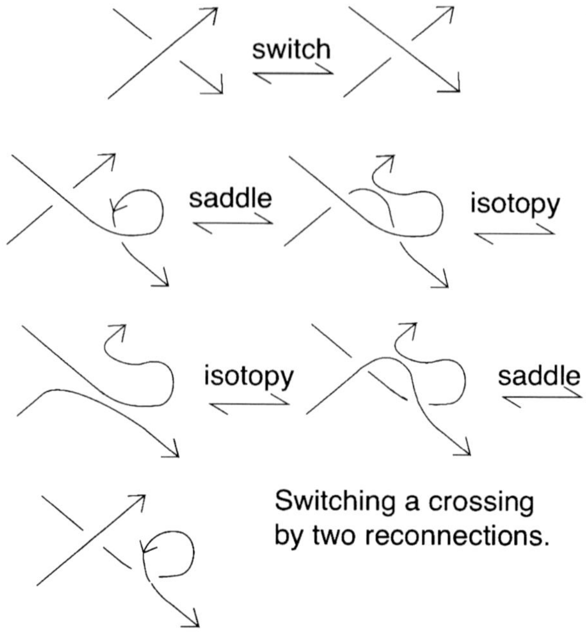

FIGURE 11. Crossing switch in two reconnections

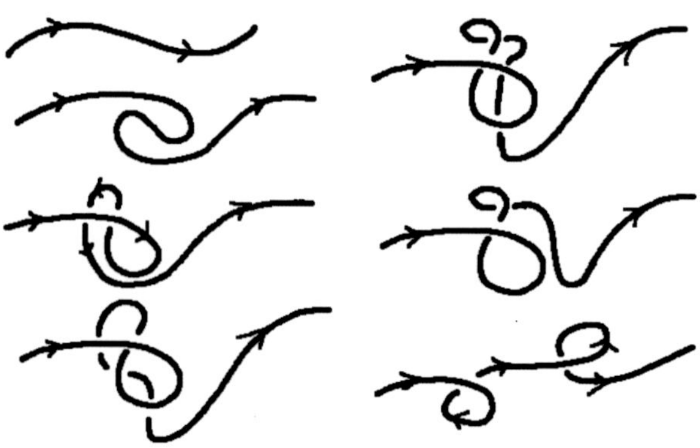

FIGURE 12. Producing opposite curls

Here is another specific example of this type. Examine Figure 14, Figure 15, Figure 16 and Figure 17. In these figures we see that two natural reconnections take the knot $K = 6_2$ to a trefoil knot. Then two more reconnections will undo the trefoil knot. Thus we see that $K = 6_2$ can be undone in 4 reconnections. This is how it did happen using the Gross-Pitaevskii simulation in [**13**]. However, 6_2 can be unknotted with one crossing switch and so we conclude that $R(6_2) = 2$ even though the physical simulation suggested that it might be 4. Figure 17 illustrates graphically how these two reconnections could occur in practice. Certainly we need to assess the probability of such sequences occurring in physical modes.

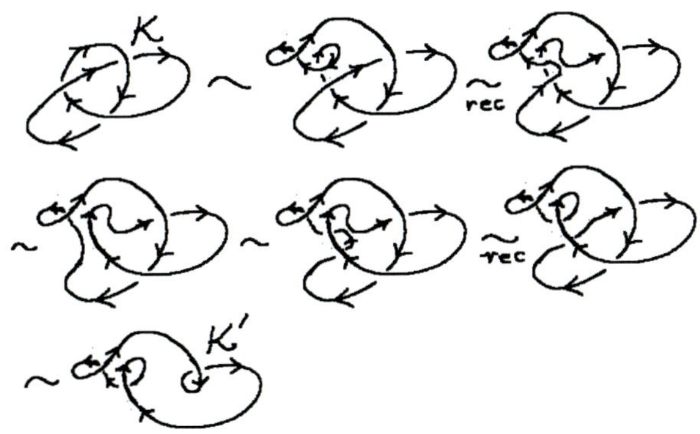

FIGURE 13. Trefoil reconnection by crossing switch

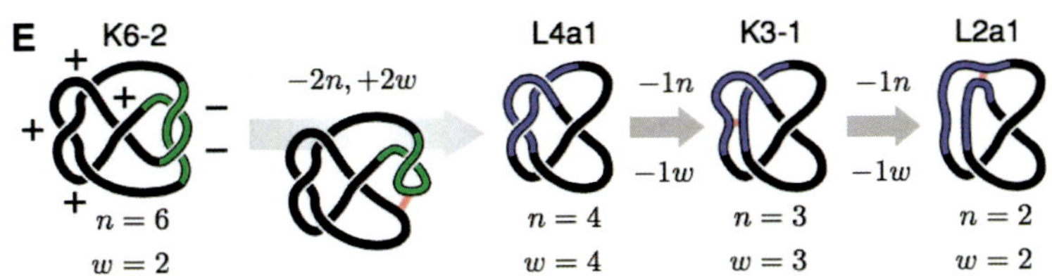

FIGURE 14. Reconnections for 6_2

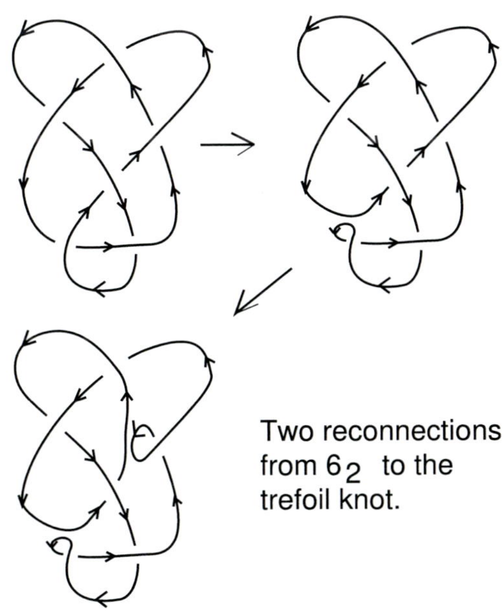

FIGURE 15. 6_2 has two reconnections to the trefoil. Hence four reconnections to an unknot.

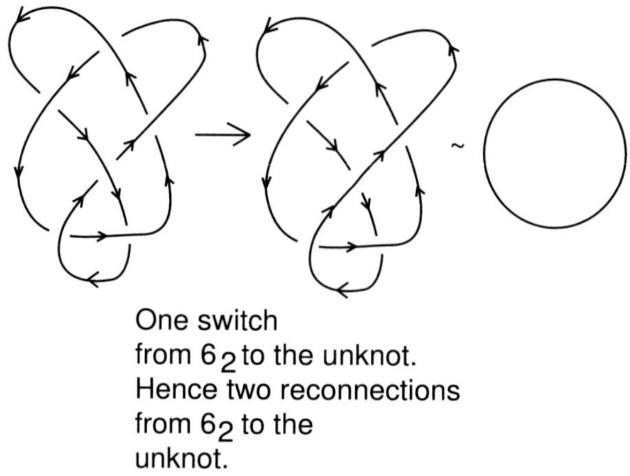

FIGURE 16. By switch 6_2 has two reconnections to the unknot.

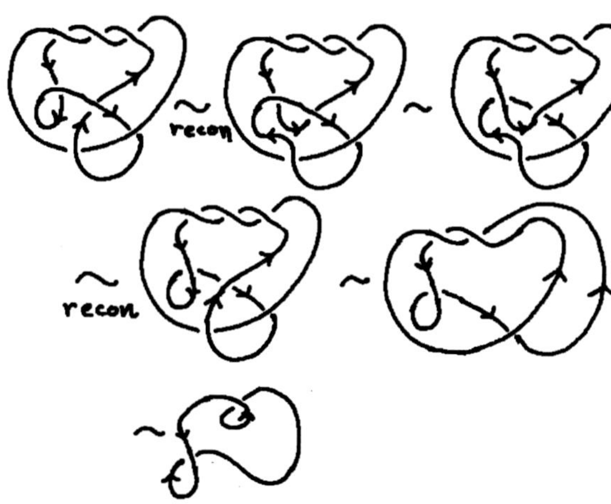

FIGURE 17. A two-fold reconnection for 6_2

EXAMPLE 3. In Figure 18 we show a positive link L of three components with $c(L) = 4, S(L) = 3$. Thus by our Reconnection Theorem we have that $R(L) = c(L) - s(L) + 1 = 4 - 3 + 1 = 2$. The figure illustrates a reconnection sequence to an unknot with three reconnections as predicted by the Theorem. Note that in this case, the Seifert surface has zero genus. In general we have, for positive links, that $R(L) = g_4(L) + \mu(L) - 1$. In this case $g_4(L) = 0$ and $R(L) = \mu(L) - 1 = 3 - 1 = 2$. We also illustrate how a different choice of initial reconnection leads to a cascade into two Hopf links and the need for three reconnections along this pathway.

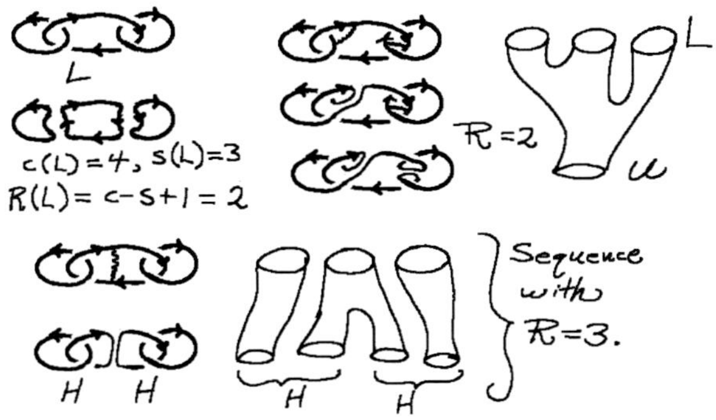

FIGURE 18. A link reconnection

EXAMPLE 4. In Figure 19 we show a link L (distinct from the previous example) that is a torus link of type $(4, 2)$ and another link L' that is, as unoriented links, the mirror image of the first link. The links L and L' are both positive but with different orientations. We find that $c(L) = c(L') = 4$ but $s(L) = 2$ while $s(L') = 4$. As a consequence $R(L) = 3$ while $R(L') = 1$. The reconnections are indicated. This example shows how the reconnection number is sensitive to orientation for links.

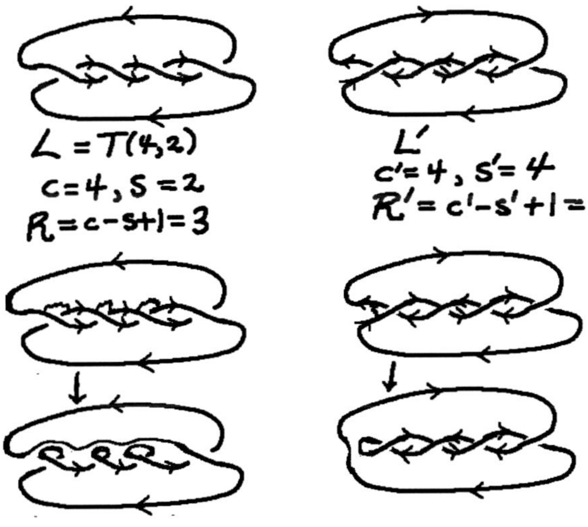

FIGURE 19. Reconnection dependent on orientation

EXAMPLE 5. Consider a torus knot $K = T(p, q)$ of type (p, q) as shown in Figure 20. The figure illustrates a (4,3) torus knot, but the pattern is the same in the general case. One closes q copies of a braid of p strands consisting of positive crossings and representing a cyclic permutation of order p so that the first strand goes over all the other strands to the last place, and each of the other strands goes up one strand-place. All crossings in the generating braid are positive. From this description it is easy to see, as in the figure, that there are q Seifert circles so that $c(K) = (q - 1)p$ and $s(K) = q$. Thus, from the Reconnection Theorem above, we

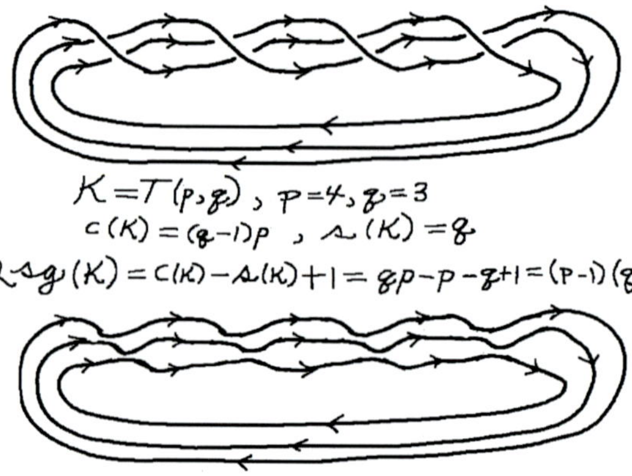

FIGURE 20. Seifert genus of torus knots

have that $R(K) = c(K) - s(K) + 1 = (q-1)p - q + 1 = pq - p - q + 1 = (p-1)(q-1)$. This example alone should become the basis for much experimental research since it is quite possible to prepare torus knots as vortices either with computer models or via the Gross-Pitaevskii equations.

EXAMPLE 6. As we have pointed out in the Reconnection Theorem, we know the reconnection number for any positive link. In Figure 21 we illustrate one more example. In this example, we have check-marked a set of crossings at which no reconnection is performed. There are $s(K) - 1$ checked crossings. By performing reconnections at the remaining $c(K) - s(K) + 1$ crossings we obtain an unknot. This is an illustration of the method of the Reconnection Theorem that produces reconnection sequences and thereby proves that he reconnection number of any positive knot is equal to $c(K) - s(K) + 1$. Since we now know the exact reconnection number of an infinite class of knots, including all the torus knots, this provides a useful ground for physical experimentation.

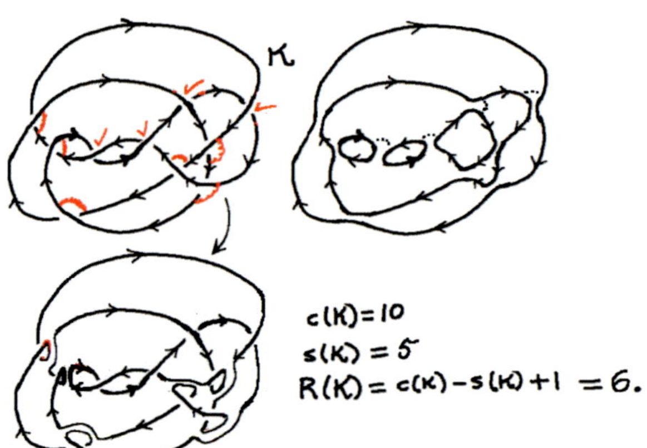

FIGURE 21. A positive knot

EXAMPLE 7. View Figure 22 where we show a knot K (the stevedore's knot) that becomes an unlinked link by a single reconnection. The stevedore's knot is indeed knotted and has unknotting number equal to one. This is the simplest example we know showing that the reconnection number can be strictly less than twice the unknotting number. There are many examples of this phenomenon.

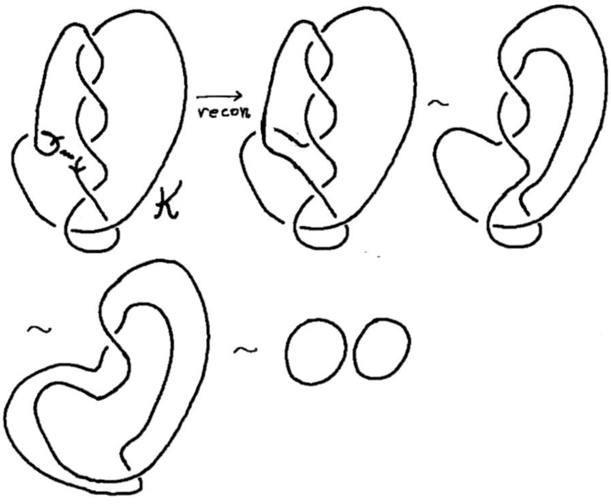

FIGURE 22. One reconnection suffices.

REMARK. Any given knot or link can be seen to give rise to various cascades of reconnection, leading to unknots and unlinks. Our formulas for the reconnection numbers of positive links give a way to navigate these pathways. It should be remarked that in [**17**, **18**] patterns of cascade in relation to Conway and Homflypt polynomials have been noted. Since it is the case [**6**] that the genus of the Seifert spanning surface for positive knots and links is seen in degrees of the Conway polynomial, some of these patterns are related to our exact calculations of the reconnection numbers. Again, more experimental work is needed in this domain.

REMARK. The results in this section all generalize essentially verbatim to virtual knot theory via our results [**10**] generalizing the Rasmussen Invariant for virtual knot theory. We can define the reconnection number $R(K)$ of a virtual knot or link diagram to be the least number of oriented saddle moves that produces an unknot or unlink in the virtual category. It follows from our results that the reconnection number for a positive virtual knot or link is given by the same formula as in the Reconnection Theorem in the classical case. Since virtual knot theory is a way to study knots and links in thickened surfaces, it is possible that experiments can be done with knotted vortices that are restricted to such three dimensional ambient spaces.

References

[1] S. V. Alekseenko, P. A. Kuibina, S. I. Shtorka, S. G. Skripkina, and M. A. Tsoya, *Vortex reconnection in a swirling flow*, JETP Letters **103** (2016), no. 7, 455–459.

[2] T. Abe and K. Tagami, *Characterization of positive links and the s-invariant for links*, Canad. J. Math. **69** (2017), no. 6, 1201–1218, DOI 10.4153/CJM-2016-030-7. MR3715009

[3] A. Beliakova and S. Wehrli, *Categorification of the colored Jones polynomial and Rasmussen invariant of links*, Canad. J. Math. **60** (2008), no. 6, 1240–1266, DOI 10.4153/CJM-2008-053-1. MR2462446

[4] A. Cavallo, *On the slice genus and some concordance invariants of links*, J. Knot Theory Ramifications **24** (2015), no. 4, 1550021, 28, DOI 10.1142/S0218216515500212. MR3346921

[5] L. H. Kauffman and L. R. Taylor, *Signature of links*, Trans. Amer. Math. Soc. **216** (1976), 351–365, DOI 10.2307/1997704. MR388373

[6] L. H. Kauffman, *Formal knot theory*, Mathematical Notes, vol. 30, Princeton University Press, Princeton, NJ, 1983. MR712133

[7] L. H. Kauffman, *On knots*, Annals of Mathematics Studies, vol. 115, Princeton University Press, Princeton, NJ, 1987. MR907872

[8] L. H. Kauffman, *Knots and physics*, 2nd ed., World Scientific Publishing Co., Inc., River Edge, NJ, 1993. MR1306280

[9] L. H. Kauffman, *An introduction to Khovanov homology*, Knot theory and its applications, Contemp. Math., vol. 670, Amer. Math. Soc., Providence, RI, 2016, pp. 105–139, DOI 10.1090/conm/670/13447. MR3568526

[10] H. A. Dye, A. Kaestner, and L. H. Kauffman, *Khovanov homology, Lee homology and a Rasmussen invariant for virtual knots*, J. Knot Theory Ramifications **26** (2017), no. 3, 1741001, 57, DOI 10.1142/S0218216517410012. MR3627701

[11] M. Khovanov, *A categorification of the Jones polynomial*, Duke Math. J. **101** (2000), no. 3, 359–426, DOI 10.1215/S0012-7094-00-10131-7. MR1740682

[12] D. Kleckner and W. T. M. Irvine, *Creation and dynamics of knotted vortices*, Nature Physics **9** (2013), 253–258.

[13] D. Kleckner, L. H. Kauffman, and W. T. M. Irvine, *How superfluid vortex knots untie*, Nature Physics **12** (2016), 650–655.

[14] P. B. Kronheimer and T. S. Mrowka, *Gauge theory for embedded surfaces. I*, Topology **32** (1993), no. 4, 773–826, DOI 10.1016/0040-9383(93)90051-V. MR1241873

[15] J. Milnor, *Singular points of complex hypersurfaces*, Annals of Mathematics Studies, No. 61, Princeton University Press, Princeton, NJ; University of Tokyo Press, Tokyo, 1968. MR239612

[16] K. Murasugi, *On a certain numerical invariant of link types*, Trans. Amer. Math. Soc. **117** (1965), 387–422, DOI 10.2307/1994215. MR171275

[17] S. Zuccher and R. L. Ricca, *Creation of quantum knots and links driven by minimal surfaces*, J. Fluid Mech. **942** (2022), Paper No. A8, 22, DOI 10.1017/jfm.2022.362. MR4421803

[18] R. Ricca and X. LIu, *Knots cascade detected by a monotonically decreasing sequence of values*, Nature - Scientific Reports, April 2016, 6:24118, DOI: 10.1038/srep24118.

[19] D. W. L. Sumners, I. I. Cruz-White, and R. L. Ricca, *Zero helicity of Seifert framed defects*, J. Phys. A **54** (2021), no. 29, Paper No. 295203, 13. MR4282966

[20] J. Rasmussen, *Khovanov homology and the slice genus*, Invent. Math. **182** (2010), no. 2, 419–447, DOI 10.1007/s00222-010-0275-6. MR2729272

Department of Mathematics, University of Illinois at Chicago, Chicago, Illinois 60607-7045

Email address: `kauffman@uic.edu`

Contemporary Mathematics
Volume **824**, 2025
https://doi.org/10.1090/conm/824/16454

Sławianowski's theory and topological invariants of manifolds

Hanna Makaruk and Robert Owczarek

Dedicated to the memory of our dear friend Professor Zbigniew Oziewicz

ABSTRACT. Inspired by Witten's construction of topological invariants of manifolds using Chern–Simons theory and suggesting independence of a quantum field theory from a metric on the manifold leads to topological invariants, we propose to consider a topological invariant, in principle in any dimension (Witten's approach worked for $3D$ manifolds only) using the action and Lagrangian built by Sławianowski in his papers from 1980's and 1990's.

1. Introduction

In 2020 our friend and mentor, Professor Oziewicz, Zbyszek, has passed. Shortly before that sad event, he talked with us on the phone and said that it is time for him to join "Jaś", by which he meant Professor Jan Jerzy Sławianowski, another our friend and mentor, who unfortunately passed just a year earlier, in 2019. Therefore, we feel memorizing Professor Oziewicz by a paper connected with scientific legacy of Professor Sławianowski is very appropriate. This way we also memorialize Professor Sławianowski and his scientific research.

The Sławianowski's theory mentioned in the title was introduced in 1985 [**1**] (see also the further developments in [**2**], and [**3**], and the recent one in [**4**]. His goal was to study an alternative formulation of general relativity. A distinguished role is played in the theory by manifolds of Lie groups, especially of semi-simple Lie groups, so the theory might miss this particular goal, since only very seldom Lie groups are solutions of Einstein equations.

2. Witten's approach to topological invariants

In his famous paper "Quantum Field Theory and the Jones Polynomial" [**5**] Ed Witten discussed not only how to describe Jones polynomial in terms of a 3D quantum field theory, but also how to define topological invariants of 3D (and other) manifolds. He formulated his philosophy as follows: "We want to formulate a generally covariant theory (in which all observables will be topological invariants), and to this aim we want to pick a Lagrangian which does not require any choice of metric." In his approach in the case of 3D manifolds this Lagrangian turned out to be the Chern–Simons Lagrangian of a gauge field.

2020 *Mathematics Subject Classification.* Primary 83D05, 83C99, 53C15, 57R22.
Key words and phrases. Differential geometry, topological invariants.

2.1. Important example: Chern–Simons theory. The gauge field is a connection on a principal G-bundle, where G is a (semi-simple compact) Lie group, is described as a 1-form with values in the Lie algebra $\mathfrak{G}$ of G,

$$A = A_i^a(x)e_a \otimes dx^i \in \mathfrak{G} \otimes C^\infty(T^*M), \tag{2.1}$$

where

$$\{e_a\}_{a=1}^{dimG} \in \mathfrak{G} \tag{2.2}$$

is a basis of the Lie algebra $\mathfrak{G}$ and

$$\{dx^i\}_{i=1}^{dimG} \in C^\infty(T^*M) \tag{2.3}$$

is the basis of 1-forms on M, in an arbitrary coordinate system x^i in M. The action (called Lagrangian by Witten) for the connection field is given by

$$S(A) = \frac{k}{4\pi}\int_M Tr(A \wedge dA + \frac{2}{3}A \wedge A \wedge A). \tag{2.4}$$

In fact, strictly speaking the notation Tr is incorrect, since taking a trace applies to endomorphisms, while here Tr is a Cartan-Killing metric, which in case of compact Lie algebras is proportional to Kronecker's delta. This action does not use any metric on the manifold M, just the smooth structure of M , as is required for construction of topological invariants.

In order to get a topological invariant of the manifold M we need to integrate over all connections A, which symbolically can be written as follows

$$\int DAexp(iS(A)) = \int DAexp\Big(i\frac{k}{4\pi}\int_M Tr\big(A \wedge dA + \frac{2}{3}A \wedge A \wedge A\big)\Big). \tag{2.5}$$

In the paper, more important was definition of the Jones polynomial as an invariant of a link in a $3D$ manifold, which resulted from a different integral

$$\int DAexp(iS(A))\prod_{i=1}^{r} W_{R_i}(C_i) \tag{2.6}$$

for a link with r components C_i where R_i denote representations of the gauge group G assigned to components C_i of the link, and $W_{R_i}(C_i)$ are Wilson loops defined as

$$W_{R_i}(C_i) = Tr_{R_i}P\oint_{C_i} A_i dx^i, \tag{2.7}$$

where P stands for the choice of orientation of C_i.

With an excellent set of tricks, Witten shows that the latter integral in the case of $M = S^3$, $G = SU(2)$, and R_i all being the fundamental representation of $G = SU(2)$, is practically the same as the Jones polynomial. Witten also analyzed the topological invariant

$$Z = \int DAexp\Big(i\frac{k}{4\pi}\int_M Tr\big(A \wedge dA + \frac{2}{3}A \wedge A \wedge A\big)\Big) \tag{2.8}$$

and connected it to several known topological invariants of manifolds, and some quantities connected to the gauge group. For example, for $M = S^2 \times S^1$, $Z = 1$. For $M = S^3$, $G = SU(2)$,

$$Z = \sqrt{\frac{2}{k+2}} sin\Big(\frac{\pi}{k+2}\Big). \tag{2.9}$$

In the analysis of the topological invariant, a crucial role play the critical points of the action, which happen to be flat connections, for which

$$F(A) = dA + A \wedge A = 0. \tag{2.10}$$

3. Sławianowski's theory

The remarkable feature of Sławianowski's theory [**1**], [**2**], [**3**], [**4**] is that it is defined on a manifold, using only its smooth structure, without introducing any metric on the manifold M, which, in contrast to the Chern–Simons theory, may be in principle of any dimension n.

Instead of a gauge field, a reper field and a dual coreper field are considered, which are sections of the principal $GL(n, \mathbb{R})$ frame and coframe bundle, respectively. Namely, let $e_A, A = 1, 2, \ldots, n$, a reper field, consisting of n linearly independent at each point vector fields of tangent vectors to the manifold M, so in arbitrary coordinates on the manifold M, $x^\mu, \mu = 1, 2, \ldots, n$ they can be expressed as $e_A = e^\mu_A \frac{\partial}{\partial x^\mu}$, while the coreper field is an n-tuple of 1-forms, $e^A, A = 1, 2, \ldots, n$, in coordinates $e^A = e^A_\mu dx^\mu$, which is dual to the reper in the sense that natural duality between vectors and convectors gives $< e^A, e_B >= \delta^A_B$, for each $A, B = 1, 2, \ldots, n$.

The action of the group $GL(n, \mathbb{R})$ on the frame bundle is given by right multiplication by the matrices,

$$e'_A = e_B M^B_A, \tag{3.1}$$

where $M^B_A \in GL(n, \mathbb{R})$, and corresponding action on the coframe bundle.

Of course, the rigorous approach is that reper and coreper fields are sections of the frame and coframe bundle, respectively. In order for this approach to topology of the manifold M to work, the sections should be global and thus the manifold M should be parallelizable. In different dimensions, this may mean more or less stringent condition on the manifold. For example, in dimension three, the orientability of the manifold is enough to be parallelizable, so it is not a very strict constraint. A family of covariant tensors of order two is defined using just the dual fields e_A and e^A

$$\gamma_{\mu\nu} = aS^\sigma_{\mu\rho}S^\rho_{\nu\sigma} + bS^\rho_{\mu\rho}S^\sigma_{\nu\sigma}, \tag{3.2}$$

where a, b are some parameters, and the tensor $S^\alpha_{\beta\gamma}$ (the torsion tensor for the teleparallelism connection, which is flat) is defined by

$$S^\alpha_{\beta\gamma} = e^\alpha_A e^A_{[\beta,\gamma]}. \tag{3.3}$$

Everywhere, the summation convention over repeating indices at lower and upper level is assumed. This is not a metric tensor, though it has some properties like one. It is a second order covariant symmetric tensor, but it is built entirely using just smooth structure of the manifold M, not imposed on the manifold. In particular, it has not fixed signature.

The action of the theory is defined as

$$S = \int_M \sqrt{det\gamma_{\mu\nu}} dx^1 dx^2 \cdots dx^n \tag{3.4}$$

so it is a "volume-like" quantity, just using the tensor built of the reper field rather than the metric. Lets us stress again, **nowhere in the construction a metric on the manifold M was used.**

It is important to state here that the equations of the motion, obtained by variational principle from the action above, which are equations for the reper field e_A, are difficult to solve in general. However, there are obvious solutions in the form of Lie algebras, in fact semi-simple Lie algebras [**2**] (the resulting Cartan-Killing metric, replacing the tensor γ is built of structure constants of the Lie algebra, and thus is also constant, but must be non-singular, thus the semi-simplicity condition). These are important in further discussion.

4. Topological invariant in Sławianowski's theory

Up to now, we related the assumption on "generally covariant" theories formulated by Witten, reminded his key example of such theory, the Chern–Simons theory, and how it leads to topological invariants of the manifolds and knots and links in it, when the manifold is three-dimensional. We then showed that the Sławianowski's theory is also an example of "generally covariant" theory in the sense of Witten, since its Lagrangian and its action is built using only smooth structure of the manifold, though in contrast to Witten's case, this is a manifold of arbitrary dimension, in principle, for which the construction works (perhaps, it should be parallelizable). Now it is time to go one step further and define the topological invariant coming from this theory. By analogy with Witten's approach, we define it formally as a path integral, a kind of quantum field theoretic partition function

$$Z = \int De\, exp(i\frac{k}{4\pi}S(e)) = \int De\, exp\Big(i\frac{k}{4\pi}\int_M \sqrt{det\gamma_{\mu\nu}} dx^1 dx^2 \cdots dx^n\Big). \tag{4.1}$$

The integral over the reper fields is as undefined as it was in the case of the Witten's path integral for the Chern–Simons theory, or mostly any other quantum field theory. We do not pretend to contribute in any way in solving this conundrum, but rather focus on the topological invariant that we possibly defined this way, inherent in the Sławianowski's theory.

We can also consider a normalized version of the partition function, which means "divide" by $\int De$ so that in case the action is zero, the result is the partition function is equal to unity, a kind of normalization.

It is not easy to figure out what are the invariants, or if they are related to known invariants. However, the critical points of the action may be important in the analysis. Interesting critical points are given by repers related to semi-simple

Lie algebras. Invariants can be also given interpretation in case of general Lie group manifolds (not only semi-simple).

- For Lie group manifolds, the tensor γ becomes the Cartan-Killing metric, which is degenerate for non-semi-simple groups, so that the "volume" integral is equal to zero.
- For semi-simple groups the metric is (up to a numerical factor) the non-degenerate Cartan-Killing metric and thus the "volume" integral becomes the volume of the group
- For compact semi-simple groups, the metric is definite, and the volume is finite (e.g., $SU(2)$ is the S^3 and the volume is the volume of a $3D$ sphere), but for non-compact groups the metric is not definite, and the volume is in general infinite

5. Summary

- Witten proposed to define topological invariants of manifolds by introducing actions depending only in smooth structures of the manifold and not on any metric that might be introduced there, and integrating over all such fields
- The well-known example is Chern–Simons theory on $3D$ manifolds, analyzed by Witten
- Sławianowski's theory is defined without using any metric, just frame and coframe fields
- It is easy to define a topological invariant á la Witten, using a (not strictly defined) integral over the reper fields
- Somewhat less easy is to interpret the invariant, so it is left to enjoy to our colleagues, mathematicians
- The definition works for parallelizable manifolds that are compact (otherwise invariants become non-interesting)
- It is difficult to go beyond manifolds that are Lie Group manifolds, at least to calculate the invariant

6. Outlook, future work

It would be desirable to calculate the value of the invariant for some manifolds that are not manifolds of Lie groups. Of course, it would be even better to connect it in general case to some known topological invariants, or prove that this invariant is new. Another interesting line of research is based on observation that there exist other classical field theories that are not introducing any metric. One such theory is affine formulation of general relativity by Kijowski [**6**], [**7**], [**8**], [**9**]. Moreover, there is an affine formulation of Sławianowski's theory, which we defined in [**10**], inspired by Kijowski's theory that may be an approach where the topological invariants defined in this paper are redefined using a connection on the frame bindle rather than frame bundle and its sections, which may proved easier to handle.

An intriguing possibility is to connect this approach with the one of Connes and Moscovici, in which a Hopf algebra structure is associated to the frame bundle of a manifold [**11**], [**12**], [**13**] and which was developed to study topological invariants associated to foliations, understood as some noncommutative spaces.

Acknowledgments

The authors appreciate discussions during the presentation of this material. They also appreciate literature hints from the anonymous referee.

References

[1] J. J. Sławianowski, *Field of linear frames as a fundamental self-interacting system*, Rep. Math. Phys. **22** (1985), no. 3, 323–371, DOI 10.1016/0034-4877(85)90028-X. MR872646

[2] J. J. Sławianowski, *Lie-algebraic solutions of affinely-invariant equations for the field of linear frames*, Rep. Math. Phys. **23** (1986), no. 2, 177–197, DOI 10.1016/0034-4877(86)90020-0. MR894911

[3] J. J. Sławianowski, $\mathrm{GL}(n, \mathbf{R})$ *as a candidate for fundamental symmetry in field theory*, Nuovo Cimento B (11) **106** (1991), no. 6, 645–668, DOI 10.1007/BF02813232. MR1118806

[4] J. J. Sławianowski, V. Kovalchuk, B. Gołubowska, A. Martens, and E. E. Rożko, *Space-time as a structured relativistic continuum*, Math. Methods Appl. Sci. **41** (2018), no. 14, 5404–5422, DOI 10.1002/mma.5087. MR3845919

[5] E. Witten, *Quantum field theory and the Jones polynomial*, Comm. Math. Phys. **121** (1989), no. 3, 351–399. MR990772

[6] J. Kijowski, *On a new variational principle in general relativity and the energy of the gravitational field*, Gen. Relativity Gravitation **9** (1978), no. 10, 857–877, DOI 10.1007/BF00759646. MR512794

[7] M. Ferraris and J. Kijowski, *On the equivalence of the relativistic theories of gravitation*, Gen. Relativity Gravitation **14** (1982), no. 2, 165–180, DOI 10.1007/BF00756921. MR649693

[8] M. Ferraris and J. Kijowski, *General relativity is a gauge type theory*, Lett. Math. Phys. **5** (1981), no. 2, 127–135, DOI 10.1007/BF00403241. MR612420

[9] J. Kijowski and R. Werpachowski, *Universality of affine formulation in general relativity*, Rep. Math. Phys. **59** (2007), no. 1, 1–31, DOI 10.1016/S0034-4877(07)80001-2. MR2308631

[10] H. Makaruk and R. Owczarek, $\mathrm{GL}(n, \mathbf{R})$*-invariant tetrad theory and its relation to the purely affine formulation of general relativity*, Proceedings of the XXVII Symposium on Mathematical Physics (Toruń, 1994), Rep. Math. Phys. **36** (1995), no. 2-3, 355–363, DOI 10.1016/0034-4877(96)83631-7. MR1382924

[11] A. Connes and H. Moscovici, *Hopf algebras, cyclic cohomology and the transverse index theorem*, Comm. Math. Phys. **198** (1998), no. 1, 199–246, DOI 10.1007/s002200050477. MR1657389

[12] H. Moscovici and B. Rangipour, *Hopf algebras of primitive Lie pseudogroups and Hopf cyclic cohomology*, Adv. Math. **220** (2009), no. 3, 706–790, DOI 10.1016/j.aim.2008.09.017. MR2483227

[13] H. Moscovici, *Geometric construction of Hopf cyclic characteristic classes*, Adv. Math. **274** (2015), 651–680, DOI 10.1016/j.aim.2015.02.003. MR3318164

P-2, Los Alamos National Laboratory, Los Alamos, New Mexico 87545
Email address: hanna_m@lanl.gov

Department of Mathematics, University of New Mexico, Los Alamos, New Mexico 87544
Email address: rowczare@unm.edu

Contemporary Mathematics
Volume **824**, 2025
https://doi.org/10.1090/conm/824/16447

A counterexample to the generalisation of Witten's conjecture

Rhea Palak Bakshi

Dedicated to the memory of Zbigniew Oziewicz

Abstract. This note provides a counterexample to a conjecture by Marché about the structure of the Kauffman bracket skein module for closed compact oriented 3-manifolds over the ring of Laurent polynomials.

1. Introduction

Classical knot and link theory is the study of embeddings of circles in the 3-sphere up to ambient isotopy. Much of this theory concerns itself with determining which knots and links are equivalent and which are not. Over the years several link invariants have been discovered, which have been reasonably successful in telling knots and links apart from one another. Link polynomials, such as the Alexander-Conway polynomial [**12**], Jones polynomial [**27**], HOMFLYPT polynomial [**16**, **45**], Kauffman 2-variable polynomial [**29**], the Kauffman bracket polynomial [**31**], and Dubrovnik polynomial [**29**], form an important class of these invariants. What these polynomials have in common is that their definitions are built upon skein relations. A skein relation is a linear relation between two or more link diagrams that are exactly the same everywhere except in the neighbourhood of a crossing where they differ. Skein modules arise naturally when one tries to find analogues of all these polynomial link invariants for links in any arbitrary 3-manifold.

Given an oriented[1] 3-manifold M and a commutative ring with unity, there are several skein modules that can be associated to M, each of which tries to capture some information about the knot theory that M admits. One reason for this plethora of definitions of skein modules is that we can consider links in M to be either oriented or unoriented, framed or unframed, or up to isotopy, homotopy, or homology, for example. What all these definitions have in common, however, is that they generalize the skein theory of all the various polynomial link invariants for links

2020 *Mathematics Subject Classification.* Primary 57K31; Secondary 57K10.

Key words and phrases. Knot, 3-manifold, link invariant, 3-manifold invariant, Kauffman bracket skein module.

The author was supported by Dr. Max Rössler, the Walter Haefner Foundation, and the ETH Zürich Foundation.

[1] In order to replace S^3 with an arbitrary 3-manifold M we consider links which are identical outside a ball B^3 in M and which appear inside B^3 as various prescribed tangles. To be able to distinguish one link from another locally inside B^3 we usually assume that M is oriented.

in S^3 to arbitrary 3-manifolds. Thus, the theory of skein modules, as introduced by Przytycki in [**39**] (and independently by Turaev in [**46**]), can be seen as the natural extension of the Alexander-Conway, HOMFLYPT, Kauffman bracket, and Jones polynomial link invariants in S^3 to links in arbitrary 3-manifolds. In fact, we allow skein relations involving arbitrary linear combinations of tangles in an oriented 3-manifold. For a brief survey of skein modules see [**3**].

The skein module based on the Kauffman bracket polynomial is by far the most extensively studied skein module of all. It has strong ties with algebraic and hyperbolic geometry via $SL(2,\mathbb{C})$ character varieties [**8**], quantum Teichmüller spaces [**5, 11, 28, 33**], and quantum cluster algebras [**38**]. It is also instrumental in the study of the AJ conjecture, which relates the coloured Jones polynomial and the A-polynomial of a knot [**17, 19, 32**]. Moreover, it is used in the combinatorial construction [**4, 35**] of the Witten-Reshetikhin-Turaev Topological Quantum Field Theories. Due to its fundamental role in this construction and its intricate ties to knot theory, geometric topology, and hyperbolic geometry, the Kauffman bracket skein module has become pivotal in the study of quantum topology.

Definition 1.1. Let M be an oriented 3-manifold, $\mathcal{L}^{fr}$ the set of ambient isotopy classes of unoriented framed links (including the empty link $\varnothing$) in M, R a commutative ring with unity, and $A \in R$ a fixed invertible element. In addition, let $R\mathcal{L}^{fr}$ be the free R-module generated by $\mathcal{L}^{fr}$ and $S_{2,\infty}^{sub}$ the submodule of $R\mathcal{L}^{fr}$ generated by all (local) skein expressions of the form:

(i) $L_+ - AL_0 - A^{-1}L_\infty$ and
(ii) $L \sqcup \bigcirc + (A^2 + A^{-2})L$,

where $\bigcirc$ denotes the trivial framed knot and the *skein triple* (L_+, L_0, L_∞) denotes three framed links in M, which are identical except in a small 3-ball in M where they differ as shown in Figure 1.1.

The Kauffman bracket skein module (KBSM) of M is defined as the quotient

$$\mathcal{S}_{2,\infty}(M; R, A) = R\mathcal{L}^{fr}/S_{2,\infty}^{sub}.$$

For simplicity, we use the notation $\mathcal{S}_{2,\infty}(M)$ when $R = \mathbb{Z}[A^{\pm 1}]$.

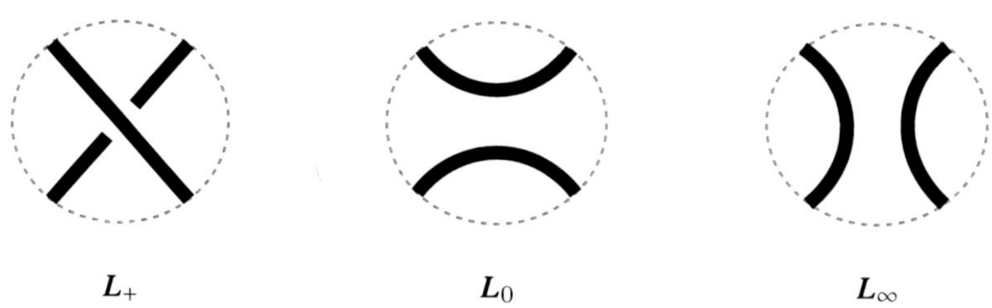

Figure 1.1. Skein triple for the Kauffman bracket skein module.

2. Marché's conjecture

For a long time the structure of the KBSM of 3-manifolds was elusive and there are few examples for which the structure is completely known. Computations are usually hard and have been carried out over various commutative rings. Over the polynomial ring $\mathbb{Z}[A^{\pm 1}]$, the KBSM has been computed for S^3, I-bundles over surfaces [**39, 42**], lens spaces $L(p, q)$ when $p \geq 1$ [**24**], $S^1 \times S^2$ [**25**], the Whitehead manifold [**26**], the product of the pair of pants with S^1 [**13**], certain classes of prism manifolds [**36**], twist knot exteriors [**6**], the exteriors of torus knots of type $(2, 2p + 1)$ [**7**], and 3-manifolds obtained by integral surgery on the right-handed trefoil knot [**9**]. Over $\mathbb{C}[A^{\pm 1}]$, the KBSM of the exteriors of 2-bridge knots and links was computed in [**32, 34**]. When the ring $\mathbb{Z}[A^{\pm 1}]$ is localised by inverting all cyclotomic polynomials, the formula for the KBSM for the quaternionic manifold is known [**20**], while over $\mathbb{Q}(A)$, the field of rational functions in the variable A, the KBSMs of T^3 [**10, 21**], trivial S^1-bundles over oriented surfaces with genus $g \geq 2$ [**15**], and some infinite families of hyperbolic 3-manifolds [**14**] are known. Sometime in 2014 or 2015, Witten conjectured that the Kauffman bracket skein module over $\mathbb{Q}(A)$ for any closed 3- manifold is always finite dimensional. This conjecture has never been mentioned in any published work by Witten himself. The first written documentation of it appears in [**10**]. Section 8 in [**22**] also briefly discusses this conjecture. Witten's conjecture comes as a surprise in the theory of skein modules since $\mathcal{S}_{2,\infty}(M, \mathbb{C}, -1)$ is related to the $SL(2, \mathbb{C})$ character variety of M, which is usually infinite dimensional.

In [**43**], Przytycki showed that if the skein modules $\mathcal{S}_{2,\infty}(M; \mathbb{Q}(A))$ and $\mathcal{S}_{2,\infty}(N; \mathbb{Q}(A))$ have free parts that are finite dimensional, then so does $\mathcal{S}_{2,\infty}(M \# N; \mathbb{Q}(A))$, thus showing that Witten's conjecture is stable under connected sums. Using factorisation algebras, the representation theory of quantum groups, and deformation quantization modules, Gunningham, Jordan, and Safronov resolved this conjecture in the affirmative.

THEOREM 2.1 ([**23**]). *The Kauffman bracket skein module of a closed oriented 3-manifold over the field $\mathbb{Q}(A)$ is finite dimensional.*

We note that the KBSM is not the only skein module that exhibits finite dimensionality over the field of rational functions. For closed oriented 3-manifolds, the q-homology skein module also possesses this property. This was proved by Przytycki in [**41**]. It must also be noted that when $R = \mathbb{Z}[A^{\pm 1}]$, the KBSM of closed oriented 3-manifolds need not be finite dimensional. The KBSM of $S^1 \times S^2$ is a quintessential example of this phenomenon. In particular, Hoste and Przytycki [**25**] showed that $\mathcal{S}_{2,\infty}(S^1 \times S^2)$ has infinitely generated torsion. Thus, much more is known about the structure of the KBSM over $\mathbb{Q}(A)$ than over $\mathbb{Z}[A^{\pm 1}]$. Marché proposed (see [**15**]) the following conjecture about the structure of the KBSM over $\mathbb{Z}[A^{\pm 1}]$.

CONJECTURE 2.2. *Let M be a closed compact oriented 3-manifold. Then there exist finitely generated $\mathbb{Z}[A^{\pm 1}]$-modules N_k and an integer $d \geq 0$ such that*

$$\mathcal{S}_{2,\infty}(M) = (\mathbb{Z}[A^{\pm 1}])^d \oplus \bigoplus_{k \geq 1} N_k,$$

where N_k is an $(A^k - A^{-k})$-torsion module for each $k \geq 1$.

This conjecture is true for all the manifolds listed at the beginning of this section. For example, for the 3-manifold $S^1 \times S^2$, $d = 1$ and $N_k = \dfrac{\mathbb{Z}[A^{\pm 1}]}{1 - A^{2k}}$ for $k \geq 2$. For the 3-manifold $F \times S^1$, where F is an oriented surface with genus $g \geq 2$, $d = 2^{2g+1} + 2g - 1$, and as a byproduct of the proof in [**15**], the torsion elements are always of $(A^k - A^{-k})$-type for some $k \geq 1$.

3. The counterexample

It turns out, however, that Conjecture 2.2 is not true in general. While writing [**44**], the author discovered a counterexample to this conjecture given by the KBSM of the connected sum of real projective spaces.

THEOREM 3.1. *Conjecture 2.2 is not true when $M = \mathbb{R}P^3 \# \mathbb{R}P^3$. In particular, $\mathcal{S}_{2,\infty}(\mathbb{R}P^3 \# \mathbb{R}P^3)$ does not split into the direct sum of free modules and torsion modules.*

We now describe the structure of the KBSM of $\mathbb{R}P^3 \# \mathbb{R}P^3$. A key ingredient in calculating $\mathcal{S}_{2,\infty}(\mathbb{R}P^3 \# \mathbb{R}P^3)$ is the concept of depicting links in $\mathbb{R}P^3 \# \mathbb{R}P^3$ using arrow diagrams, which we briefly discuss.

3.1. Arrow diagrams in $\mathbb{R}P^2 \hat{\times} S^1$. In [**13**], nice diagrammatic representations of unoriented links in trivial circle bundles over orientable surfaces were introduced by Dąbkowski and Mroczkowski.[2] These representations are known as arrow diagrams. These arrow diagrams were used to compute the KBSM of the trivial S^1-bundle over the pair of pants. In [**37**], Mroczkowski extended these representations to links in twisted circle bundles over unorientable surfaces (see also [**18**]). Note that $\mathbb{R}P^3 \setminus B^3 \cong \mathbb{R}P^2 \hat{\times} I$, where $\mathbb{R}P^2 \hat{\times} I$ is the twisted I-bundle over the real projective plane. Thus, $\mathbb{R}P^3 \# \mathbb{R}P^3 \cong \mathbb{R}P^2 \hat{\times} S^1$, where $\mathbb{R}P^2 \hat{\times} S^1$ is the twisted circle bundle over $\mathbb{R}P^2$.

Let F be an orientable surface and consider an unoriented link $L \hookrightarrow F \times S^1$. Cut $F \times S^1$ along $F \times \{1\}$, where $1 \in S^1$. In doing so, we obtain the manifold $F \times [0, 1]$ and a collection of circles and arcs that come from L. Denote this collection by L'. By a general position argument we may assume that L intersects $F \times \{1\}$ transversely (but not orthogonally) in a finite number of points. Consider the projection $\pi : F \times [0, 1] \longrightarrow F \times \{0\}$. Projecting the arcs onto $F \times \{0\}$ gives us a set of closed curves in F. By another general position argument, we may assume that $\pi(L')$ contains transverse double points and, under π, the endpoints of the arcs in L' are projected onto points that are distinct from these double points. The closed curves in $\pi(L')$ are equipped with some additional information. Besides the usual undercrossing and overcrossing information based on the appearance of the curves in $F \times [0, 1]$, we keep track of where the arcs intersect $F \times \{0\}$. This is done by placing at the points of intersection a dot and an arrow pointing in the direction of increasing height in $[0, 1]$ just before and after the cut was made. Thus, at the points that are projections of the endpoints $(x, 0)$ and $(x, 1)$ of arcs in L', an arrow is placed in the direction in which the height drops by 1 in L' when the first coordinate crosses x. This means that if we travel along L in the direction of the arrow and cross it, we have essentially travelled through the roof, $F \times \{1\}$, and resurfaced at the floor, $F \times \{0\}$. See Figure 3.1.

[2] A precursor to arrow diagrams was introduced by Turaev in [**47**].

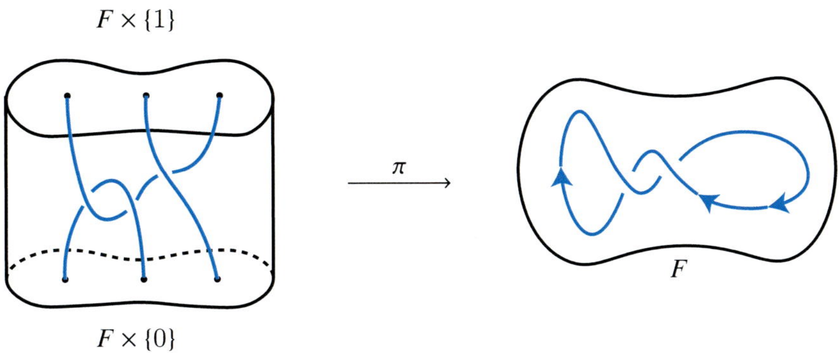

FIGURE 3.1. A link and its arrow diagram in $F \times S^1$.

In summary, an arrow diagram D of a link L in $F \times S^1$ is a link diagram in F with crossing information at double points and some arrows. Links in $F \times S^1$ are ambient isotopic to each other if their arrow diagrams are related by a finite sequence of the usual Reidemeister moves and the two additional moves illustrated in Figure 3.2.

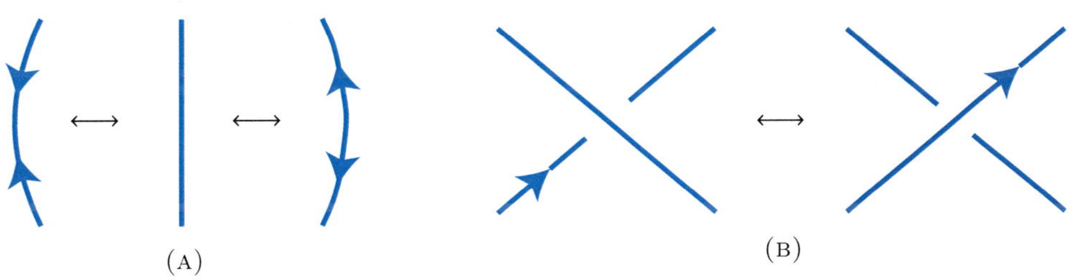

FIGURE 3.2. Isotopy moves for arrow diagrams.

We can generalise this construction to unoriented links in $\mathbb{R}P^2 \hat{\times} S^1$ in the following way. The unorientable surface $\mathbb{R}P^2$ is obtained by identifying the boundary of D^2 using the antipodal map. Thus, $\mathbb{R}P^2 \hat{\times} S^1$ is obtained from $D^2 \times S^1$ by identifying $(x, y) \in \partial D^2 \times S^1$ to $(-x, \overline{y})$, where $\overline{y}$ denotes the complex conjugate of y. Consider a link $L \hookrightarrow \mathbb{R}P^2 \hat{\times} S^1$. In $D^2 \times S^1$, L becomes a collection of closed curves and arcs whose endpoints come in antipodal pairs, (x, y) and $(x, \overline{y})$, in $\partial D^2 \times S^1$. As before, we construct arrow diagrams for L' in $D^2 \times S^1$ with the difference that now the arcs have endpoints that appear in antipodal pairs. Thus, arrow diagrams in $\mathbb{R}P^2 \hat{\times} S^1$ consist of closed curves and arcs in D^2, with the endpoints of the arcs coming in antipodal pairs on ∂D^2. Any two links in $\mathbb{R}P^2 \hat{\times} S^1$ are ambient isotopic to each other if their arrow diagrams are related by a finite sequence of the five isotopy moves listed above and the three additional moves illustrated in Figure 3.3.

Let x denote the arrow diagram  and $t = -A^{-3}x$, that is, t represents x with a negative full-twist. Then we have the following result about the structure of the KBSM of $\mathbb{R}P^3 \# \mathbb{R}P^3$ over the ring $\mathbb{Z}[A^{\pm 1}]$.

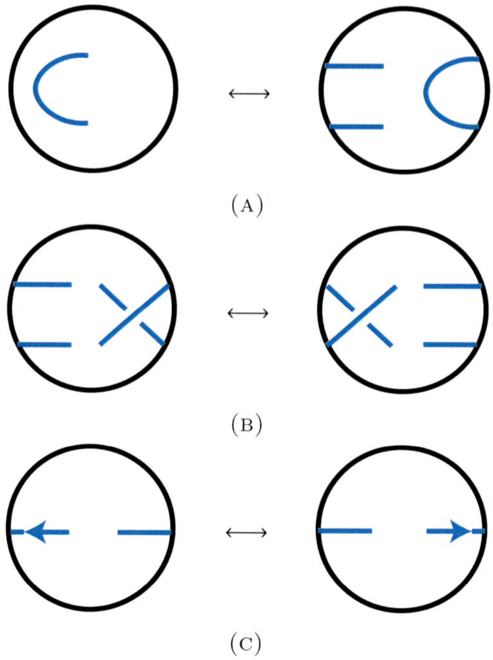

FIGURE 3.3. Additional isotopy moves for arrow diagrams in $\mathbb{R}P^2 \hat{\times} S^1$.

THEOREM 3.2 ([**37**]). *$\mathcal{S}_{2,\infty}(\mathbb{R}P^3 \# \mathbb{R}P^3) = \mathbb{Z}[A^{\pm 1}] \oplus \mathbb{Z}[A^{\pm 1}] \oplus \mathbb{Z}[A^{\pm 1}][t]/S$, where S is a submodule of $\mathbb{Z}[A^{\pm 1}][t]$ generated by the following two relations:*

(i) $(A^{n+1} + A^{-(n+1)})(S_n(t) - 1) - 2(A + A^{-1}) \sum_{k=1}^{n/2} A^{n+2-4k}$, *for $n \geq 2$ even, and*

(ii) $(A^{n+1} + A^{-(n+1)})(S_n(t) - t) - 2t \sum_{k=1}^{(n-1)/2} A^{n+1-4k}$, *for $n \geq 3$ odd.*

Here $S_n(t)$ denotes the Chebyshev polynomial of the second kind, which is defined recursively by the equation $S_{n+1}(t) = t \cdot S_n(t) - S_{n-1}(t)$, with the initial conditions $S_0(t) = 1$ and $S_1(t) = t$. The generators of the free part are the knots K and K' illustrated in Figure 3.4. K is a knot whose diagram consists of an arc with two antipodal endpoints and no crossings, while K' is a knot whose diagram has an arc with an arrow.

REMARK 3.3. $\mathcal{S}_{2,\infty}(\mathbb{R}P^3 \# \mathbb{R}P^3)$ does not split into the sum of free and torsion parts. See Proposition 4.19 in [**37**]. This is a direct contradiction to Conjecture 2.2, which states that the KBSM of a closed compact oriented 3-manifold over $\mathbb{Z}[A^{\pm 1}]$ splits into the direct sums of free modules and torsion modules.

The KBSM of $\mathbb{R}P^3$ over any commutative ring R with unity has two generators, namely the empty link and the noncontractible curve in $\mathbb{R}P^2$. An implication of Theorem 1.2 in [**43**] about the KBSM over $\mathbb{Q}(A)$ of the connected sum of two compact oriented 3-manifolds is the following:

$$\mathcal{S}_{2,\infty}(\mathbb{R}P^3 \# \mathbb{R}P^3; \mathbb{Q}(A)) = \mathcal{S}_{2,\infty}(\mathbb{R}P^3; \mathbb{Q}(A)) \otimes \mathcal{S}_{2,\infty}(\mathbb{R}P^3; \mathbb{Q}(A)).$$

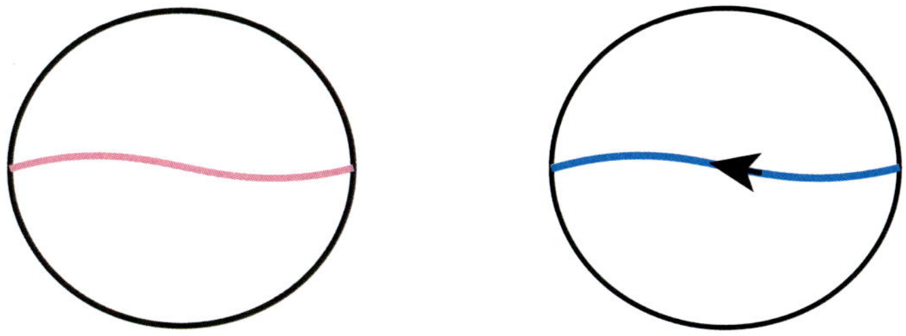

(A) K (B) K'

FIGURE 3.4. The generators of the free part of $\mathcal{S}_{2,\infty}(\mathbb{R}P^3 \# \mathbb{R}P^3)$.

Thus, the dimension of the KBSM of $\mathbb{R}P^3 \# \mathbb{R}P^3$ over $\mathbb{Q}(A)$ is four. Hence, the rank of the KBSM over $\mathbb{Z}[A^{\pm 1}]$ of $\mathbb{R}P^3 \# \mathbb{R}P^3$ is also four. Moreover, $\mathcal{S}_{2,\infty}(\mathbb{R}P^3 \# \mathbb{R}P^3)$ contains torsion elements as seen in [**37**]. Compare with Theorem 4.4 in [**40**].

4. Future directions

While Marché's conjecture is true for all of the examples listed in Section 1, the 3-manifold $\mathbb{R}P^3 \# \mathbb{R}P^3$ is an outlier. Thus, to understand the structure of the Kauffman bracket skein module of an oriented 3-manifold over the ring $\mathbb{Z}[A^{\pm 1}]$, one should study that of the connected sums of oriented 3-manifolds. The first step in doing this would be determining the structure of the Kauffman bracket skein module of the connected sums of handlebodies. In [**43**], a theorem about the structure of the KBSM of the connected sums of two handlebodies $H_n \# H_m$, $n, m \in \mathbb{N}$, was stated. A counterexample to the theorem was provided in [**2**] for $n + m \geq 3$. In soon to be published work, joint with Lê and Przytycki, we proved the following result about the KBSM of the connected sum of two solid tori, that is, when $n = m = 1$.

THEOREM 4.1 ([**1**]). *Let H_n denote a genus n handlebody and $F_{0,n+1}$ be a disc with n holes so that $H_n = F_{0,n+1} \times I$. Then,*

$$\mathcal{S}_{2,\infty}(H_1 \# H_1) = \mathcal{S}_{2,\infty}(H_2)/\mathcal{I}.$$

More precisely, the induced natural epimorphism $i_* : \mathcal{S}_{2,\infty}(H_2)/\mathcal{I} \longrightarrow \mathcal{S}_{2,\infty}(H_1 \# H_1)$, due to the embedding $i_* : H_2 \hookrightarrow H_1 \# H_1$, is an isomorphism. Here, $\mathcal{I}$ is the submodule generated by the expressions $z_k - A^6 u(z_k)$, for any even $k \geq 2$. Here $z_k \in B_k(F_{0,3})$, where $B_k(F_{0,3})$ is a subset of a basis of $\mathcal{S}_{2,\infty}(F_{0,3} \times I)$ composed of links that have no contractible components and have geometric intersection number k with a disc D that separates H_n and H_m. $u(z_k)$ is a modification of z_k in the neighbourhood of D, as shown in Figure 4.1b. *The relation $z_k = A^6 u(z_k)$, is a result of the sliding relation $z_k = sl_{\partial D}(z_k)$ illustrated in Figure* 4.1a.

In future work, the author plans to work on generalising this result to the KBSMs of $L(p, q) \# H_1$ and $L(p, q) \# L(p', q')$, for lens spaces $L(p, q)$ and $L(p', q')$. We also note that the counterexample provided to Marché's conjecture involves the connected sum of oriented 3-manifolds. We conjecture the following, where part (i) is stronger than (ii) and (ii) is in turn stronger than (iii):

(A) Sliding the top string of z_k along the 2-handle results in $sl_{\partial D}(z_k)$.

(B) After resolving the two crossings we get $u(z_k)$.

FIGURE 4.1. Illustration of the handle sliding relation that generates $\mathcal{I}$.

CONJECTURE 4.2.

(i) *The KBSM of any closed, prime, oriented* 3*-manifold can be decomposed into the direct sum of free modules and torsion modules.*

(ii) *The KBSM of any closed, irreducible, atoroidal, oriented* 3*-manifold can be decomposed into the direct sum of free modules and torsion modules.*

(iii) *The KBSM of any closed, oriented, irreducible, non-Haken*[3] 3*-manifold does not contain torsion and is free. This conjecture follows from a conjecture of Przytycki in* [**30**] *and from Theorem* 2.1.

Acknowledgments

The author would like to thank Hanna Makaruk and Robert Owczarek for organising several AMS Special Sessions on Inverse Problems over the years and for their endless support. It was at a session they had organised in Portland in 2018 that the author first had the opportunity to meet Zbigniew Oziewicz.

References

[1] R. P. Bakshi, T. T. Q. Lê, and J. H. Przytycki, *The Kauffman bracket skein module of the connected sum of two solid tori*, 2025, draft available on request.

[2] R. P. Bakshi and J. H. Przytycki, *Kauffman bracket skein module of the connected sum of handlebodies: a counterexample*, Manuscripta Math. **167** (2022), no. 3-4, 809–820, DOI 10.1007/s00229-021-01288-5. MR4385392

[3] R. P. Bakshi, J. H. Przytycki, and H. Wong, Chapter 63: Skein modules of 3-manifolds, *Encyclopedia of Knot Theory*, Chapman and Hall/CRC Press, 2020, pp. 617–623, ISBN 9781138297845.

[3] A non-Haken closed, oriented 3-manifold is an irreducible manifold that contains no embedded oriented, incompressible surfaces.

[4] C. Blanchet, N. Habegger, G. Masbaum, and P. Vogel, *Three-manifold invariants derived from the Kauffman bracket*, Topology **31** (1992), no. 4, 685–699, DOI 10.1016/0040-9383(92)90002-Y. MR1191373

[5] F. Bonahon and H. Wong, *Quantum traces for representations of surface groups in* $\mathrm{SL}_2(\mathbb{C})$, Geom. Topol. **15** (2011), no. 3, 1569–1615, DOI 10.2140/gt.2011.15.1569. MR2851072

[6] D. Bullock and W. Lo Faro, *The Kauffman bracket skein module of a twist knot exterior*, Algebr. Geom. Topol. **5** (2005), 107–118, DOI 10.2140/agt.2005.5.107. MR2135547

[7] D. Bullock, *The* $(2,\infty)$*-skein module of the complement of a* $(2,2p+1)$ *torus knot*, J. Knot Theory Ramifications **4** (1995), no. 4, 619–632, DOI 10.1142/S0218216595000260. MR1361084

[8] D. Bullock, *Rings of* $\mathrm{SL}_2(\mathbf{C})$*-characters and the Kauffman bracket skein module*, Comment. Math. Helv. **72** (1997), no. 4, 521–542, DOI 10.1007/s000140050032. MR1600138

[9] D. Bullock, *On the Kauffman bracket skein module of surgery on a trefoil*, Pacific J. Math. **178** (1997), no. 1, 37–51, DOI 10.2140/pjm.1997.178.37. MR1447403

[10] A. Carrega, *Nine generators of the skein space of the 3-torus*, Algebr. Geom. Topol. **17** (2017), no. 6, 3449–3460, DOI 10.2140/agt.2017.17.3449. MR3709652

[11] V. V. Fok and L. O. Chekhov, *Quantum Teichmüller spaces* (Russian, with Russian summary), Teoret. Mat. Fiz. **120** (1999), no. 3, 511–528, DOI 10.1007/BF02557246; English transl., Theoret. and Math. Phys. **120** (1999), no. 3, 1245–1259. MR1737362

[12] J. H. Conway, *An enumeration of knots and links, and some of their algebraic properties*, Computational Problems in Abstract Algebra (Proc. Conf., Oxford, 1967), Pergamon, Oxford-New York-Toronto, Ont., 1970, pp. 329–358. MR258014

[13] M. Mroczkowski and M. K. Dabkowski, *KBSM of the product of a disk with two holes and* S^1, Topology Appl. **156** (2009), no. 10, 1831–1849, DOI 10.1016/j.topol.2009.03.045. MR2519219

[14] R. Detcherry, *Infinite families of hyperbolic 3-manifolds with finite-dimensional skein modules*, J. Lond. Math. Soc. (2) **103** (2021), no. 4, 1363–1376, DOI 10.1112/jlms.12410. MR4273472

[15] R. Detcherry and M. Wolff, *A basis for the Kauffman skein module of the product of a surface and a circle*, Algebr. Geom. Topol. **21** (2021), no. 6, 2959–2993, DOI 10.2140/agt.2021.21.2959. MR4344875

[16] P. Freyd, D. Yetter, J. Hoste, W. B. R. Lickorish, K. Millett, and A. Ocneanu, *A new polynomial invariant of knots and links*, Bull. Amer. Math. Soc. (N.S.) **12** (1985), no. 2, 239–246, DOI 10.1090/S0273-0979-1985-15361-3. MR776477

[17] C. Frohman, R. Gelca, and W. Lofaro, *The A-polynomial from the noncommutative viewpoint*, Trans. Amer. Math. Soc. **354** (2002), no. 2, 735–747, DOI 10.1090/S0002-9947-01-02889-6. MR1862565

[18] B. Gabrovšek and M. Mroczkowski, *Link diagrams in Seifert manifolds and applications to skein modules*, Algebraic modeling of topological and computational structures and applications, Springer Proc. Math. Stat., vol. 219, Springer, Cham, 2017, pp. 117–141, DOI 10.1007/978-3-319-68103-0_6. MR3748893

[19] S. Garoufalidis, *On the characteristic and deformation varieties of a knot*, Proceedings of the Casson Fest, Geom. Topol. Monogr., vol. 7, Geom. Topol. Publ., Coventry, 2004, pp. 291–309, DOI 10.2140/gtm.2004.7.291. MR2172488

[20] P. M. Gilmer and J. M. Harris, *On the Kauffman bracket skein module of the quaternionic manifold*, J. Knot Theory Ramifications **16** (2007), no. 1, 103–125, DOI 10.1142/S0218216507005208. MR2300430

[21] P. M. Gilmer, *On the Kauffman bracket skein module of the 3-torus*, Indiana Univ. Math. J. **67** (2018), no. 3, 993–998, DOI 10.1512/iumj.2018.67.7327. MR3820238

[22] P. M. Gilmer and G. Masbaum, *On the skein module of the product of a surface and a circle*, Proc. Amer. Math. Soc. **147** (2019), no. 9, 4091–4106, DOI 10.1090/proc/14553. MR3993800

[23] S. Gunningham, D. Jordan, and P. Safronov, *The finiteness conjecture for skein modules*, Invent. Math. **232** (2023), no. 1, 301–363, DOI 10.1007/s00222-022-01167-0. MR4557403

[24] J. Hoste and J. H. Przytycki, *The* $(2,\infty)$*-skein module of lens spaces; a generalization of the Jones polynomial*, J. Knot Theory Ramifications **2** (1993), no. 3, 321–333, DOI 10.1142/S0218216593000180. MR1238877

[25] J. Hoste and J. H. Przytycki, *The Kauffman bracket skein module of* $S^1 \times S^2$, Math. Z. **220** (1995), no. 1, 65–73, DOI 10.1007/BF02572603. MR1347158

[26] J. Hoste and J. H. Przytycki, *The* $(2,\infty)$*-skein module of Whitehead manifolds*, J. Knot Theory Ramifications **4** (1995), no. 3, 411–427, DOI 10.1142/S021821659500020X. MR1347362

[27] V. F. R. Jones, *Hecke algebra representations of braid groups and link polynomials*, Ann. of Math. (2) **126** (1987), no. 2, 335–388, DOI 10.2307/1971403. MR908150

[28] R. M. Kashaev, *Quantization of Teichmüller spaces and the quantum dilogarithm*, Lett. Math. Phys. **43** (1998), no. 2, 105–115, DOI 10.1023/A:1007460128279. MR1607296

[29] L. H. Kauffman, *An invariant of regular isotopy*, Trans. Amer. Math. Soc. **318** (1990), no. 2, 417–471, DOI 10.2307/2001315. MR958895

[30] R. Kirby, *Problems in low dimensional manifold theory*, Algebraic and geometric topology (Proc. Sympos. Pure Math., Stanford Univ., Stanford, Calif., 1976), Proc. Sympos. Pure Math., XXXII, Amer. Math. Soc., Providence, RI, 1978, pp. 273–312. MR520548

[31] L. H. Kauffman, *State models and the Jones polynomial*, Topology **26** (1987), no. 3, 395–407, DOI 10.1016/0040-9383(87)90009-7. MR899057

[32] T. T. Q. Lê, *The colored Jones polynomial and the A-polynomial of knots*, Adv. Math. **207** (2006), no. 2, 782–804, DOI 10.1016/j.aim.2006.01.006. MR2271986

[33] T. T. Q. Lê, *Quantum Teichmüller spaces and quantum trace map*, J. Inst. Math. Jussieu **18** (2019), no. 2, 249–291, DOI 10.1017/s1474748017000068. MR3915288

[34] T. T. Q. Le and A. T. Tran, *The Kauffman bracket skein module of two-bridge links*, Proc. Amer. Math. Soc. **142** (2014), no. 3, 1045–1056, DOI 10.1090/S0002-9939-2013-11835-6. MR3148538

[35] W. B. R. Lickorish, *Invariants for 3-manifolds from the combinatorics of the Jones polynomial*, Pacific J. Math. **149** (1991), no. 2, 337–347. MR1105702

[36] M. Mroczkowski, *Kauffman bracket skein module of a family of prism manifolds*, J. Knot Theory Ramifications **20** (2011), no. 1, 159–170, DOI 10.1142/S0218216511008681. MR2777022

[37] M. Mroczkowski, *Kauffman bracket skein module of the connected sum of two projective spaces*, J. Knot Theory Ramifications **20** (2011), no. 5, 651–675, DOI 10.1142/S0218216511008905. MR2806337

[38] G. Muller, *Skein and cluster algebras of marked surfaces*, Quantum Topol. **7** (2016), no. 3, 435–503, DOI 10.4171/QT/79. MR3551171

[39] J. H. Przytycki, *Skein modules of 3-manifolds*, Bull. Polish Acad. Sci. Math. **39** (1991), no. 1-2, 91–100. MR1194712

[40] J. H. Przytycki, *Algebraic topology based on knots: an introduction*, KNOTS '96 (Tokyo), World Sci. Publ., River Edge, NJ, 1997, pp. 279–297. MR1664968

[41] J. H. Przytycki, *A q-analogue of the first homology group of a 3-manifold*, Perspectives on quantization (South Hadley, MA, 1996), Contemp. Math., vol. 214, Amer. Math. Soc., Providence, RI, 1998, pp. 135–144, DOI 10.1090/conm/214/02910. MR1601241

[42] J. H. Przytycki, *Fundamentals of Kauffman bracket skein modules*, Kobe J. Math. **16** (1999), no. 1, 45–66. MR1723531

[43] J. H. Przytycki, *Kauffman bracket skein module of a connected sum of 3-manifolds*, Manuscripta Math. **101** (2000), no. 2, 199–207, DOI 10.1007/s002290050014. MR1742248

[44] J. H. Przytycki, R. P. Bakshi, D. Ibarra, G. Montoya-Vega, and D. Weeks, *Lectures in knot theory—an exploration of contemporary topics*, Universitext, Springer, Cham, [2024] ©2024, DOI 10.1007/978-3-031-40044-5. MR4738299

[45] J. H. Przytycki and P. Traczyk, *Invariants of links of Conway type*, Kobe J. Math. **4** (1988), no. 2, 115–139. MR945888

[46] V. G. Turaev, *The Conway and Kauffman modules of a solid torus* (Russian, with English summary), Zap. Nauchn. Sem. Leningrad. Otdel. Mat. Inst. Steklov. (LOMI) **167** (1988), no. Issled. Topol. 6, 79–89, 190, DOI 10.1007/BF01099241; English transl., J. Soviet Math. **52** (1990), no. 1, 2799–2805. MR964255

[47] V. G. Turaev, *Shadow links and face models of statistical mechanics*, J. Differential Geom. **36** (1992), no. 1, 35–74. MR1168981

Department of Mathematics, University of California, Santa Barbara, CA 93117
Email address: rheapalak@math.ucsb.edu | rheapalakbakshi@gmail.com

Contemporary Mathematics
Volume **824**, 2025
https://doi.org/10.1090/conm/824/16451

A note on the Gram determinant of type Mb

Rhea Palak Bakshi, Dionne Ibarra, Sujoy Mukherjee,
and Józef H. Przytycki

Dedicated to the memory of Zbigniew Oziewicz

ABSTRACT. A closed formula for the determinant of a matrix given by a bilinear form on crossingless curves in the Möbius band with 2n boundary points was conjectured by Qi Chen in 2009. This determinant is called the Gram determinant of type Mb. The construction of this determinant is related to the Gram determinant of Type A, generalized Type A, and Type B. In this paper, we continue the study of the Gram determinant of type Mb and its conjectured closed formula. We will conclude with a result supporting the conjecture.

E. Witten's speculation on the existence of a 3-manifold invariant related to the Jones Polynomial in [**Wit**] sparked interest in Gram determinants in knot theory. N. Reshetikhin and V. Turaev constructed the invariant in [**RT**] using the theory of representations of $SU(2)$. After their paper was announced, W. B. R. Lickorish in [**Lic1**] thought about a simpler approach to constructing the Witten-Reshetikhin-Turaev invariant. The Gram determinant of type A was formulated in Lickorish's construction. It was later suggested by V. F. R. Jones to calculate the value of the determinant by using the Jones-Wenzl idempotents.

R. Simion investigated bilinear forms of type B in her work on chromatic joins (see [**Sim, Sch, CP1**]). M. Dabkowski and the fourth author hypothesized that the Gram determinant is related to Chebyshev polynomials of the first kind, and G. Barad conjectured the complete closed formula for the determinant. This formula was proven by Q. Chen and the fourth author in [**CP2**]. The proof in [**CP2**] uses basic combinatorics and properties of Jones-Wenzl idempotents.[1]

2020 *Mathematics Subject Classification.* Primary 57K10; Secondary 57K31, 05A19.
Key words and phrases. Knot theory, skein module, Gram determinant.
The first author was supported by Dr. Max Rössler, the Walter Haefner Foundation, and the ETH Zürich Foundation. The second author was supported by the Australian Research Council grant DP210103136. The third author was supported by the American Mathematical Society and the Simons Foundation through the AMS-Simons Travel Grant. The fourth author was partially supported by the Simons Foundation Collaboration Grant for Mathematicians - 637794 and the CCAS Dean's Research Chair award.

[1] The closure of the Jones-Wenzl idempotents lead to Chebyshev polynomials when the variable is the core of the annulus [**Lic2, PBIMW**].

In 2008, via personal communications with Qi Chen [**Prz2**], the fourth author proposed a new Gram determinant called the Gram determinant of type Mb. This Gram determinant utilizes crossingless connections between $2n$ points on the boundary of a Möbius band and the bilinear form is defined by identifying two Möbius bands along their boundaries producing a Klein bottle. The Klein bottle has five homotopically distinct curves (including the trivial curve). Therefore, the Gram matrix has entries in the five variables that denote these curves. In a personal email to the fourth author [**Che**], Chen proposed a conjecture for a closed formula of the Gram determinant of type Mb. First, we will give a formal definition of the Gram determinant of type Mb along with examples. Then we will prove a supporting result to Chen's conjecture. This paper serves as a continuation of [**BIMP**].

We start by showing how many crossingless connections are possible between $2n$ points (connected pairwise) on the boundary of the Möbius band. This number gives the size of the Gram matrix of type Mb.

LEMMA 0.1. *The number of crossingless connections between $2n$ points in the boundary of the Möbius band, up to isotopy, is equal to*

$$|Mb_n| = \sum_{k=0}^{n} \binom{2n}{k} = 2^{2n-1} + \frac{1}{2}\binom{2n}{n},$$

where k denotes the number of arcs that do not intersect the crosscap.

PROOF. Consider the crossingless connections in which $n-k$ arcs, out of n arcs, intersect the crosscap. We modify the proof of Corollary 3.12 in [**Pzr1**] by replacing the table with a crosscap. In doing so, we see that the number of crossingless connections between k arcs which do not intersect the crosscap is equal to $\binom{2n}{k}$. By summing over k, where $0 \leq k \leq n$, we obtain all possible crossingless connections, so that $|Mb_n| = \sum_{k=0}^{n} \binom{2n}{k}$. By using the two identities $\sum_{k=0}^{2n} \binom{2n}{k} = 2^{2n}$ and $\binom{2n}{k} = \binom{2n}{2n-k}$ we obtain

$$\begin{aligned} |Mb_n| &= \sum_{k=0}^{n} \binom{2n}{k} \\ &= \frac{1}{2}\left(\sum_{k=0}^{n} \binom{2n}{k} + \sum_{k=0}^{n} \binom{2n}{2n-k}\right) \\ &= \frac{1}{2}\left(\binom{2n}{n} + \sum_{k=0}^{2n} \binom{2n}{k}\right) \\ &= 2^{2n-1} + \frac{1}{2}\binom{2n}{n}. \end{aligned} \tag{0.1}$$

□

EXAMPLE 0.2. For $n = 1$, we have

$M_1 = \{$ 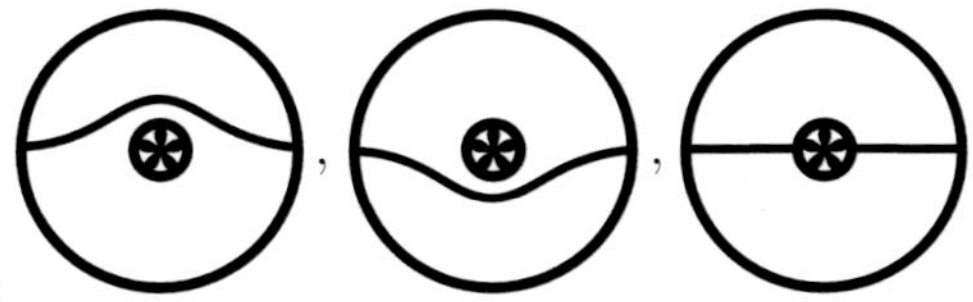 $\}.$

DEFINITION 0.3. The **Gram determinant of type *Mb*** is defined as follows. Consider the set $Mb_n = \{m_i\}_{i=1}^{2^{2n-1}+\frac{1}{2}\binom{2n}{n}}$ of crossingless connections, up to ambient isotopy, between $2n$ labelled framed points fixed in the boundary of the Möbius band Mb. The bilinear form

$$(0.2) \qquad \langle \cdot, \cdot \rangle : \mathcal{S}_{2,\infty}(Mb\hat{\times}I, \{x_i\}_{i=1}^{2n}) \times \mathcal{S}_{2,\infty}(Mb\hat{\times}I, \{x_i\}_{i=1}^{2n}) \to \mathbb{Z}[d, w, x, y, z]$$

is defined using the relative Kauffman bracket skein module of the Möbius band [**Pzr1**, **PBIMW**], $\mathcal{S}_{2,\infty}(Mb\hat{\times}I, \{x_i\}_{i=1}^{2n})$, as explained below:

Given $m_i, m_j \in Mb_n$, for $\langle m_i, m_j \rangle$, we identify the boundary of m_i with the boundary of the inversion of m_j, denoted by $\overline{m}_j$, while identifying the framed points according to their labeling. The result is a collection of five homotopically distinct simple closed curves in the Klein bottle which we denote by d, w, x, y, and z (see Figure 1). We define

$$(0.3) \qquad \langle m_i, m_j \rangle := d^k w^l x^m y^n z^h,$$

where k, l, m, n, and h are the number of each homotopically distinct simple closed curves. The Gram matrix of Type Mb is defined by $G_n^{Mb} = (\langle m_i, m_j \rangle)_{1 \leq i,j, \leq |Mb_n|}$. Thus, the Gram determinant of type Mb is $D_n^{Mb}(d, w, x, y, z) = \det(G_n^{Mb})$.

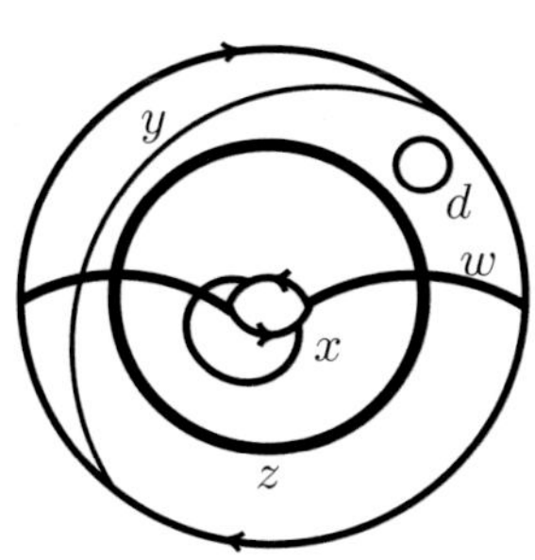

FIGURE 1. The Klein bottle with the five homotopically distinct simple closed curves denoted by d, w, x, y, and z.

Conjecture 0.4 (Chen). *A closed formula for the Gram determinant of type* Mb *for* $n \geq 1$ *is*

$$D_n^{Mb}(d,w,x,y,z) = \left(\prod_{i=1}^{n} D_{n,i}\right) \cdot \left(\prod_{k=1}^{n} (T_k(d) + (-1)^k z)^{\binom{2n}{n-k}}\right) \cdot \left(\prod_{\substack{k=1 \\ k \text{ odd}}}^{n} ((T_k(d) - (-1)^k z)T_k(w) - 2xy)^{\binom{2n}{n-k}}\right) \cdot \left(\prod_{\substack{k=1 \\ k \text{ even}}}^{n} ((T_k(d) - (-1)^k z)T_k(w) - 2(2-z))^{\binom{2n}{n-k}}\right), \tag{0.4}$$

where $D_{n,i} = \prod_{k=1+i}^{n} (T_{2k}(d) - 2)^{\binom{2n}{n-k}}$, *i represents the number of curves passing through the crosscap, and* T_n *represents the* n^{th} *Chebyshev polynomial of the first kind.*[2]

Remark 0.5. The last three factors of equation (0.4) are related to the Gram determinant of Type B [**CP2**].

Example 0.6. The Gram matrix of Type Mb for $n = 1$ is shown in Table 1 and the determinant is equal to $D_1^{Mb}(d,w,x,y,z) = d^2w - 2xyd - wz^2 + 2xyz$. This can be written in terms of Chebyshev polynomials of the first kind:

$$D_1^{Mb} = (T_1(d) - z)((T_1(d) + z)T_1(w) - 2xy). \tag{0.5}$$

Table 1. The Gram matrix G_1^{Mb}.

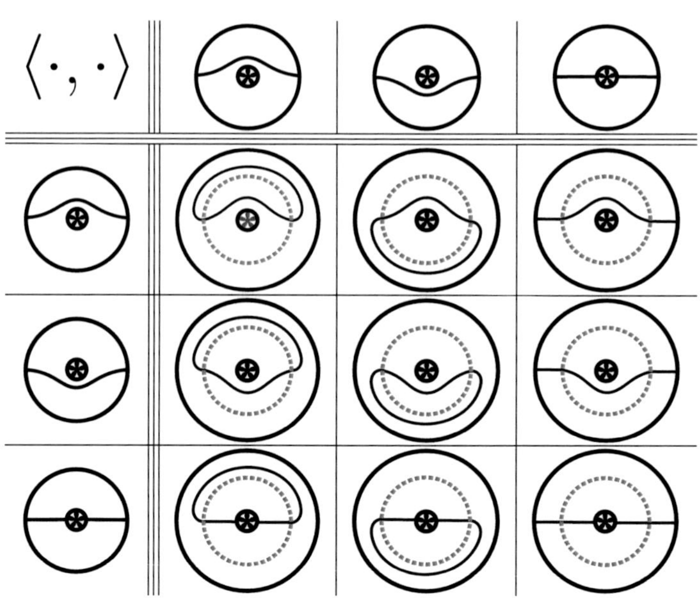

[2] Chebyshev polynomials of the first kind in variable d are defined by the initial conditions $T_0(d) = 2$, $T_1(d) = d$, and the recursive relation $T_n(d) = dT_{n-1}(d) - T_{n-2}(d)$.

PROPOSITION 0.7. *$Det(G_n^{Mb})$ is divisible by $(d-z)^{\binom{2n}{n-1}}$.*

PROOF. For crossingless connections in the disk with $2n$ points in the boundary, the arcs connecting the points separate the disk into $n+1$ regions. This implies that for each crossingless connection there are $n+1$ ways to place a table. Thus, for each crossingless connection in a disk there is a class of $n+1$ crossingless connections in the annulus (D^2 – table) and we can move from one element in the class to another in the same class by moving the table from one region to another (See Figure 2).

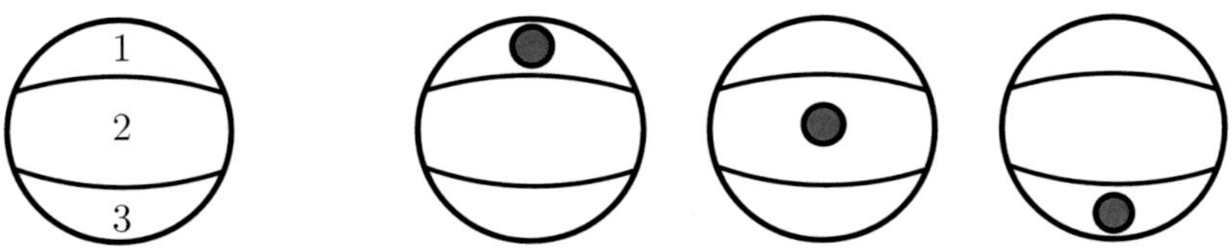

(A) A crossingless connection in the disk.

(B) The class of three crossingless connections in the annulus that corresponds to Figure 2a by placing a table in each region.

FIGURE 2. A crossingless connection in the disk with its corresponding class of crossingless connections in the annulus.

For each class there are n different moves that will change a z to d; these moves are between one element in the class to any other element in the same class. If we now replace the table with a crosscap then the same move in the Möbius band changes a z to d while leaving w, x, and y unchanged. In fact, w will not be in any row of the Gram matrix that corresponds to the class because we are only considering the class of crossingless connections with no curves intersecting the crosscap. x will also not appear in those rows because it is not possible to have curves intersecting the inner crosscap. Thus, the row obtained by taking the difference of the two rows in the class associated to the move is divisible by $d-z$. We conclude that $(d-z)^n$ is a factor of $Det(G_n^{Mb})$ and since there are C_n number of classes (where C_n is the nth Catalan number), $(d-z)^{nC_n}$ is a factor of $Det(G_n^{Mb})$. Finally, by equation (0.6), we obtain our desired result:

$$\begin{aligned} nC_n &= \frac{n}{n+1}\binom{2n}{n} = \frac{n}{n+1}\left(\frac{2n!}{(2n-n)!n!}\right) \\ &= \frac{2n!}{(n+1)!(n-1)!} \\ &= \binom{2n}{n-1}. \end{aligned} \tag{0.6}$$

□

FIGURE 3. Participants of the Special Session on Inverse Problems at the AMS Spring Western Sectional Meeting #1137, Apr. 2018, PSU, Portland, OR. From left to right: Jerzy Kocik, Józef H. Przytycki, Robert Owczarek, Zbigniew Oziewicz, Hilda Colin, Rhea Palak Bakshi, Sujoy Mukherjee, Larissa V. Sbitneva. Picture taken by Dionne Ibarra.

FIGURE 4. Participants of the Special Session on Inverse Problems at the AMS Fall Western Sectional Meeting #1153, Nov. 2019, UC Riverside. Picture taken by Rhea Palak Bakshi.

Acknowledgments

The authors would like to thank Hanna Makaruk and Robert Owczarek for organizing the AMS Special Sessions on Inverse Problems for many years and giving us the opportunity to participate and learn from Zbigniew Oziewicz. Figure 3 (taken in 2018) and Figure 4 (taken in 2019) show some of the participants of the special sessions on Inverse Problems.

References

[BIMP] R. P. Bakshi, D. Ibarra, S. Mukherjee, and J. H. Przytycki, *A generalization of the Gram determinant of type A*, Topology Appl. **295** (2021), Paper No. 107663, 15, DOI 10.1016/j.topol.2021.107663. MR4241642

[Che] Q. Chen, Personal communication (email) with J. H. Przytycki, April 3, 2009.

[CP1] Q. Chen and J. H. Przytycki, *The Gram matrix of a Temperley-Lieb algebra is similar to the matrix of chromatic joins*, Commun. Contemp. Math. **10** (2008), no. suppl. 1, 849–855, DOI 10.1142/S021919970800306X. MR2468368

[CP2] Q. Chen and J. H. Przytycki, *The Gram determinant of the type B Temperley-Lieb algebra*, Adv. in Appl. Math. **43** (2009), no. 2, 156–161, DOI 10.1016/j.aam.2008.12.002. MR2531918

[Lic1] W. B. R. Lickorish, *Invariants for 3-manifolds from the combinatorics of the Jones polynomial*, Pacific J. Math. **149** (1991), no. 2, 337–347. MR1105702

[Lic2] W. B. R. Lickorish, *An introduction to knot theory*, Graduate Texts in Mathematics, vol. 175, Springer-Verlag, New York, 1997, DOI 10.1007/978-1-4612-0691-0. MR1472978

[Pzr1] J. H. Przytycki, *Fundamentals of Kauffman bracket skein modules*, Kobe J. Math. **16** (1999), no. 1, 45–66. MR1723531

[Prz2] J. H. Przytycki, Personal communication (email) with Q. Chen, November 21, 2008.

[PBIMW] J. H. Przytycki, R. P. Bakshi, D. Ibarra, G. Montoya-Vega, and D. Weeks, *Lectures in knot theory—an exploration of contemporary topics*, Universitext, Springer, Cham, 2024, DOI 10.1007/978-3-031-40044-5. MR4738299

[RT] N. Reshetikhin and V. G. Turaev, *Invariants of 3-manifolds via link polynomials and quantum groups*, Invent. Math. **103** (1991), no. 3, 547–597, DOI 10.1007/BF01239527. MR1091619

[Sch] F. Schmidt, *Problems related to type-A and type-B matrices of chromatic joins*, Adv. in Appl. Math. **32** (2004), no. 1-2, 380–390, DOI 10.1016/S0196-8858(03)00090-3. Special issue on the Tutte polynomial. MR2037637

[Sim] R. Simion, *A type-B associahedron*, Adv. in Appl. Math. **30** (2003), no. 1-2, 2–25, DOI 10.1016/S0196-8858(02)00522-5. Formal power series and algebraic combinatorics (Scottsdale, AZ, 2001). MR1979780

[Wit] E. Witten, *Quantum field theory and the Jones polynomial*, Comm. Math. Phys. **121** (1989), no. 3, 351–399. MR990772

Institute for Theoretical Studies, ETH Zürich, Zürich 8006, Switzerland
Email address: `rheapalakbakshi@gmail.com`

School of Mathematics, Monash University, Clayton, VIC 3800, Australia
Email address: `dionne.ibarra@monash.edu`

Department of Mathematics, University of Denver, Denver, Colorado 80208
Email address: `sujoymukherjee.math@gmail.com`

Department of Mathematics, The George Washington University, Washington DC, USA and Department of Mathematics, University of Gdańsk, Gdańsk 80-309, Poland
Email address: `przytyck@gwu.edu`

Contemporary Mathematics
Volume **824**, 2025
https://doi.org/10.1090/conm/824/16455

Hyperbolic knots and torsion in Khovanov homology

Micah Chrisman and Sujoy Mukherjee

Dedicated to the memory of Professor Zbigniew Oziewicz

ABSTRACT. In this note, we show that if there is a knot in S^3 having $\mathbb{Z}_m$ torsion in its Khovanov homology, then there are infinitely many hyperbolic knots and infinitely many prime satellite knots having $\mathbb{Z}_m$ torsion in their Khovanov homology. As an application, we give the first known examples of hyperbolic knots and links with odd (and other non-$\mathbb{Z}_2$) torsion in their Khovanov homology.

1. Introduction

Let R be a commutative ring with identity and let L be an oriented link in S^3. Khovanov's homology theory [**18**] associates a bigraded chain complex over R to L, whose homology is an invariant of the link. The cohomology at bigrading (i, j) is denoted by $\mathrm{Kh}_R^{i,j}(L)$ and the collection of these groups is denoted by $\mathrm{Kh}_R(L)$. For $R = \mathbb{Z}$, we write $\mathrm{Kh}(L)$ instead of $\mathrm{Kh}_{\mathbb{Z}}(L)$. In this case, the torsion part of the abelian group $\mathrm{Kh}^{i,j}(L)$ is the direct sum of finite cyclic groups $\mathbb{Z}_m$. If $\mathbb{Z}_m$ appears as a direct summand of $\mathrm{Kh}^{i,j}(L)$ for some bigrading (i, j), we will say that $\mathrm{Kh}(L)$ has $\mathbb{Z}_m$ torsion.

Torsion is a regular occurrence in integral Khovanov homology. Applications of Khovanov homology, however, typically employ its free part only. Famously, the graded Euler characteristic of Khovanov homology gives the Jones polynomial of L. It is in this sense that $\mathrm{Kh}(L)$ categorifies the Jones polynomial. Another application is the Rasmussen invariant [**28**], which gives a lower bound on the smooth slice genus. This invariant is defined only for Khovanov homology with rational coefficients. The near ubiquity of torsion in Khovanov homology suggests that torsion has further geometric properties which have yet to be fully explored.

2020 *Mathematics Subject Classification.* Primary 57K18, 57K32; Secondary 57K10.

Key words and phrases. Khovanov homology, hyperbolic knots and links, ribbon concordance, torsion.

The first author was supported by the Ohio State University, Marion Campus, who provided funds and release time. The second author was supported by the American Mathematical Society and the Simons Foundation through the AMS-Simons Travel Grant.

The most commonly observed torsion in Khovanov homology is $\mathbb{Z}_2$ torsion. Every nonsplit alternating link has either no torsion or only $\mathbb{Z}_2$ torsion. For knots, it is conjectured that $\mathbb{Z}_2$ torsion is an unknot detector (see Shumakovitch [**30, 31**]). Przytycki and Sazdanović have suggested that the size of torsion groups in the Khovanov homology of a closed braid is bounded by the braid index [**27**]. In particular, they have conjectured that the Khovanov homology of a closed 3-braid also has only $\mathbb{Z}_2$ torsion. This was proved for 3-strand torus links by Chandler, Lowrance, Sazdanović, and Summers, who also established the result for a larger infinite family of closed 3-braids [**11**]. Far less common is $\mathbb{Z}_{2^s}$ torsion in Khovanov homology for $s > 1$. Even so, $\mathbb{Z}_{2^s}$ torsion of large order is known to exist. The second author, together with Przytycki, Silvero, Wang, and Yang [**23**], found infinitely many links having $\mathbb{Z}_{2^s}$ torsion for $1 \leq s \leq 23$.

Also less common than $\mathbb{Z}_2$ torsion is torsion of odd order. Until recently, only a finite number of links were known to have $\mathbb{Z}_m$ torsion for m odd. For example, Bar-Natan [**6**] found that the $T(5,6)$ torus knot has $\mathbb{Z}_3$ and $\mathbb{Z}_5$ torsion in its Khovanov homology. Infinite families of links containing $\mathbb{Z}_3$, $\mathbb{Z}_5$, and $\mathbb{Z}_7$ torsion were constructed in [**23**]. In [**24**], Schütz and the second author proved that there are links having $\mathbb{Z}_{p^k}$ torsion for all $k \geq 1$ and $p = 3, 5, 7$. The second author showed in [**25**] that the connected sum $T(5,6)\#T(5,6)$ has $\mathbb{Z}_9$ torsion while $T(5,6)\#T(5,6)\#T(5,6)$ has $\mathbb{Z}_{27}$ torsion. Schütz [**29**] has also shown that the torus knots $T(9,10)$ and $T(9,11)$ have $\mathbb{Z}_9$ and $\mathbb{Z}_{27}$ torsion.

Now, if L is a nonsplit prime alternating link that is not a torus link, then L is hyperbolic (Menasco [**21**]). As discussed above, nonsplit alternating links can only have $\mathbb{Z}_2$ torsion in their Khovanov homology. On the other hand, the examples of non-$\mathbb{Z}_2$ torsion previously mentioned are constructed from torus links, connected sums of torus links, and certain generalized twisted torus links.[1] The third type of example is obtained from the braid representation $(\sigma_{m-1}\sigma_{m-2}\cdots\sigma_1)^n$ of $T(m,n)$ by multiplying on the right by σ_1^k. Here σ_i denotes the i-th standard generator of the braid group on m strings. Twisted torus links can often be shown to be satellite links (see, e.g., [**23**], Figure 3.5). This leads to the natural question:

> "Are there hyperbolic links having non-$\mathbb{Z}_2$ torsion in their Khovanov homology?"

The purpose of this note is to answer the above question in the affirmative. We prove that there are infinitely many inequivalent hyperbolic knots and links having non-$\mathbb{Z}_2$ torsion in Khovanov homology. Likewise, we show that there are infinitely many inequivalent prime satellite knots having non-$\mathbb{Z}_2$ torsion in their Khovanov homology. Our main tool will be the result of Levine and Zemke [**20**], that the Khovanov homology functor applied to a ribbon concordance is injective. The results then follow from some well-known facts about knot and link concordance.

This paper is organized as follows. In Section 2, we give a brief review of Khovanov homology (Section 2.1) and ribbon concordance (Section 2.2). Our main results are proved in Section 3.

[1] This terminology is due to Birman and Koffman [**2**].

2. Preliminaries

2.1. A brief description of Khovanov homology. For a comprehensive description of Khovanov homology, the reader is directed to [**4**, **35**, **36**]. Our description follows Viro's approach [**36**]. Start with an oriented link diagram D. Next, forgetting the orientation, associate to the unoriented link diagram $|D|$, a bigraded chain complex whose homology is an invariant of the link under regular isotopy. Hence, we first categorify the unreduced Kauffman bracket polynomial [**17**]. Next, the correspondence between this version and the standard version is established by introducing an orientation on the link diagram and calculating its writhe.

Choose a diagram $|D|$ for an unoriented link $|L|$, and let $C(|D|)$ denote its set of crossings. Let

$$ks : C(|D|) \longrightarrow \{A, B\}$$

be a function defined by associating either an A marker or a B marker to each crossing of $|D|$. For each function ks, following the conventions in Figure 1, smooth the crossings of $|D|$ to obtain a system of circles on the plane. This is called a *Kauffman state*. Therefore, there is a bijective correspondence between the functions ks and the Kauffman states of $|L|$. Assuming $|D|$ has n crossings, there are 2^n Kauffman states and functions ks. Finally, let $\sigma(ks) = |ks^{-1}(A)| - |ks^{-1}(B)|$.

Let $c(ks)$ denote the set of circles of the Kauffman state associated to ks. Let

$$eks : c(ks) \longrightarrow \{1, -1\}$$

be a function, called an *enhanced Kauffman state*, defined by associating either a 1 or a -1 to each circle of the Kauffman state. Furthermore, let

$$\tau(eks) = \sum_{x \in c(ks)} eks(x).$$

We are now ready to define the chain modules and the boundary maps. First, let $EKS(|D|)$ denote the set of enhanced Kauffman states of $|D|$. For $eks \in EKS(|D|)$, let ks be the function corresponding to the Kauffman state obtained from eks by forgetting the enhancements. Define

$$a(eks) = \sigma(ks) \quad \text{and} \quad b(eks) = \sigma(ks) + 2 \cdot \tau(eks).$$

Next, let $C_{a,b}(|D|)$ be the free R-module with generators

$$\{eks \in EKS(|D|) \mid a(eks) = a \text{ and } b(eks) = b\}.$$

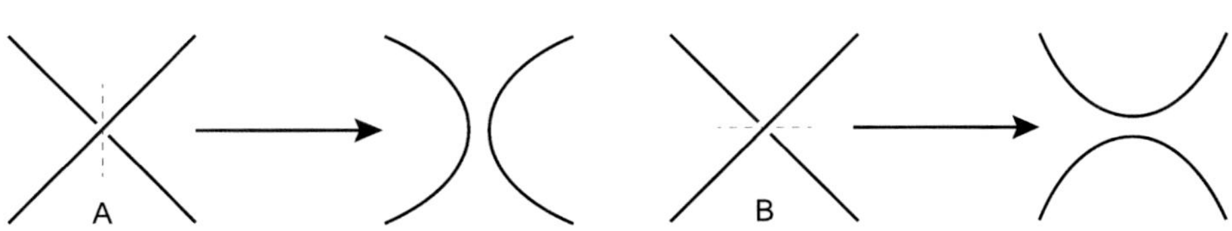

FIGURE 1. Smoothing conventions.

Before defining the boundary maps, we need the notion of adjacency of enhanced Kauffman states. Let $eks, eks' \in EKS(|D|)$. Then, eks' is said to be *adjacent* to eks if $b(eks) = b(eks')$, $a(eks') = a(eks) - 2$, and their corresponding Kauffman states, ks and ks', differ only at one crossing $y \in C(|D|)$ with $ks(y) = A$ and $ks'(y) = B$.

After ordering the crossings of $|D|$, for $b \in \mathbb{Z}$ (that is, for a fixed quantum grading), the boundary maps $\partial_{a,b} : C_{a,b}(|D|) \longrightarrow C_{a-2,b}(|D|)$ are defined as

$$\partial_{a,b}(eks) = \sum_{eks' \in EKS(|D|)} (eks : eks')eks'.$$

Here, if eks' is not adjacent to eks, $(eks : eks') = 0$; else, $(eks : eks') = (-1)^{\alpha}$, where α is the number of B-labeled crossings that occur in eks after the only crossing at which ks and ks' differ. Finally, the homology groups are defined in the usual way, $\mathrm{Kh}^R_{a,b}(|D|) = \frac{ker(\partial_{a,b})}{im(\partial_{a+2,b})}$. So far, these groups are invariant under only Reidemeister moves 2 and 3.

The path to the standard version of Khovanov cohomology is now quite simple. After reintroducing the given orientation on $|D|$, and denoting it by D,

$$\mathrm{Kh}^{i,j}_R(D) = \mathrm{Kh}^R_{w(D)-2i,3w(D)-2j}(|D|),$$

where $w(D) = \#\text{positive crossings} - \#\text{negative crossings}$, is called the *writhe* of D.

2.2. Ribbon concordance. Let K_0, K_1 be oriented knots in S^3. A *concordance from K_0 to K_1* is a smoothly embedded annulus $A \subset S^3 \times I$ such that $\partial A = K_1 \times \{1\} \sqcup -K_0 \times \{0\}$. A concordance of an m-component link $L_0 = K_{0,1} \cup \cdots \cup K_{0,m}$ to a m-component link $L_1 = K_{1,1} \cup \cdots \cup K_{1,m}$ is a collection $A = A_1 \cup \cdots \cup A_m$ of concordances A_i from $K_{0,i}$ to $K_{1,i}$ for $1 \leq i \leq m$ such that $A_i \cap A_j = \varnothing$ for all $i \neq j$. Since a concordance A between two knots is required to be smooth, it may be assumed that the projection $S^3 \times [0,1] \longrightarrow [0,1]$ restricts to a Morse function on A. Each time slice of A corresponding to a regular value of the Morse function is either a knot or a link. The index 0 critical points, i.e., those of the form $x^2 + y^2$, correspond to *births*. This move is of the form $L \longrightarrow L \cup \bigcirc$, where an unknotted component $\bigcirc$ is added to the link diagram L so that it is unlinked from L. Index 1 critical points (i.e., those of the form $x^2 - y^2$) correspond to saddle moves. Index 2 critical points (i.e., those of the form $-x^2 - y^2$) are called *deaths*. They are the inverse to births. Concordance is an equivalence relation on oriented knots.

A concordance from K_0 to K_1 is called a *ribbon concordance*[2] if the Morse function has only critical points of index 0 and 1. In this case, we will write $K_0 \leq K_1$. Figure 2 shows a movie of a ribbon concordance with one birth and one saddle. The annulus A is depicted schematically in Figure 2, right. Note that for any ribbon knot K, there is a ribbon concordance from the unknot to K. Ribbon concordance was first studied by Gordon [**15**], who conjectured that ribbon concordance defines a partial ordering on knots in S^3. The relation is easily seen to be reflexive and transitive. A proof that $\leq$ is antisymmetric was only recently announced by Agol [**1**]. In the interim, many authors investigated the effect of ribbon concordance on various knot invariants. See, for example, Gordon [**15**], Gilmer [**14**], Miyazaki [**22**], and Silver [**32**].

[2]Our concordances will go from K_0 to K_1, as in Zemke [**37**], rather than from K_1 to K_0, as in Gordon [**15**].

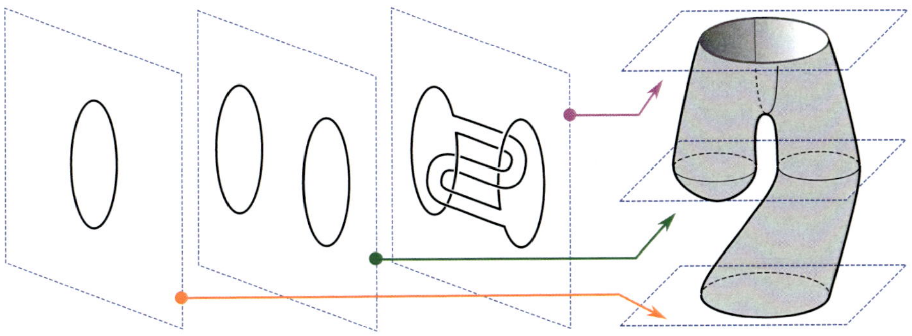

FIGURE 2. A ribbon concordance consisting of a birth $t = \frac{1}{4}$ and a saddle $t = \frac{3}{4}$.

Given an arbitrary concordance A of links from L_0 to L_1, there is a linear map $\mathrm{Kh}_R(A) : \mathrm{Kh}_R(L_0) \longrightarrow \mathrm{Kh}_R(L_1)$. The map $\mathrm{Kh}_R(A)$ is a functor on the cobordism category of links. For more details, see Bar-Natan [**5**], Caprau [**7**], and Clark-Morrison-Walker [**12**]. When A is a ribbon concordance from L_0 to L_1, the following result of Levine and Zemke shows that $\mathrm{Kh}_R(A)$ is an embedding.

THEOREM 2.1 (Levine-Zemke [**20**, Theorem 1]). *If A is a ribbon concordance from L_0 to L_1, then $\mathrm{Kh}_R(A) : \mathrm{Kh}_R(L_0) \longrightarrow \mathrm{Kh}_R(L_1)$ is an injective and grading-preserving linear map. Furthermore, for any bigrading (i, j), $\mathrm{Kh}_R^{i,j}(L_0)$ is a direct summand of $\mathrm{Kh}_R^{i,j}(L_1)$.*

Theorem 2.1 has been shown to hold for many link homology theories. It was first proved for knot Floer homology by Zemke [**37**]. In particular, a ribbon concordance A from a knot K_0 to a knot K_1 gives an injective linear map $F_A : \widehat{HFK}(K_0) \longrightarrow \widehat{HFK}(K_1)$ that preserves the grading. Here homology is calculated over $\mathbb{Z}_2$. In [**13**], Daemi et al. prove that any flavor of Heegard-Floer homology applied to the 2-fold branch cover of a ribbon concordance also induces an injection. Caprau et al. [**9**] showed that Theorem 2.1 holds for the $\mathfrak{sl}(n)$-homology functor for all $n \geq 2$. An axiomatic TQFT approach was used by Kang [**16**] to prove Theorem 2.1 for a wide class of link homology theories.

3. Main results

In [**26**], Myers proved that every link is concordant to a hyperbolic link. We will modify this argument to prove our main result. Here we consider Khovanov homology over a fixed ring R.

LEMMA 3.1 (Hyperbolic Subway Coupon Lemma). *Let M be an R-module. If there is an oriented link L with M a submodule of $\mathrm{Kh}_R^{i,j}(L)$ for some grading (i, j), then there are infinitely many distinct hyperbolic links L' such that M is a submodule of $\mathrm{Kh}_R^{i,j}(L')$.*

PROOF. Suppose first that $L = K$ is a knot. Let J_0 be a ribbon knot having Alexander polynomial not equal to 1. For example, take J_0 to be the stevedore's knot 6_1. Let $n > 0$ be an integer. Since J_0 is ribbon, there is a knot K_n such that $K \leq K_n$ and $K_n = K\#J_0\#J_0\#\cdots\#J_0$ is a connected sum of K with n copies of J_0. The Alexander polynomial of K_n is $\Delta_{K_n}(t) = \Delta_K(t) \cdot (\Delta_{J_0}(t))^n$ and hence there is an infinite family of knots K_n such that $K \leq K_n$ but K_n is not equivalent to K.

Next we modify K_n to a hyperbolic knot K'_n that is ribbon concordant to K and has the same Alexander polynomial as K_n. Let X_n be the exterior of K_n in S^3. Let a_1, a_2 be disjoint arcs of K_n. By Myers [**26**], Proposition 6.1, there is a properly embedded arc $\tilde{J}$ in X_n such that $\overline{X_n \smallsetminus N(\tilde{J})}$ is a simple Haken 3-manifold, where $N(\tilde{J})$ is a regular neighborhood of $\tilde{J}$ in X_n. The arc $\tilde{J}$ may be chosen so that one component of $\partial\tilde{J}$ lies in the boundary of a regular neighborhood N_1 of a_1 while the other component lies in the boundary of a regular neighborhood N_2 of a_2. Note that $B = N(\tilde{J}) \cup N_1 \cup N_2$ is a 3-ball and that $(B, a_1 \cup a_2)$ is a trivial tangle. See Figure 3. Let (T_n, t_n) be the 2-tangle formed by deleting the interior of B from S^3 and the interior of the arcs a_1, a_2 from K_n.

Let $KT = (B', t')$ denote the Kinoshita-Terasaka tangle (see Figure 4, bottom right). The KT tangle is prime (Bleiler [**3**]) and atoroidal (Soma [**33**]). Then it follows from Myers [**26**], Lemma 2.5, that the knot K'_n formed from the join $t_n \vee t'$ has simple Haken exterior. Hence, by Thurston [**34**], it is hyperbolic. The reason for choosing the Kinoshita-Terasaka tangle is that the join $K'_n = t_n \vee t'$ has the same Alexander polynomial as K_n (see Bleiler [**3**], Theorem 2.2). This implies that $\{K'_0, K'_1, \ldots\}$ is an infinite set of distinct hyperbolic knots.

The movie shown in Figure 4 proves that there is a ribbon concordance from K_n to $K'_n = t_n \vee t'$. Hence for each K'_n, there is a ribbon concordance A_n from K to K'_n. By Theorem 2.1, $\mathrm{Kh}(A_n)$ is injective. Hence, the linear map $\mathrm{Kh}_R(A_n)$ embeds the given submodule M into $\mathrm{Kh}_R(K'_n)$.

Lastly, suppose that L is a multi-component link and fix one of its components K. Perform the construction above on K to obtain knots K_n that are components of a link L_n. Myers [**26**], Proposition 6.1, again guarantees that we may choose an arc $\tilde{J}$ beginning and ending on K_n so that the deleting $N(\tilde{J})$ from the link exterior

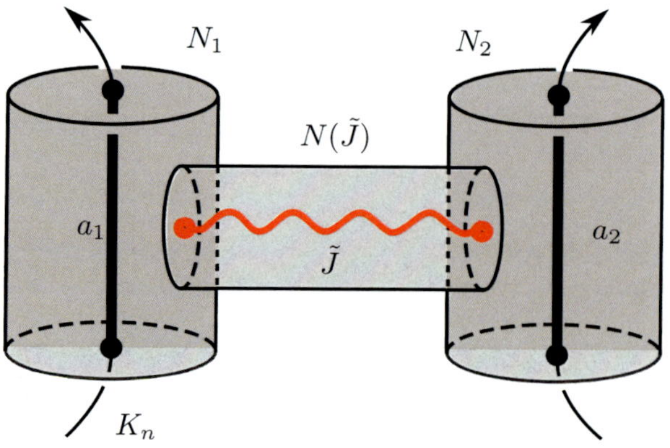

FIGURE 3. Construction used in the proof of Lemma 3.1.

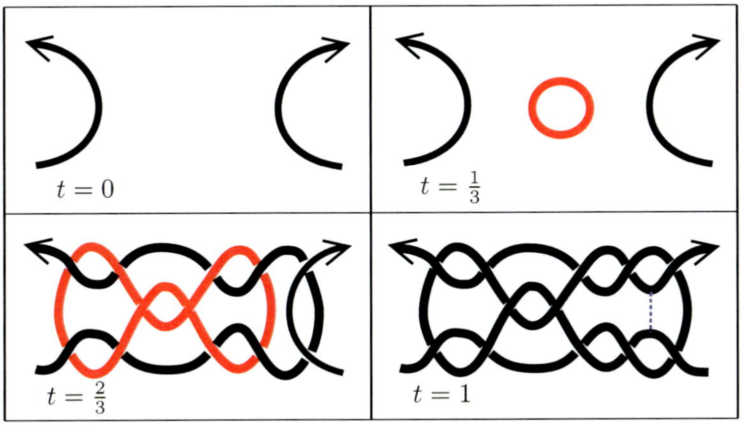

FIGURE 4. A ribbon concordance from the trivial tangle ($t = 0$) to the KT-tangle ($t = 1$). There is one birth (red unknot) and one saddle (blue dashed arc).

X_n gives a simple Haken 3-manifold. As before, replace the trivial tangle $(B, a_1 \cup a_2)$ with the KT-tangle to obtain a hyperbolic link L'_n. As in the previous case, only one component K'_n has been modified. The set of links $\{L'_0, L'_1, \ldots\}$ is infinite because the components $K'_0, K'_1, \ldots$ have distinct Alexander polynomials. □

In [**19**], Livingston proved that every knot is concordant to a prime satellite knot. As in Lemma 3.1, this argument can be modified to show that corresponding to any knot with torsion in Khovanov homology, there is a prime satellite knot having the same torsion.

LEMMA 3.2 (Satellite Subway Coupon Lemma). *Let M be an R-module. If there is a knot K with M a submodule of $\mathrm{Kh}_R^{i,j}(K)$ for some grading (i, j), then there are infinitely many distinct prime satellite knots K' such that M is a submodule of $\mathrm{Kh}_R^{i,j}(K')$.*

PROOF. Let K_n be as in the proof of Lemma 3.1, so that K is ribbon concordant to K_n and K, K_n are inequivalent. The knot K_n will be the companion of a satellite knot K'_n with the pattern P as shown in Figure 5, $t = 1$. By Livingston [**19**], Theorem 5.1, this pattern P is a nontrivial knot in the solid torus and has wrapping number greater than 1. Then any satellite knot $P(K_n)$ is therefore a prime knot in S^3 (Livingston [**19**], Theorem 4.2). The movie shown in Figure 5 shows that there is a ribbon concordance from K_n to $P(K_n)$ and hence from K to $P(K_n)$. Then Theorem 2.1 implies that M is a submodule of $\mathrm{Kh}_R^{i,j}(P(K_n))$. Using the skein relation for the Alexander polynomial, it is easy to prove that $P(K_n)$ and K_n have the same Alexander polynomial (see, e.g., Chrisman [**10**], Section 5.2). Therefore, the set $\{P(K_0), P(K_1), \ldots\}$ satisfies the conclusion of the theorem. □

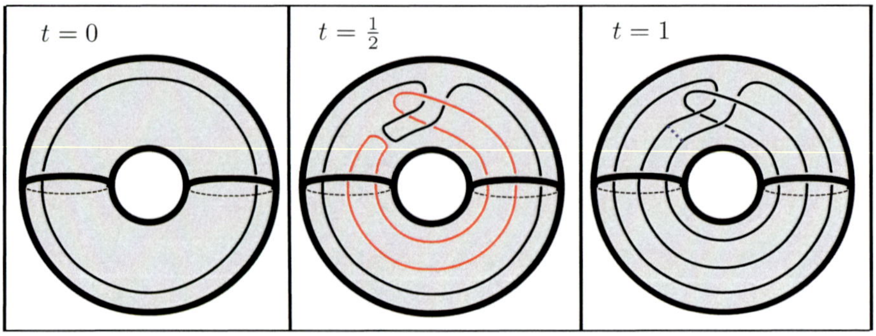

FIGURE 5. A ribbon concordance from the core circle ($t = 0$) to Livingston's pattern ($t = 1$). There is one birth (red unknot) and one saddle (blue dashed arc).

REMARK 3.3. The constructions used in Lemmas 3.1 and 3.2 apply to any link homology theory for which Theorem 2.1 is true. Hence, they hold for knot Floer homology [**37**] and $\mathfrak{sl}(n)$-homology [**9**].

THEOREM 3.4. *There are infinitely many hyperbolic links and infinitely many prime satellite knots having non-*$\mathbb{Z}_2$ *torsion in their Khovanov homology. In particular, the following are true.*

(1) *For* $m = 3, 4, 5, 7, 8, 9, 27$, *there are infinitely many hyperbolic knots and infinitely many prime satellite knots having* $\mathbb{Z}_m$ *torsion in their Khovanov homology.*
(2) *For* $1 \leq k$, $1 \leq s \leq 23$, *and* $m = 2^s, 3^k, 5^k, 7^k$, *there are infinitely many multi-component hyperbolic links having* $\mathbb{Z}_m$ *torsion in their Khovanov homology.*

PROOF. By Lemmas 3.1 and 3.2, it suffices to show that there exists some link with the desired torsion. For (1), knots with these torsion groups can be found in [**6**], [**25**], [**23**], and [**29**]. For (2), multi-component links with these torsion groups can be found in [**23**] or [**24**]. □

EXAMPLE 3.5. An example of a hyperbolic knot K with odd torsion in its Khovanov homology is given in Figure 6, right. According to SnapPy [**8**], the knot is hyperbolic with volume ≈ 20.9424081371. The knot K can be obtained from the torus knot $T(5, 6)$ (Figure 6, left) using the construction of Lemma 3.1. The red arc in Figure 6 is the arc $\widetilde{J}$, as in Figure 3. The saddle occurring in the ribbon concordance (see Figure 4) is shown as a blue dashed arc in Figure 6, right. The Khovanov homology of $T(5, 6)$ is given in Figure 7 while that of K is given in Figure 8. In both tables, we write A_b in position (i, j) of the table if $\mathbb{Z}_b^A = \mathbb{Z}_b \oplus \cdots \oplus \mathbb{Z}_b$ (A-times) is a direct summand of $\text{Kh}^{i,j}(L)$. For $\mathbb{Z}^A$, we simply write A. The shaded entries in Figure 8 correspond to the nontrivial bigradings in the Khovanov homology of $T(5, 6)$, so that the conclusion of Theorem 2.1 is evident. Observe that K has both $\mathbb{Z}_3$ and $\mathbb{Z}_5$ torsion and hence the odd torsion is preserved.

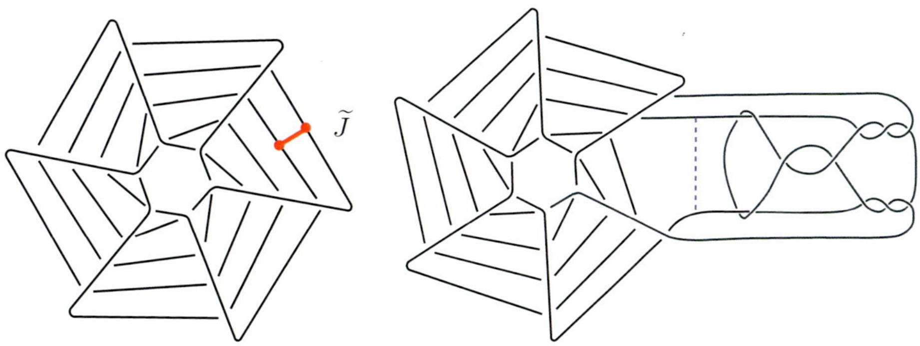

FIGURE 6. $T(5,6)$ (left) is ribbon concordant to the hyperbolic knot K (right).

j | i	0	1	2	3	4	5	6	7	8	9	10	11	12	13	14
43															1_3
41													1	1	
39													$1_2, 1_5$	1	
37											1_2	2	1		
35										2	1_2	$1_2, 1_5$	1		
33								1		$1, 1_2$	1				
31						1		$1, 1_2$	2						
29						1	1	1_2	1						
27				1	1		1								
25				1_2	1										
23			1												
21	1														
19	1														

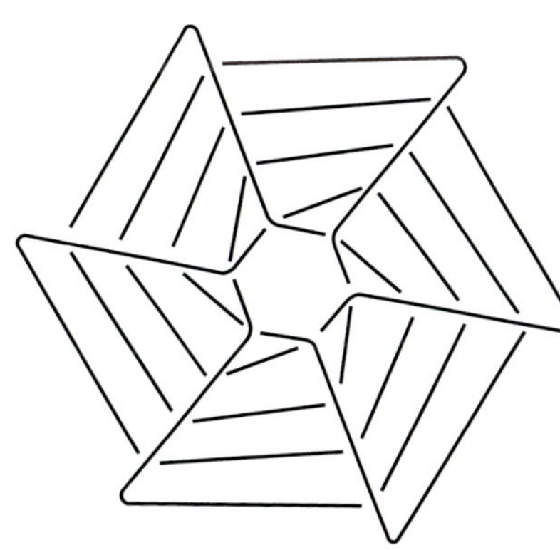

FIGURE 7. The Khovanov homology (left) of the $T(5,6)$ torus knot (right).

j \ i	−4	−3	−2	−1	0	1	2	3	4	5	6	7	8	9	10	11	12	13	14	15	16	17	18	19	20
55																									1
53																								$1,1_3$	1_2
51																						1	$2,1_3$	$2,1_2$	
49																					4	$4,1_2,1_3$	$2,2_2,1_3$	1_2	
47																				$5,1_2,1_3$	$4,4_2,2_3$	$3,4_2,1_3$	$1,1_2$		
45																		1	$8,3_2,1_3$	$1,6_2,1_3$	$6,3_2,1_3$	$1,1_2$			
43																	4	$12,5_2,1_3$	$11,11_2,3_3$	$4,6_2,2_3$	$1,2_2$				
41															1	$6,1_2$	$12,9_2,2_3$	$13,16_2,2_3$	$7,6_2,1_3$	$2,1_2$					
39														2	$8,4_2$	$16,13_2,1_3$	$18,16_2,1_3,1_5$	$8,4_2,1_3$	$1,1_2$						
37													4	$12,6_2$	$17,17_2,1_3$	$15,18_2,2_3$	$5,6_2,1_3$	$1,1_2$							
35											1	7	$12,9_2$	$16,19_2,1_3$	$14,17_2,1_3$	$6,4_2,1_5$	2								
33										2	$7,1_2$	$14,13_2$	$17,18_2$	$14,11_2,1_3$	$4,2_2$										
31									3	$1,2_2$	$15,12_2$	$14,17_2$	$8,9_2,1_3$	$2,2_2$											
29								5	$1,3_2$	$12,12_2$	$11,13_2$	$6,5_2$	3												
27							5	$9,5_2$	$11,11_2$	$9,8_2$	$4,1_2$														
25						5	$9,5_2$	$9,7_2$	$5,5_2$	$1,1_2$															
23					5	$7,5_2$	$6,4_2$	$2,2_2$	1																
21				3	$6,5_2$	$4,2_2$	1																		
19			2	$5,3_2$	3																				
17		1	$3,2_2$																						
15		$2,1_2$																							
13	1																								

FIGURE 8. The Khovanov homology of the knot from Figure 6.

Acknowledgments

Lemmas 3.1 and 3.2 were appropriately discovered while the authors were walking to redeem a buy-one-get-one-free coupon from their nearest (now closed) *Subway* restaurant (1952 N High St, Columbus, OH). The authors would like to thank *Subway* for their generosity.

The computational data in this note was obtained using JavaKh-v2 written by Scott Morrison. It is an update of Jeremy Green's JavaKh-v1 written under the supervision of Dror Bar-Natan.

The authors would like to thank Rhea Palak Bakshi, Carmen Caprau, and Józef H. Przytycki for their valuable insights and suggestions. The second author is grateful to Hanna Makaruk and Robert Owczarek for organizing the AMS Special Sessions on Inverse Problems for many years and giving him the opportunity to participate.

References

[1] I. Agol, *Ribbon concordance of knots is a partial ordering*, Commun. Am. Math. Soc. **2** (2022), 374–379, DOI 10.1090/cams/15. MR4520779

[2] J. Birman and I. Kofman, *A new twist on Lorenz links*, J. Topol. **2** (2009), no. 2, 227–248, DOI 10.1112/jtopol/jtp007. MR2529294

[3] S. A. Bleiler, *Realizing concordant polynomials with prime knots*, Pacific J. Math. **100** (1982), no. 2, 249–257. MR669325

[4] D. Bar-Natan, *On Khovanov's categorification of the Jones polynomial*, Algebr. Geom. Topol. **2** (2002), 337–370, DOI 10.2140/agt.2002.2.337. MR1917056

[5] D. Bar-Natan, *Khovanov's homology for tangles and cobordisms*, Geom. Topol. **9** (2005), 1443–1499, DOI 10.2140/gt.2005.9.1443. MR2174270

[6] D. Bar-Natan, *Fast Khovanov homology computations*, J. Knot Theory Ramifications **16** (2007), no. 3, 243–255, DOI 10.1142/S0218216507005294. MR2320156

[7] C. L. Caprau, sl(2) *tangle homology with a parameter and singular cobordisms*, Algebr. Geom. Topol. **8** (2008), no. 2, 729–756, DOI 10.2140/agt.2008.8.729. MR2443094

[8] M. Culler, N. M. Dunfield, M. Goerner, and J. R. Weeks, *SnapPy, a computer program for studying the geometry and topology of* 3*-manifolds*, Available at `http://snappy.computop.org`.

[9] C. Caprau, N. González, C. R. S. Lee, A. M. Lowrance, R. Sazdanović, and M. Zhang, *On Khovanov homology and related invariants*, Research directions in symplectic and contact geometry and topology, Assoc. Women Math. Ser., vol. 27, Springer, Cham, [2021] ©2021, pp. 273–292, DOI 10.1007/978-3-030-80979-9_6. MR4417719

[10] M. Chrisman, *Concordances to prime hyperbolic virtual knots*, Geom. Dedicata **212** (2021), 379–414, DOI 10.1007/s10711-020-00563-1. MR4251678

[11] A. Chandler, A. M. Lowrance, R. Sazdanović, and V. Summers, *Torsion in thin regions of Khovanov homology*, Canad. J. Math. **74** (2022), no. 3, 630–654, DOI 10.4153/s0008414x21000043. MR4430925

[12] D. Clark, S. Morrison, and K. Walker, *Fixing the functoriality of Khovanov homology*, Geom. Topol. **13** (2009), no. 3, 1499–1582, DOI 10.2140/gt.2009.13.1499. MR2496052

[13] A. Daemi, T. Lidman, D. S. Vela-Vick, and C.-M. M. Wong, *Ribbon homology cobordisms*. part B, Adv. Math. **408** (2022), no. part B, Paper No. 108580, 68, DOI 10.1016/j.aim.2022.108580. MR4467148

[14] P. M. Gilmer, *Ribbon concordance and a partial order on S-equivalence classes*, Topology Appl. **18** (1984), no. 2-3, 313–324, DOI 10.1016/0166-8641(84)90016-6. MR769297

[15] C. McA. Gordon, *Ribbon concordance of knots in the* 3*-sphere*, Math. Ann. **257** (1981), no. 2, 157–170, DOI 10.1007/BF01458281. MR634459

[16] S. Kang, *Link homology theories and ribbon concordances*, Quantum Topol. **13** (2022), no. 1, 183–205, DOI 10.4171/qt/162. MR4404800

[17] L. H. Kauffman, *State models and the Jones polynomial*, Topology **26** (1987), no. 3, 395–407, DOI 10.1016/0040-9383(87)90009-7. MR899057

[18] M. Khovanov, *A categorification of the Jones polynomial*, Duke Math. J. **101** (2000), no. 3, 359–426, DOI 10.1215/S0012-7094-00-10131-7. MR1740682
[19] C. Livingston, *Homology cobordisms of 3-manifolds, knot concordances, and prime knots*, Pacific J. Math. **94** (1981), no. 1, 193–206. MR625818
[20] A. S. Levine and I. Zemke, *Khovanov homology and ribbon concordances*, Bull. Lond. Math. Soc. **51** (2019), no. 6, 1099–1103, DOI 10.1112/blms.12303. MR4041014
[21] W. Menasco, *Closed incompressible surfaces in alternating knot and link complements*, Topology **23** (1984), no. 1, 37–44, DOI 10.1016/0040-9383(84)90023-5. MR721450
[22] K. Miyazaki, *Ribbon concordance does not imply a degree one map*, Proc. Amer. Math. Soc. **108** (1990), no. 4, 1055–1058, DOI 10.2307/2047967. MR994781
[23] S. Mukherjee, J. H. Przytycki, M. Silvero, X. Wang, and S. Y. Yang, *Search for torsion in Khovanov homology*, Exp. Math. **27** (2018), no. 4, 488–497, DOI 10.1080/10586458.2017.1320242. MR3894728
[24] S. Mukherjee and D. Schütz, *Arbitrarily large torsion in Khovanov cohomology*, Quantum Topol. **12** (2021), no. 2, 243–264, DOI 10.4171/qt/149. MR4261658
[25] S. Mukherjee, *On odd torsion in even Khovanov homology*, Exp. Math. **31** (2022), no. 3, 1039–1045, DOI 10.1080/10586458.2020.1743215. MR4477421
[26] R. Myers, *Homology cobordisms, link concordances, and hyperbolic 3-manifolds*, Trans. Amer. Math. Soc. **278** (1983), no. 1, 271–288, DOI 10.2307/1999315. MR697074
[27] J. H. Przytycki and R. Sazdanović, *Torsion in Khovanov homology of semi-adequate links*, Fund. Math. **225** (2014), no. 1, 277–304, DOI 10.4064/fm225-1-13. MR3205574
[28] J. Rasmussen, *Khovanov homology and the slice genus*, Invent. Math. **182** (2010), no. 2, 419–447, DOI 10.1007/s00222-010-0275-6. MR2729272
[29] D. Schütz, *Torsion calculations in Khovanov cohomology*, J. Knot Theory Ramifications **29** (2020), no. 8, 2071001, 7, DOI 10.1142/S0218216520710017. MR4143698
[30] A. N. Shumakovitch, *Torsion of Khovanov homology*, Fund. Math. **225** (2014), no. 1, 343–364, DOI 10.4064/fm225-1-16. MR3205577
[31] T. Sano, *Fixing the functoriality of Khovanov homology: a simple approach*, J. Knot Theory Ramifications **30** (2021), no. 11, Paper No. 2150074, 12, DOI 10.1142/S0218216521500747. MR4376719
[32] D. S. Silver, *On knot-like groups and ribbon concordance*, J. Pure Appl. Algebra **82** (1992), no. 1, 99–105, DOI 10.1016/0022-4049(92)90013-6. MR1181096
[33] T. Soma, *Simple links and tangles*, Tokyo J. Math. **6** (1983), no. 1, 65–73, DOI 10.3836/tjm/1270214326. MR707839
[34] W. P. Thurston, *Three-dimensional manifolds, Kleinian groups and hyperbolic geometry*, Bull. Amer. Math. Soc. (N.S.) **6** (1982), no. 3, 357–381, DOI 10.1090/S0273-0979-1982-15003-0. MR648524
[35] P. Turner, *Five lectures on Khovanov homology*, J. Knot Theory Ramifications **26** (2017), no. 3, 1741009, 41, DOI 10.1142/S0218216517410097. MR3627709
[36] O. Viro, *Khovanov homology, its definitions and ramifications*, Fund. Math. **184** (2004), 317–342, DOI 10.4064/fm184-0-18. MR2128056
[37] I. Zemke, *Knot Floer homology obstructs ribbon concordance*, Ann. of Math. (2) **190** (2019), no. 3, 931–947, DOI 10.4007/annals.2019.190.3.5. MR4024565

Department of Mathematics, The Ohio State University, Columbus, Ohio 43210
Email address: `chrisman.76@osu.edu`

Department of Mathematics, The Ohio State University, Columbus, Ohio 43210
Current address: Department of Mathematics, University of Denver, Denver, Colorado 80208
Email address: `sujoymukherjee.math@gmail.com | sujoy.mukherjee@du.edu`

Contemporary Mathematics
Volume **824**, 2025
https://doi.org/10.1090/conm/824/16449

Coefficients of Catalan states of lattice crossing I: Θ_A-state expansion

Mieczyslaw K. Dabkowski and Cheyu Wu

Abstract. Plucking polynomial for plane rooted trees was introduced by J.H. Przytycki in 2014. As it was shown later, this polynomial can be used to find coefficients $C(A)$ of Catalan states C of $m \times n$-lattice crossing $L(m, n)$ without returns on one side. In this paper, we show that $C(A)$ for any C can be found by using Θ_A-state expansion which represents $C(A)$ as a linear combination of coefficients of Catalan states with no top returns over $\mathbb{Q}(A)$. We also provide an algorithm for finding Θ_A-state expansions and examples of its applications. Finally, an example of a Catalan state with non-unimodal coefficient is given.

1. Introduction

Kauffman Bracket Skein Module (KBSM)[1] for a product of an oriented surface of genus g with n boundary components $F_{g,n}$ and an interval $I = [0, 1]$ is an algebra[2] called the Kauffman Bracket Skein Algebra (KBSA) of $F_{g,n}$. In 2000, D. Bullock and J.H. Przytycki studied these algebras for $(g, n) \in \{(1, 0), (1, 1), (1, 2), (0, 4)\}$ and found their presentations (see [**3**]). Furthermore, an elegant and important *"product-to-sum formula"* for multiplication of curves in KBSA of $F_{1,0}$ was given by C. Frohman and R. Gelca in [**13**] and later used in [**16, 17, 20**]. Unfortunately, formulas for product in KBSA of $F_{g,n}$ with $(g, n) \in \{(1, 1), (1, 2), (0, 4)\}$ appear to be much more complex and difficult to find. Some partial results have been though obtained. For instance, formulas for product of some special curves in KBSA of $F_{1,1}$ were given in [**7, 8**] and for product of some classes of curves in KBSA of $F_{0,4}$ were obtained in [**19**] and [**1**].

Since the algebraic structure of KBSA of $F_{g,n}$ is a subject of recent research interests (see [**2, 14, 15, 18, 23, 26, 27**]), determining formulas for products of curves on $F_{g,n}$ is an important problem. To understand such products locally, in [**9**] authors considered lattice crossing $L(m, n)$ (see Figure 1.1) and asked if closed-form formulas for coefficients $C(A)$ of Catalan states C of $L(m, n)$ in the Relative Kauffman Bracket Skein Module (RKBSM) of a cylinder with $2(m + n)$ points fixed on its boundary can be found. The problem of finding these coefficients led to some new and interesting developments. For instance, plucking polynomial of

2020 *Mathematics Subject Classification.* Primary 57K31, 57K10.

[1]Skein modules were introduced by J.H. Przytycki [**21**] in 1987 and later also defined by V. Turaev [**28**] in 1988.

[2]Multiplication of links L_1 and L_2 in KBSM of $F_{g,n} \times I$ is defined by placing L_1 above L_2.

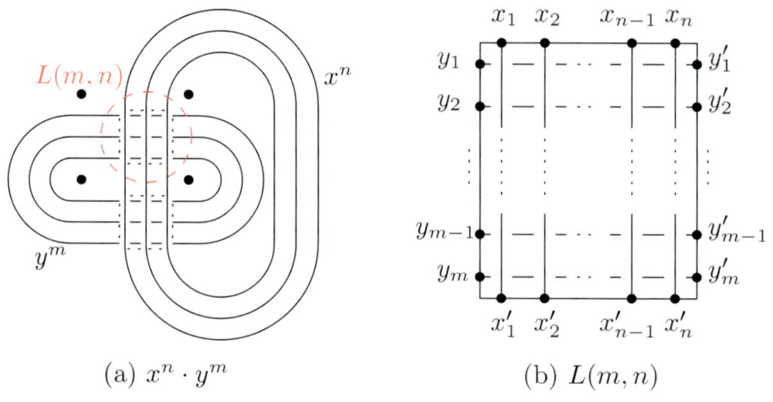

FIGURE 1.1. Product of curves in $F_{0,4}$ and lattice crossing $L(m,n)$

plane rooted trees was introduced by J.H. Przytycki in [**24, 25**][3] and studied in [**4–6**]. Using a version of this polynomial (for weighted plane rooted trees), a formula for coefficients of Catalan states of $L(m,n)$ with no returns on one side was given in [**11**]. Finally, formulas for $C(A)$ for some families of Catalan states with returns on all sides were found in [**10**] and the problem of determining coefficients of Catalan states of B-type lattice crossing was studied in [**12**]. In this paper, we show that $C(A)$ for any Catalan state C of $L(m,n)$ can be expressed as a linear combination over $\mathbb{Q}(A)$ of coefficients of Catalan states with no top returns. This fact combined with results of [**11**] gives an efficient method for finding $C(A)$.

The paper is organized as follows. In Section 2, we introduce some new terminology describing parts of a Catalan state, such as a *roof state*, a *middle state*, a *floor state*, etc. (see Definition 2.1), and give a short summary of results from [**9**] and [**11**]. In Section 3, we derive formula (3.6) (called the *first-row expansion*) in Proposition 3.11 which plays an important role in the next section. We also show that Laurent polynomials $C(A)$ have non-negative integer coefficients (Corollary 3.10) and $C(A) \neq 0$ if and only if the Catalan state C is realizable (Theorem 3.13). In Section 4, we define Θ_A-*state expansion* (Definition 4.1) and show that it exists for any roof state R (Theorem 4.7). The proof of Theorem 4.7 yields an algorithm (Algorithm 4.1) for finding Θ_A-state expansion for R which allows us to compute $C(A)$ for any Catalan state C of $L(m,n)$ with R as its roof state. We conclude this paper by presenting applications of Algorithm 4.1 (see Example 4.9 and Corollary 4.10) and providing an example of a Catalan state C in $L(9,5)$ with non-unimodal coefficient $C(A)$ (see Example 4.11).

[3]Plucking polynomial was introduced by J.H. Przytycki during his talk *"q-polynomial invariant of rooted trees"*, Knots in Washington XXXVIII: *30 years of the Jones polynomial*, GWU, May 9-11, 2014.

2. Preliminaries

Let $\mathrm{R}^2_{m,n,2k-n}$ be a rectangle with $2(m+k)$ points fixed on its boundary as shown in Figure 2.1(a). A *crossingless connection* C in $\mathrm{R}^2_{m,n,2k-n}$ (or simply *crossingless connection*) consists of $(m+k)$ arcs embedded in the rectangle that join points that are fixed on its boundary. For a crossingless connection C in $\mathrm{R}^2_{m,n,2k-n}$, we will refer to points fixed on the top side (respectively, bottom, left, and right side) of $\mathrm{R}^2_{m,n,2k-n}$ as its top-boundary points (respectively, bottom-, left-, and right-boundary points) and we denoted by $\mathrm{ht}(C), n_t(C), n_b(C)$ the number of left-, top-, bottom-boundary points of C. We label these points using the labels of corresponding points (x_i, x_i', y_j, y_j') fixed on the boundary of $\mathrm{R}^2_{m,n,2k-n}$.

An arc of C that joins a pair of top-boundary points is called a *top return*. Analogously, we can define *bottom, left,* and *right returns.*

DEFINITION 2.1. A *roof state* (respectively *floor state*) is a crossingless connection with no bottom (respectively top) returns (see Figure 2.1(b) and (c)). A roof state with no top returns is called a *middle state* (see Figure 2.1(d)). A roof state R with $\mathrm{ht}(R) = 0$ is called a *top state*, likewise, a floor state F with $\mathrm{ht}(F) = 0$ is called a *bottom state.*

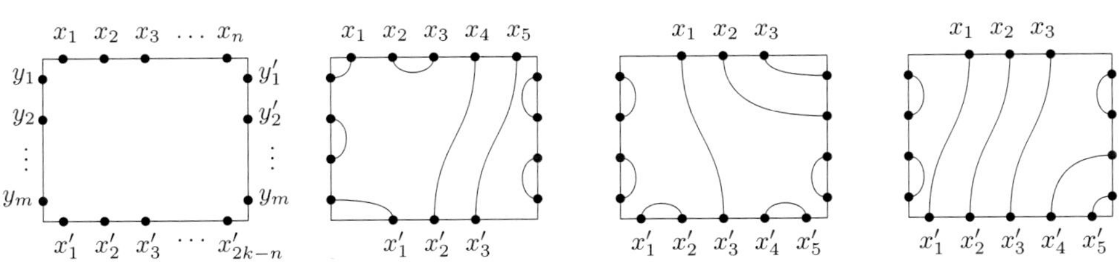

(a) $\mathrm{R}^2_{m,n,2k-n}$ (b) R (c) F (d) M

FIGURE 2.1. Rectangle $\mathrm{R}^2_{m,n,2k-n}$, roof state R, floor state F, and middle state M

For a pair of tangles T_1 and T_2, their *vertical product* $T_1 *_v T_2$ is the tangle shown in Figure 2.2 whenever the numbers of bottom-boundary points of T_1 and top-boundary points of T_2 are equal. Otherwise $T_1 *_v T_2$ is not a tangle and in such a case we set $T_1 *_v T_2 = K_0$. Thus, naturally we define

$$K_0 *_v T_2 = T_1 *_v K_0 = K_0 *_v K_0 = K_0.$$

Horizontal product $T_1 *_h T_2$ of tangles T_1 and T_2 can be defined in an analogous way to the vertical product (see Figure 2.2). Moreover, as it could easily be seen,

$$(T_{11} *_h T_{12}) *_v (T_{21} *_h T_{22}) = (T_{11} *_v T_{21}) *_h (T_{12} *_v T_{22})$$

for any tangles $T_{11}, T_{12}, T_{21}, T_{22}$ provided that all vertical and horizontal products above are not K_0. Furthermore, for crossingless connections C_1, C_2 and a floor state F, if $C_1 *_v F = C_2 *_v F \neq K_0$, then $C_1 = C_2$. Analogously, if $R *_v C_1 = R *_v C_2 \neq K_0$ for a roof state R, then also $C_1 = C_2$.

A crossingless connection C is called a *Catalan state* if $n_t(C) = n_b(C)$, and we denote by $\mathrm{Cat}(m, n)$ the set of all such states in $\mathrm{R}^2_{m,n,n}$.

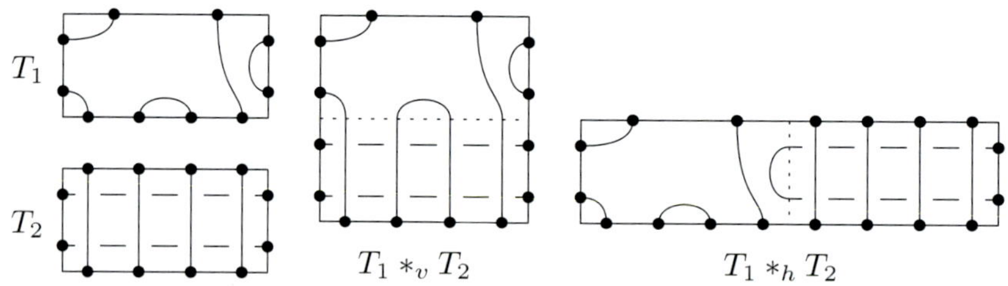

FIGURE 2.2. Vertical and horizontal products of T_1 and T_2

Recall after [**9**], an $m \times n$-*lattice crossing* $L(m,n)$ is a $2(m+n)$-tangle obtained by placing n parallel vertical line segments above m parallel horizontal line segments (see Figure 1.1(b)). A Kauffman state s of $L(m,n)$ is an assignment of positive and negative markers to all of its crossings.[4] Let $\mathcal{K}(m,n)$ be the set of all Kauffman states of $L(m,n)$. For $s \in \mathcal{K}(m,n)$, we denote by D_s the diagram obtained from $L(m,n)$ after smoothing all of its crossings according to the rule shown in Figure 2.3. Let C_s be the Catalan state obtained from D_s after removing all of its closed components. We say that a Catalan state C is *realizable* if $C = C_s$ for some $s \in \mathcal{K}(m,n)$.

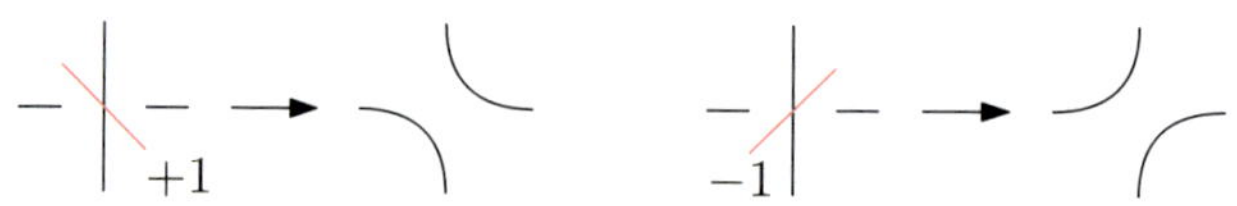

FIGURE 2.3. Positive and negative markers

As it was shown in [**22**], $\mathrm{Cat}(m,n)$ is a basis for RKBSM of a cylinder with $2(m+n)$ points fixed on its boundary. Therefore,

$$L(m,n) = \sum_{C \in \mathrm{Cat}(m,n)} C(A)\, C. \tag{2.1}$$

In the linear combination above, the Laurent polynomials $C(A)$ are called *coefficients of Catalan states C of lattice crossing $L(m,n)$*. Let $|D_s|$ be the number of closed components of D_s, and let $p(s)$ and $n(s)$ be the number of positive and negative markers of s, respectively. Denote by $\mathcal{K}(C)$ the set of all $s \in \mathcal{K}(m,n)$ such that $C_s = C$. Then $C(A)$ is given by

$$C(A) = \sum_{s \in \mathcal{K}(C)} A^{p(s)-n(s)}(-A^2 - A^{-2})^{|D_s|}. \tag{2.2}$$

We extend the definition of the coefficient of a Catalan state of a lattice crossing as follows. Let C be a crossingless connection or $C = K_0$, define $[[C]]_A$ by

$$[[C]]_A = \begin{cases} C(A), & \text{if } C \text{ is a Catalan state,} \\ 0, & \text{otherwise.} \end{cases}$$

For a crossingless connection C, denote by $\overline{C}$ its reflection about a vertical line and by C^* its π-rotation. We also put $\overline{K_0} = K_0$ and $K_0^* = K_0$.

---

[4] A Kauffman state s of $L(m,n)$ can also be viewed as an $m \times n$ matrix with entries ± 1.

REMARK 2.2. As one can check, for any crossingless connection C or $C = K_0$,

(i) $[[\overline{C}]]_A = [[C]]_{A^{-1}}$ and
(ii) $[[C^*]]_A = [[C]]_A$.

Let l_i^h and l_j^v be horizontal and vertical lines in $\mathrm{R}^2_{m,n,n}$ that are shown in Figure 2.4(a). Note that, horizontal lines l_i^h are also well-defined in $\mathrm{R}^2_{m,n,2k-n}$ when $k \neq n$. Given a crossingless connection C and a horizontal line $l = l_i^h$, using a regular isotopy (that keeps boundary points fixed) we deform C so that l intersects C at the minimal number of points and let us denote this number by $\#(C \cap l)$.

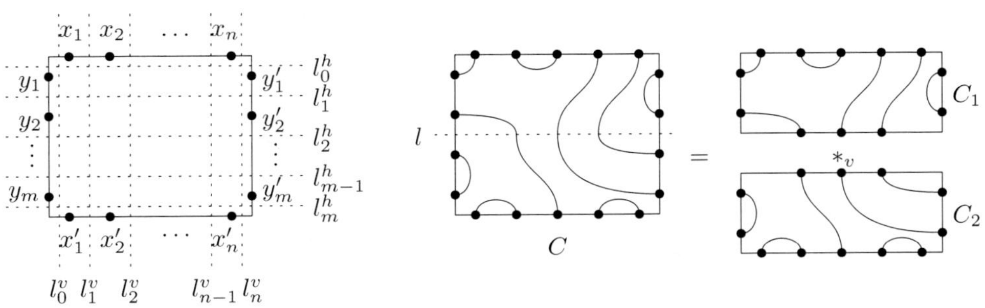

(a) l_i^h and l_j^v

(b) $C/l = (C_1, C_2)$

FIGURE 2.4. Horizontal lines l_i^h and vertical lines l_j^v; splitting C/l of C along l

THEOREM 2.3 ([**9**], Theorem 2.5). *A Catalan state $C \in \mathrm{Cat}(m,n)$ is realizable if and only if $\#(C \cap l_i^h) \leq n$ for $i = 1, 2, \ldots, m-1$ and $\#(C \cap l_j^v) \leq m$ for $j = 1, 2, \ldots, n-1$.*

Define *splitting* C/l of a crossingless connection C along a horizontal line l to be a pair (C_1, C_2) of crossingless connections, such that $C = C_1 *_v C_2$ and $n_b(C_1) = \#(C \cap l)$ (see Figure 2.4(b)).

THEOREM 2.4 ([**11**], Theorem 5.1). *Given a crossingless connection C and a horizontal line l with $\#(C \cap l) \geq \min\{n_t(C), n_b(C)\}$, let (C_1, C_2) be the splitting C/l of C along l. Then*

$$[[C]]_A = [[C_1]]_A \, [[C_2]]_A.$$

REMARK 2.5. An analogous result to Theorem 2.4 holds for a Catalan state C and a vertical line $l = l_j^v$ with $\#(C \cap l) = \mathrm{ht}(C)$.

Let $C \in \mathrm{Cat}(m,n)$ be a Catalan state with no bottom returns. Following the construction given in [**11**], one defines a dual plane rooted tree $T(C)$ with root v_0 and a weight function α (called *delay*)[5] as shown in Figure 2.5.

[5]The delay is defined from the set of leaves of $T(C)$ (not including the root) to $\mathbb{N}$ by $\alpha(v) = k$ if v corresponds to a left or right return with its lower end y_k or y'_k and $\alpha(v) = 1$ otherwise.

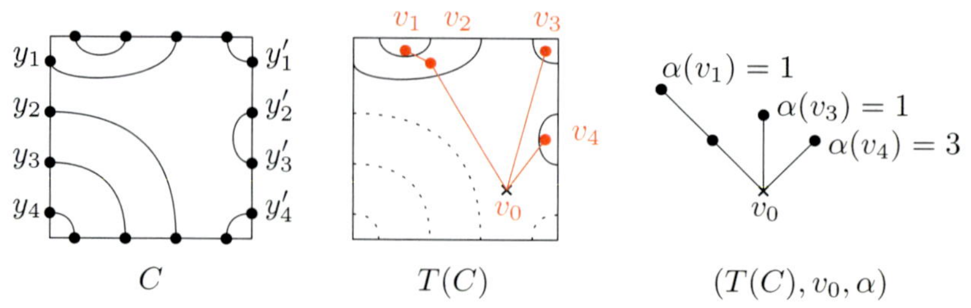

FIGURE 2.5. Plane rooted tree $(T(C), v_0)$ with delay α

DEFINITION 2.6 ([**11**], Definition 3.1). Let (T, v_0) be a plane rooted tree T with root v_0 and let α be a weight function from the set of leaves of T (not including v_0) to $\mathbb{N}$. Denote by $L_1(T)$ the set of all leaves v of T with $\alpha(v) = 1$. The *plucking polynomial* Q_q of (T, v_0, α) is a polynomial in variable q defined as follows: For T with no edges $Q_q(T, v_0, \alpha) = 1$, and otherwise

$$Q_q(T, v_0, \alpha) = \sum_{v \in L_1(T)} q^{r(T, v_0, v)} Q_q(T - v, v_0, \alpha_v),$$

where $r(T, v_0, v)$ is the number of vertices of T to the right of the unique path connecting v_0 and v, and α_v is the weight function defined by $\alpha_v(u) = \max\{1, \alpha(u) - 1\}$ if u is a leaf of T and $\alpha_v(u) = 1$ if u is a new leaf of $T - v$.

As shown in [**11**], coefficients of realizable Catalan states of $L(m, n)$ with no bottom returns can be found using Kauffman states with rows r_i, $1 \leq i \leq m$, in the form

$$r_i = (\underbrace{1, 1, \ldots, 1}_{b_i}, \underbrace{-1, -1, \ldots, -1}_{n - b_i}),$$

where $0 \leq b_i \leq n$. Hence, these states are uniquely determined by sequences $\mathbf{b} = (b_1, b_2, \ldots, b_m)$. Denote by $\mathfrak{b}(C)$ the set of all such sequences that realize C and let $\beta(C) = \max\{\|\mathbf{b}\| : \mathbf{b} \in \mathfrak{b}(C)\}$, where $\|\mathbf{b}\| = b_1 + b_2 + \cdots + b_m$.

THEOREM 2.7 ([**11**], Theorem 3.4). *Let $C \in \mathrm{Cat}(m, n)$ be a realizable Catalan state with no bottom returns, and let $T = T(C)$ be its corresponding plane rooted tree with root v_0 and delay α. Then*

$$C(A) = A^{2\beta(C) - mn} Q^*_{A^{-4}}(T, v_0, \alpha),$$

where $Q^*_q(T, v_0, \alpha) = q^{-\min \deg_q(Q_q(T, v_0, \alpha))} Q_q(T, v_0, \alpha)$.

3. First-row expansion

Let $\mathcal{B}_n$ be the set of all bottom states F with $n_b(F) = n$.[6] Define a map φ_n from $\mathcal{B}_n$ to the set $\mathrm{Fin}(\mathbb{N})$ of all finite subsets of $\mathbb{N}$ by $\varphi_n(F) = \{i_1, i_2, \ldots, i_k\}$, where $i_1 < i_2 < \cdots < i_k$ are the indices of the left ends of bottom returns of F. For example, $\varphi_7(F) = \{3, 4\}$ for the bottom state F shown in Figure 3.1.

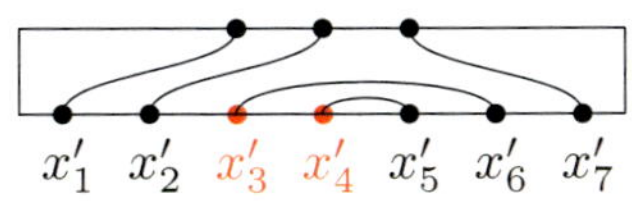

FIGURE 3.1. Indices of bottom returns of F

Let $\mathcal{L}_n = \varphi_n(\mathcal{B}_n)$, then clearly $\varphi_n : \mathcal{B}_n \to \mathcal{L}_n$ is a bijection. Thus, for each $I \in \mathcal{L}_n$ there is a corresponding bottom state $\varphi_n^{-1}(I)$. Clearly, $\varphi_n^{-1}(\emptyset) = L(0, n)$ and $\varphi_n^{-1}(\{i\})$ for $1 \le i \le n-1$ is given in Figure 3.2. We show how to find $\varphi_n^{-1}(I)$ for an arbitrary $I \in \mathcal{L}_n$ in Remark 3.3.

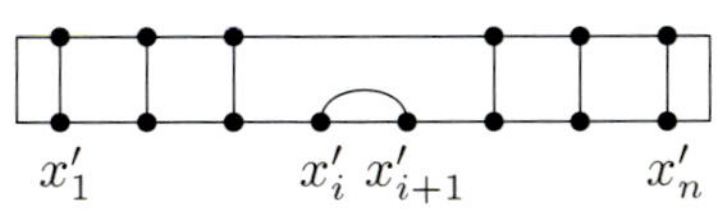

FIGURE 3.2. $\varphi_n^{-1}(\{i\})$ for $1 \le i \le n-1$

Let C be a crossingless connection with $n_b(C) = n$ or $C = K_0$, and $I \in \mathrm{Fin}(\mathbb{N})$. If $C \neq K_0$ and $I \in \mathcal{L}_n$, define C_I as the unique crossingless connection C' such that

$$C = C' *_v \varphi_n^{-1}(I)$$

provided that C' exists, and set $C_I = K_0$ otherwise. If $C = K_0$ or $I \notin \mathcal{L}_n$, we also set $C_I = K_0$. Geometrically, C_I is simply a crossingless connection obtained from C after removing its all bottom returns indexed by I together with their ends (see Figure 3.3).

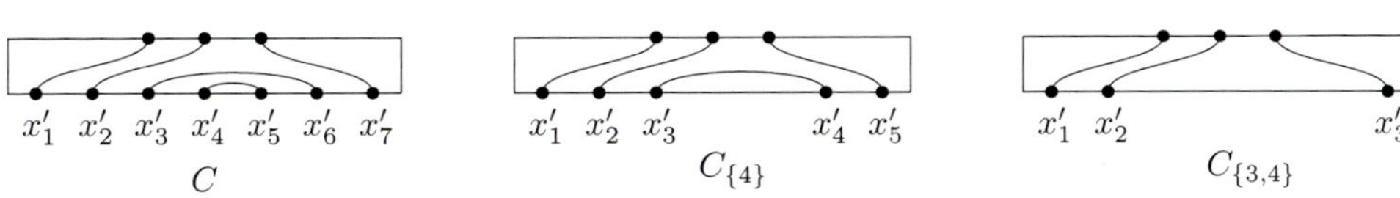

FIGURE 3.3. Crossingless connections C and C_I for $I = \{4\}, \{3, 4\}$

Define a partial order on $\mathcal{L}_n$ as follows: for $I, J \in \mathcal{L}_n$, $J \preceq_n I$ if $(\varphi_n^{-1}(I))_J \neq K_0$ (see Figure 3.4). Notice that, if $J \preceq_n I$ then

$$\begin{aligned}
\varphi_{n+1}^{-1}(I) = \varphi_n^{-1}(I) *_h L(0,1) &= ((\varphi_n^{-1}(I))_J *_v \varphi_n^{-1}(J)) *_h (L(0,1) *_v L(0,1)) \\
&= ((\varphi_n^{-1}(I))_J *_h L(0,1)) *_v (\varphi_n^{-1}(J) *_h L(0,1)) \\
&= ((\varphi_n^{-1}(I))_J *_h L(0,1)) *_v \varphi_{n+1}^{-1}(J).
\end{aligned}$$

It follows that, if $J \preceq_n I$ in $\mathcal{L}_n$ then $J \preceq_{n+1} I$ in $\mathcal{L}_{n+1}$. Therefore, $(\mathcal{L}_n, \preceq_n)$ is a subposet of $(\mathcal{L}_{n+1}, \preceq_{n+1})$ and thus one defines a partial order $\preceq$ on $\mathrm{Fin}(\mathbb{N})$ by: $I \preceq J$ if $I \preceq_n J$ for some n.

[6] One can show that the number of such states is $|\mathcal{B}_n| = \binom{n}{\lfloor \frac{n}{2} \rfloor}$.

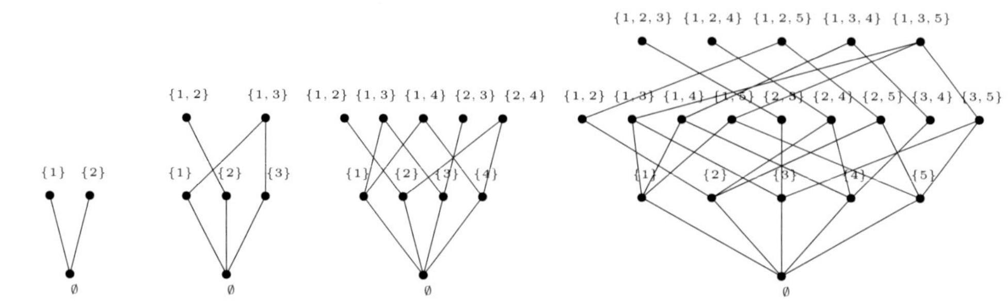

(a) $(\mathcal{L}_3, \preceq_3)$ (b) $(\mathcal{L}_4, \preceq_4)$ (c) $(\mathcal{L}_5, \preceq_5)$ (d) $(\mathcal{L}_6, \preceq_6)$

FIGURE 3.4. Posets $(\mathcal{L}_n, \preceq_n)$ for $n = 3, 4, 5, 6$

Clearly, $I = \{i\} \in \mathcal{L}_{i+1}$, and if $I \in \mathcal{L}_m$ for some m and $I \neq \emptyset$ then $I \cup \{j\} \in \mathcal{L}_{m+1}$ for any $j < \min I$. Therefore, given a non-empty finite subset I of $\mathbb{N}$, $I \in \mathcal{L}_{\max I + |I|}$. Hence, there is a non-negative integer n_I such that, $I \in \mathcal{L}_n$ for $n \geq n_I$ and $I \notin \mathcal{L}_n$ for $n < n_I$.

For $I, J \in \mathrm{Fin}(\mathbb{N})$, let $n^* = \max\{n_I + 2|J|, n_J\}$. Define an operation $I \oplus J$ by

$$I \oplus J = \varphi_{n^*}(\varphi^{-1}_{n^*-2|J|}(I) *_v \varphi^{-1}_{n^*}(J)).$$

The geometric interpretation of $I \oplus J$ is given in Figure 3.5.

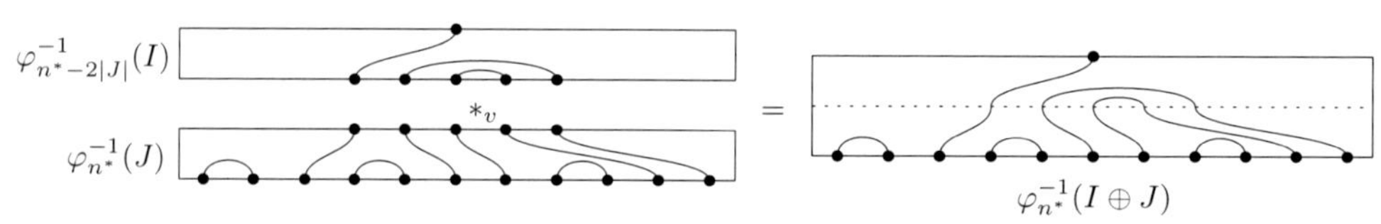

FIGURE 3.5. Geometric interpretation of $I \oplus J$

REMARK 3.1. Based on the geometric interpretation of $I \oplus J$ given above, it can easily be seen that

$$\{i\} \oplus \{j\} = \begin{cases} \{i, j\}, & \text{if } i < j, \\ \{j, i+2\}, & \text{otherwise,} \end{cases} = \begin{cases} \{j-2\} \oplus \{i\}, & \text{if } i \leq j-2, \\ \{j\} \oplus \{i+2\}, & \text{if } i \geq j. \end{cases}$$

PROPOSITION 3.2. *Let $I, J \in \mathrm{Fin}(\mathbb{N})$.*

(i) $I \oplus J = I \cup J$ *if either* $I = \emptyset$ *or* $J = \emptyset$, *or both* I *and* J *are non-empty and* $\max I < \min J$.

(ii) $I \oplus J \in \mathcal{L}_n$ *if and only if* $n \geq \max\{n_I + 2|J|, n_J\}$, *i.e.,*

$$n_{I \oplus J} = \max\{n_I + 2|J|, n_J\}.$$

(iii) *If* $n \geq n_{I \oplus J}$, *then* $\varphi_n^{-1}(I \oplus J) = \varphi^{-1}_{n-2|J|}(I) *_v \varphi_n^{-1}(J)$.

(iv) *If* $I \in \mathcal{L}_{n-2|J|}$ *and* $J \in \mathcal{L}_n$, *then* $I \oplus J \in \mathcal{L}_n$.

PROOF. Let $n^* = \max\{n_I + 2|J|, n_J\}$. For (i), if either $I = \emptyset$ or $J = \emptyset$, then clearly $I \oplus J = I \cup J$. Thus, assume that I and J are non-empty and notice that J is a subset of indices of bottom returns of $F = \varphi^{-1}_{n^*-2|J|}(I) *_v \varphi^{-1}_{n^*}(J)$. Since for each $i = 1, 2, \dots, \min J - 1$ there is an arc joining x_i and x'_i in $\varphi^{-1}_{n^*}(J)$ and $\max I \le \min J - 1$, it follows that indices of bottom returns of $\varphi^{-1}_{n^*-2|J|}(I)$ are among the indices of bottom returns of F. Therefore, $I \cup J \subseteq I \oplus J$. Since $|I \oplus J| = |I| + |J| = |I \cup J|$, we see that $I \oplus J = I \cup J$.

For (ii), by definition $I \oplus J \in \mathcal{L}_{n^*}$, so $I \oplus J \in \mathcal{L}_n$ for any $n \ge n^*$. To prove (ii), it suffices to show that if $n < n^*$ then $I \oplus J \notin \mathcal{L}_n$. Indeed, either $n^* - 2|J| = n_I$ or $n^* = n_J$. Therefore, $x'_{n^*-2|J|}$ is a right end of a bottom return in $\varphi^{-1}_{n^*-2|J|}(I)$ or x'_{n^*} is a right end of a bottom return in $\varphi^{-1}_{n^*}(J)$. Hence, x'_{n^*} is a right end of a bottom return in $\varphi^{-1}_{n^*}(I \oplus J)$. It follows that $I \oplus J \notin \mathcal{L}_n$ for any $n < n^*$.

For (iii), we see that by (ii) $n_{I \oplus J} = n^*$, so $I \in \mathcal{L}_{n^*-2|J|} \subseteq \mathcal{L}_{n-2|J|}$ and $J \in \mathcal{L}_{n^*} \subseteq \mathcal{L}_n$ for $n \ge n_{I \oplus J}$, and therefore $\varphi^{-1}_{n-2|J|}(I)$ and $\varphi^{-1}_n(J)$ are bottom states. Consequently,

$$\begin{aligned}
\varphi^{-1}_n(I \oplus J) &= \varphi^{-1}_{n^*}(I \oplus J) *_h L(0, n - n^*) \\
&= (\varphi^{-1}_{n^*-2|J|}(I) *_v \varphi^{-1}_{n^*}(J)) *_h (L(0, n - n^*) *_v L(0, n - n^*)) \\
&= (\varphi^{-1}_{n^*-2|J|}(I) *_h L(0, n - n^*)) *_v (\varphi^{-1}_{n^*}(J) *_h L(0, n - n^*)) \\
&= \varphi^{-1}_{n-2|J|}(I) *_v \varphi^{-1}_n(J).
\end{aligned}$$

For (iv), we notice that, since $I \in \mathcal{L}_{n-2|J|}$ and $J \in \mathcal{L}_n$, by definition of n_I and n_J, $n - 2|J| \ge n_I$ and $n \ge n_J$, i.e., $n \ge \max\{n_I + 2|J|, n_J\}$. Therefore, using (ii) we conclude that $I \oplus J \in \mathcal{L}_n$. □

REMARK 3.3. Let $I = \{i_1, i_2, \dots, i_t\} \in \mathcal{L}_n$ with $I \neq \emptyset$, then by Proposition 3.2(i) and (iii), we see that

$$\begin{aligned}
\varphi^{-1}_n(I) &= \varphi^{-1}_n(\{i_1\} \oplus \{i_2\} \oplus \cdots \oplus \{i_t\}) \\
&= \varphi^{-1}_{n-2t+2}(\{i_1\}) *_v \varphi^{-1}_{n-2t+4}(\{i_2\}) *_v \cdots *_v \varphi^{-1}_n(\{i_t\}).
\end{aligned}$$

Thus, $\varphi^{-1}_n(I)$ can be found for all $I \in \mathcal{L}_n$. Furthermore, since $n_I = \max I'$, where I' is the set of indices of right ends of the bottom returns of $\varphi^{-1}_{i_t+|I|}(I)$, we can also find n_I.

LEMMA 3.4. *For every roof state R, floor state F, crossingless connection C, and $I, J \in \mathrm{Fin}(\mathbb{N})$,*

$$(R *_v F)_I = R *_v F_I \tag{3.1}$$

and

$$(C_J)_I = C_{I \oplus J}. \tag{3.2}$$

PROOF. Let $n = n_b(F)$. If $R *_v F = K_0$ then $n_b(R) \neq n_t(F)$. Since either $F_I = K_0$ or F_I is a floor state with $n_t(F_I) = n_t(F)$, it follows that $R *_v F_I = K_0$. If $R *_v F$ is a crossingless connection, $R *_v F$ and F have same bottom returns and then one argues that $(R *_v F)_I = K_0$ if and only if $F_I = K_0$. Therefore, it suffices to show (3.1) when $R *_v F \neq K_0$ and $F_I \neq K_0$. Indeed, in such a case,

$$R *_v F = R *_v F_I *_v \varphi^{-1}_n(I),$$

so $(R *_v F)_I = R *_v F_I$.

To prove (3.2), we see that by using (3.1) it suffices to assume that $C = F$ is a bottom state. Notice that, $(\varphi_{n'}^{-1}(I \oplus J))_J = \varphi_{n'-2|J|}^{-1}(I) \neq K_0$ for $n' = n_{I\oplus J}$, so $J \preceq_{n'} I \oplus J$ and consequently $J \preceq I \oplus J$. Hence, if $J \not\preceq \varphi_n(F)$ then $I \oplus J \not\preceq \varphi_n(F)$, and therefore

$$(F_J)_I = (K_0)_I = K_0 = F_{I\oplus J}.$$

Assume that $J \preceq \varphi_n(F)$. If $I \not\preceq \varphi_{n-2|J|}(F_J)$, we show that $I \oplus J \not\preceq \varphi_n(F)$, so consequently

$$(F_J)_I = K_0 = F_{I\oplus J}.$$

Suppose that $I \oplus J \preceq \varphi_n(F)$, then by Proposition 3.2(ii), $n \geq n_{I\oplus J}$. Since by the definition and Proposition 3.2(iii)

$$F_J *_v \varphi_n^{-1}(J) = F = F_{I\oplus J} *_v \varphi_n^{-1}(I \oplus J) = F_{I\oplus J} *_v \varphi_{n-2|J|}^{-1}(I) *_v \varphi_n^{-1}(J),$$

it follows that

$$F_J = F_{I\oplus J} *_v \varphi_{n-2|J|}^{-1}(I).$$

Therefore, $(F_J)_I = F_{I\oplus J} \neq K_0$, i.e., $I \preceq \varphi_{n-2|J|}(F_J)$, a contradiction. Finally, if $I \preceq \varphi_{n-2|J|}(F_J)$, then $n \geq n_{I\oplus J}$ and by Proposition 3.2(iii),

$$F = F_J *_v \varphi_n^{-1}(J) = (F_J)_I *_v \varphi_{n-2|J|}^{-1}(I) *_v \varphi_n^{-1}(J) = (F_J)_I *_v \varphi_n^{-1}(I \oplus J),$$

hence, $F_{I\oplus J} = (F_J)_I$. □

For $I, J \in \mathrm{Fin}(\mathbb{N})$ with $J \preceq I$, define

$$I \ominus J = \varphi_{n_I-2|J|}((\varphi_{n_I}^{-1}(I))_J).$$

The geometric interpretation of $I \ominus J$ is given in Figure 3.6.

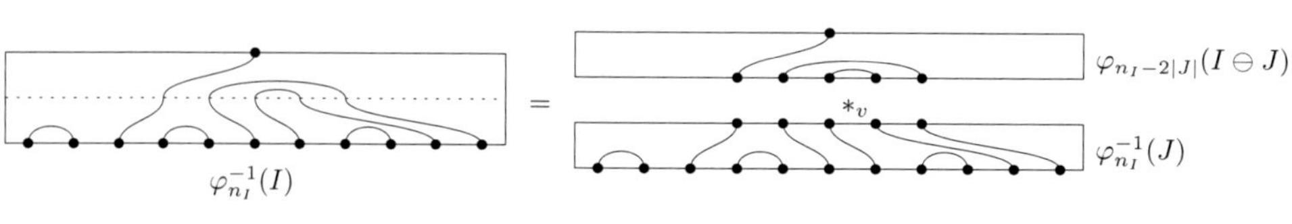

FIGURE 3.6. Geometric interpretation of $I \ominus J$

PROPOSITION 3.5. *Let $I, J, K \in \mathrm{Fin}(\mathbb{N})$ and $J \preceq I$.*

(i) *If $I \in \mathcal{L}_n$, then $I \ominus J \in \mathcal{L}_{n-2|J|}$, so consequently*

$$\varphi_{n-2|J|}^{-1}(I \ominus J) = (\varphi_n^{-1}(I))_J.$$

(ii) *$J \preceq K \oplus I$ and $(K \oplus I) \ominus J = K \oplus (I \ominus J)$. In particular, $(K \oplus I) \ominus I = K$.*
(iii) *$(I \ominus J) \oplus J = I$.*
(iv) *$J \oplus K \preceq I \oplus K$ and $(I \oplus K) \ominus (J \oplus K) = I \ominus J$.*

PROOF. For (i), since $n \geq n_I$, $I \ominus J \in \mathcal{L}_{n_I-2|J|} \subseteq \mathcal{L}_{n-2|J|}$. Consequently,

$$\begin{aligned}\varphi_{n-2|J|}^{-1}(I \ominus J) &= \varphi_{n_I-2|J|}^{-1}(I \ominus J) *_h L(0, n-n_I) = (\varphi_{n_I}^{-1}(I))_J *_h L(0, n-n_I) \\ &= (\varphi_{n_I}^{-1}(I) *_h L(0, n-n_I))_J \\ &= (\varphi_n^{-1}(I))_J.\end{aligned}$$

For (ii), let $n = n_{K\oplus I}$ then clearly $(\varphi_n^{-1}(K \oplus I))_I = \varphi_{n-2|I|}^{-1}(K) \neq K_0$. Therefore, $I \preceq_n K \oplus I$ and consequently $J \preceq I \preceq K \oplus I$. Moreover,

$$\begin{aligned}(\varphi_n^{-1}(K \oplus I))_J *_v \varphi_n^{-1}(J) = \varphi_n^{-1}(K \oplus I) &= \varphi_{n-2|I|}^{-1}(K) *_v \varphi_n^{-1}(I) \\ &= \varphi_{n-2|I|}^{-1}(K) *_v (\varphi_n^{-1}(I))_J *_v \varphi_n^{-1}(J),\end{aligned}$$

so $(\varphi_n^{-1}(K \oplus I))_J = \varphi_{n-2|I|}^{-1}(K) *_v (\varphi_n^{-1}(I))_J$. Hence,

$$\begin{aligned}\varphi_{n-2|J|}^{-1}((K \oplus I) \ominus J) = (\varphi_n^{-1}(K \oplus I))_J &= \varphi_{n-2|I|}^{-1}(K) *_v (\varphi_n^{-1}(I))_J \\ &= \varphi_{n-2|I|}^{-1}(K) *_v \varphi_{n-2|J|}^{-1}(I \ominus J) \\ &= \varphi_{n-2|J|}^{-1}(K \oplus (I \ominus J)),\end{aligned}$$

and consequently $(K \oplus I) \ominus J = K \oplus (I \ominus J)$. In particular, since $I \ominus I = \emptyset$ and $K \oplus \emptyset = K$, taking $J = I$ yields $(K \oplus I) \ominus I = K$.

For (iii), since $\varphi_{n_I-2|J|}^{-1}(I \ominus J) = (\varphi_{n_I}^{-1}(I))_J$,

$$\varphi_{n_I}^{-1}((I \ominus J) \oplus J) = \varphi_{n_I-2|J|}^{-1}(I \ominus J) *_v \varphi_{n_I}^{-1}(J) = (\varphi_{n_I}^{-1}(I))_J *_v \varphi_{n_I}^{-1}(J) = \varphi_{n_I}^{-1}(I).$$

Therefore, $(I \ominus J) \oplus J = I$.

For (iv), since $J \preceq I$, by (iii) and Proposition 3.2(iii),

$$\varphi_n^{-1}(I) = \varphi_n^{-1}((I \ominus J) \oplus J) = \varphi_{n-2|J|}^{-1}(I \ominus J) *_v \varphi_n^{-1}(J)$$

for all $n \geq n_I$. Since $n_I \geq n_J$, using Proposition 3.2(ii) one shows that

$$\begin{aligned}n_{I\oplus K} &= \max\{n_I + 2|K|, n_K\} \geq \max\{n_I + 2|K|, n_J + 2|K|, n_K\} \\ &= \max\{n_I + 2|K|, n_{J\oplus K}\}.\end{aligned}$$

Therefore, for $n = n_{I\oplus K}$,

$$\begin{aligned}\varphi_n^{-1}(I \oplus K) &= \varphi_{n-2|K|}^{-1}(I) *_v \varphi_n^{-1}(K) \\ &= \varphi_{n-2|K|-2|J|}^{-1}(I \ominus J) *_v \varphi_{n-2|K|}^{-1}(J) *_v \varphi_n^{-1}(K) \\ &= \varphi_{n-2|J\oplus K|}^{-1}(I \ominus J) *_v \varphi_n^{-1}(J \oplus K).\end{aligned}$$

It follows that $J \oplus K \preceq I \oplus K$. Moreover,

$$\varphi_{n-2|J\oplus K|}^{-1}((I \oplus K) \ominus (J \oplus K)) = (\varphi_n^{-1}(I \oplus K))_{J\oplus K} = \varphi_{n-2|J\oplus K|}^{-1}(I \ominus J),$$

so $(I \oplus K) \ominus (J \oplus K) = I \ominus J$. □

Let $\mathcal{W}$ be the set of all pairs (R, I), where R is a roof state and $I \in \text{Fin}(\mathbb{N})$. Define a relation $\trianglelefteq$ on $\mathcal{W}$ as follows: $(R, I) \trianglelefteq (R', I')$ if

(a) $I' \ominus I \in \mathcal{L}_{n_t(R)}$,
(b) $n_t(R) + 2|I| = n_t(R') + 2|I'|$, and
(c) $n_b(R) = n_b(R')$.

We also use $(R', I') \trianglerighteq (R, I)$ when $(R, I) \trianglelefteq (R', I')$ and, for any $\mathcal{P}' \subset \mathcal{W}$, we write $\mathcal{P}' \trianglerighteq (R, I)$ if $(R', I') \trianglerighteq (R, I)$ for all $(R', I') \in \mathcal{P}'$. The relation $\trianglelefteq$ is transitive as shown in the following lemma.

LEMMA 3.6. *If $(R, I) \trianglelefteq (R', I')$ and $(R', I') \trianglelefteq (R'', I'')$, then $(R, I) \trianglelefteq (R'', I'')$.*

PROOF. For (R, I) and (R'', I''), conditions (b) and (c) given in the definition of $\trianglelefteq$ are clearly satisfied. Notice that $I \preceq I' \preceq I''$, so $I'' \ominus I$ is defined. Since $I'' \ominus I' \in \mathcal{L}_{n_t(R')} = \mathcal{L}_{n+2|I|-2|I'|} = \mathcal{L}_{n-2|I'\ominus I|}$ and $I' \ominus I \in \mathcal{L}_n$, where $n = n_t(R)$, it follows by Proposition 3.5(iii), (ii), and Proposition 3.2(iv) that

$$I'' \ominus I = ((I'' \ominus I') \oplus I') \ominus I = (I'' \ominus I') \oplus (I' \ominus I) \in \mathcal{L}_n,$$

i.e., condition (a) also holds. Therefore, $(R, I) \trianglelefteq (R'', I'')$. □

Given $(R, I) \in \mathcal{W}$, let $\Theta_A(R, I; \cdot)$ be a function of variable F defined by

$$\Theta_A(R, I; F) = [[R *_v F_I]]_A, \tag{3.3}$$

where F is a floor state or $F = K_0$.

PROPOSITION 3.7. *Let $(R, I) \in \mathcal{W}$ with $n_t(R) = n$ and $n_b(R) = k$. Assume that there is a finite collection $\mathcal{P} \subset \mathcal{W}$ of pairs (R', I') with $I \preceq I'$, and rational functions $Q_{R',I'}(A) \in \mathbb{Q}(A)$ for each $(R', I') \in \mathcal{P}$, such that*

$$\Theta_A(R, I; \cdot) = \sum_{(R',I')\in\mathcal{P}} Q_{R',I'}(A)\, \Theta_A(R', I'; \cdot).$$

Then

$$\Theta_A(R, I; \cdot) = \sum_{(R',I')\in\mathcal{P}'} Q_{R',I'}(A)\, \Theta_A(R', I'; \cdot), \tag{3.4}$$

where $\mathcal{P}'$ is the set of all pairs $(R', I') \in \mathcal{P}$ such that $(R', I') \trianglerighteq (R, I)$. Furthermore, the following are true:

(i) *Let $J \in \text{Fin}(\mathbb{N})$, then $(R, I \oplus J) \trianglelefteq (R', I' \oplus J)$ for all $(R', I') \in \mathcal{P}'$ and*

$$\Theta_A(R, I \oplus J; \cdot) = \sum_{(R',I')\in\mathcal{P}'} Q_{R',I'}(A)\, \Theta_A(R', I' \oplus J; \cdot).$$

(ii) *Let M be a middle state with $n_t(M) = k$, then $(R *_v M, I) \trianglelefteq (R' *_v M, I')$ for all $(R', I') \in \mathcal{P}'$ and*

$$\Theta_A(R *_v M, I; \cdot) = \sum_{(R',I')\in\mathcal{P}'} Q_{R',I'}(A)\, \Theta_A(R' *_v M, I'; \cdot).$$

(iii) *If $I = \emptyset$ in (3.4) then $(\overline{R}, \emptyset) \trianglelefteq (\overline{R'}, \overline{I'})$ for all $(R', I') \in \mathcal{P}'$ and*

$$\Theta_A(\overline{R}, \emptyset; \cdot) = \sum_{(R',I')\in\mathcal{P}'} Q_{R',I'}(A^{-1})\, \Theta_A(\overline{R'}, \overline{I'}; \cdot),$$

where $\overline{R}$ and $\overline{R'}$ are reflections of R and R' about a vertical line, and $\overline{I'} = \varphi_n(\overline{\varphi_n^{-1}(I')})$.

PROOF. Given a floor state F either $R *_v F_I$ is a Catalan state or $R *_v F_I$ is not a Catalan state. In the former case, for each $(R', I') \in \mathcal{P} \setminus \mathcal{P}'$ either $F_{I'} = K_0$ or $F_{I'}$ is a floor state. If $F_{I'} = K_0$, then $\Theta_A(R', I'; F) = 0$. Otherwise, $n_b(F_{I'}) = n_b(F) - 2|I'| = n + 2|I| - 2|I'|$ and $n_t(F_{I'}) = n_t(F) = k$. Since $I \preceq I' \in \mathcal{L}_{n+2|I|}$, $I' \ominus I \in \mathcal{L}_n$ by Proposition 3.5(i). Therefore, it must be $n_t(R') \neq n + 2|I| - 2|I'|$ or $n_b(R') \neq k$, i.e., $R' *_v F_{I'}$ is not a Catalan state, so $\Theta_A(R', I'; F) = 0$. Consequently,

$$\Theta_A(R, I; F) = \sum_{(R', I') \in \mathcal{P}} Q_{R', I'}(A)\, \Theta_A(R', I'; F) = \sum_{(R', I') \in \mathcal{P}'} Q_{R', I'}(A)\, \Theta_A(R', I'; F).$$

In the latter case, for each $(R', I') \in \mathcal{P}'$ either $F_{I'}$ is a floor state with $n_t(F_{I'}) \neq k$ or $n_b(F_{I'}) \neq n + 2|I| - 2|I'|$, or $F_{I'} = K_0$. Thus, $\Theta_A(R', I'; F) = 0$. It follows that

$$\Theta_A(R, I; F) = 0 = \sum_{(R', I') \in \mathcal{P}'} Q_{R', I'}(A)\, \Theta_A(R', I'; F).$$

For (i), given $(R', I') \in \mathcal{P}'$ we see that conditions (b) and (c) in the definition of $\trianglelefteq$ for $(R, I \oplus J)$ and $(R', I' \oplus J)$ are clearly satisfied. Moreover, notice that $I \oplus J \preceq I' \oplus J$ and $(I' \oplus J) \ominus (I \oplus J) = I' \ominus I \in \mathcal{L}_n$ by Proposition 3.5(iv). Thus, condition (a) in the definition of $\trianglelefteq$ also holds. Therefore, $(R, I \oplus J) \trianglelefteq (R', I' \oplus J)$. Furthermore, by (3.3) and Lemma 3.4,

$$\Theta_A(R, I \oplus J; F) = [[R *_v F_{I \oplus J}]]_A = [[R *_v (F_J)_I]]_A = \Theta_A(R, I; F_J)$$

and

$$\Theta_A(R, I' \oplus J; F) = [[R *_v F_{I' \oplus J}]]_A = [[R *_v (F_J)_{I'}]]_A = \Theta_A(R, I'; F_J).$$

Since for any F and $J \in \mathrm{Fin}(\mathbb{N})$,

$$\Theta_A(R, I; F_J) = \sum_{(R', I') \in \mathcal{P}'} Q_{R', I'}(A)\, \Theta_A(R', I'; F_J)$$

by (3.4), it follows that

$$\Theta_A(R, I \oplus J; F) = \sum_{(R', I') \in \mathcal{P}'} Q_{R', I'}(A)\, \Theta_A(R', I' \oplus J; F).$$

The first part of statement (ii) is obvious. For the second part we see that, by (3.3) and Lemma 3.4,

$$\Theta_A(R, I; M *_v F) = [[R *_v (M *_v F)_I]]_A = [[R *_v M *_v F_I]]_A = \Theta_A(R *_v M, I; F)$$

and

$$\Theta_A(R', I'; M *_v F) = [[R' *_v (M *_v F)_{I'}]]_A = [[R' *_v M *_v F_{I'}]]_A = \Theta_A(R' *_v M, I'; F).$$

Since for any F and a middle state M with $n_t(M) = k$,

$$\Theta_A(R, I; M *_v F) = \sum_{(R', I') \in \mathcal{P}'} Q_{R', I'}(A)\, \Theta_A(R', I'; M *_v F)$$

by (3.4), it follows that

$$\Theta_A(R *_v M, I; F) = \sum_{(R', I') \in \mathcal{P}'} Q_{R', I'}(A)\, \Theta_A(R' *_v M, I'; F).$$

The first part of statement (iii) is obvious since $I = \emptyset$. For the second part, we start by showing that $[[\overline{R' *_v F'}]]_A = [[\overline{R' *_v F''}]]_A$, where $F' = (\overline{F})_{I'}$ and $F'' = F_{\overline{I'}}$. Consider two cases $n_b(F) \neq n$ or $n_b(F) = n$.

In the former case, for the left-hand side of the equality, we see that either $F' = K_0$ or F' is a floor state with $n_b(F') = n_b(\overline{F}) - 2|I'| \neq n - 2|I'|$. Since in both situations $\overline{R' *_v F'}$ is not a Catalan state, $[[\overline{R' *_v F'}]]_A = 0$. Analogously, for the right-hand side of the equality, either $F'' = K_0$ or F'' is a floor state with $n_b(F'') = n_b(F) - 2|\overline{I'}| \neq n - 2|I'|$. Thus, in both cases $\overline{R' *_v F''}$ is not a Catalan state and consequently, $[[\overline{R' *_v F''}]]_A = 0$.

In the latter case, if $F' \neq K_0$, then $\overline{F} = F' *_v \varphi_n^{-1}(I')$ and consequently $F = \overline{F'} *_v \overline{\varphi_n^{-1}(I')} = \overline{F'} *_v \varphi_n^{-1}(\overline{I'})$. It follows that $F'' = F_{\overline{I'}} = \overline{F'}$ and therefore $\overline{R' *_v F'} = \overline{R'} *_v \overline{F'} = \overline{R'} *_v F''$. If $F' = K_0$, we show that $F'' = K_0$. Suppose that $F'' \neq K_0$, then $F = F'' *_v \varphi_n^{-1}(\overline{I'})$ and consequently

$$\overline{F} = \overline{F''} *_v \overline{\varphi_n^{-1}(\overline{I'})} = \overline{F''} *_v \varphi_n^{-1}(I').$$

It follows that $F' = (\overline{F})_{I'} = \overline{F''} \neq K_0$, a contradiction. Therefore,

$$\overline{R' *_v F'} = K_0 = \overline{R'} *_v F''.$$

As we see from the above,

$$[[\overline{R' *_v (\overline{F})_{I'}}]]_A = [[\overline{R' *_v F'}]]_A = [[\overline{R'} *_v F'']]_A = [[\overline{R'} *_v F_{\overline{I'}}]]_A$$

for all F. Therefore, by (3.3) and Remark 2.2(i),

$$\Theta_{A^{-1}}(R, \emptyset; \overline{F}) = [[R *_v \overline{F}]]_{A^{-1}} = [[\overline{\overline{R} *_v F}]]_{A^{-1}} = [[\overline{R} *_v F]]_A = \Theta_A(\overline{R}, \emptyset; F)$$

and

$$\Theta_{A^{-1}}(R', I'; \overline{F}) = [[R' *_v (\overline{F})_{I'}]]_{A^{-1}} = [[\overline{R' *_v (\overline{F})_{I'}}]]_A = [[\overline{R'} *_v F_{\overline{I'}}]]_A = \Theta_A(\overline{R'}, \overline{I'}; F).$$

Since by (3.4),

$$\Theta_{A^{-1}}(R, \emptyset; \overline{F}) = \sum_{(R',I') \in \mathcal{P}'} Q_{R',I'}(A^{-1})\, \Theta_{A^{-1}}(R', I'; \overline{F})$$

for all F, it follows that

$$\Theta_A(\overline{R}, \emptyset; F) = \sum_{(R',I') \in \mathcal{P}'} Q_{R',I'}(A^{-1})\, \Theta_A(\overline{R'}, \overline{I'}; F).$$

□

Consider rectangle $\mathrm{R}^2_{m,n,2k-n}$ as shown in Figure 2.1(a) with $m \geq 1$. We label points y_1 by x_0 and y_1' by x_{n+1} and denote by e_j an arc that joins x_j and x_{j+1} for $0 \leq j \leq n$ (see Figure 3.7(a)).[7] For $\mathrm{R}^2_{0,n,2k-n}$, we define e_j for $1 \leq j \leq n-1$ in an analogous way.

Given a crossingless connection C, we write $e_j \in C$ if C has an arc that is regularly isotopic to e_j and define

$$\mathcal{J}(C) = \{j \mid e_j \in C\}.$$

For example, $\mathcal{J}(R) = \{0, 4, 6\}$ for the roof state R shown in Figure 3.7(b).

[7]If $n = 0$ and $m \geq 1$, e_0 is an arc with ends y_1, y_1'.

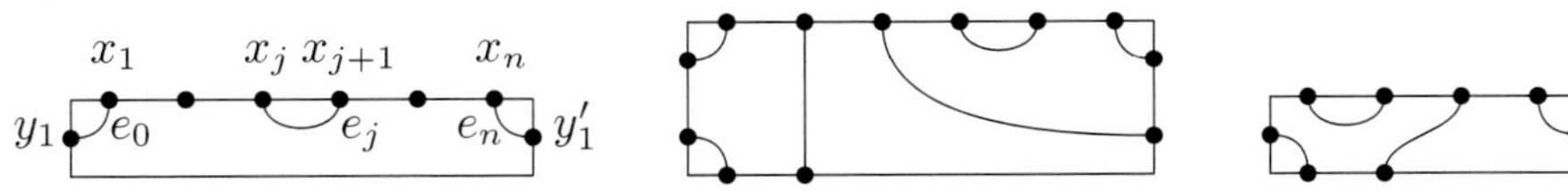

(a) e_0, e_j, e_n (b) R (c) $R^{\{4,6\}}$

FIGURE 3.7. Arcs e_0, e_j, e_n, roof state R, and R^J for $J = \{4, 6\}$

Let $\mathcal{D}_n$ be the set of all pairs (J, I) of finite subsets of $\mathbb{Z}_{\geq 0}$ such that $J = \{j_1, j_2, \ldots, j_{t+1}\}$ and $I = \{i_1, i_2, \ldots, i_t\}$, where $t \geq 0$ and $0 \leq j_k < i_k < j_{k+1} \leq n$ for all k. There is a bijection between $\mathcal{D}_n$ and the set $\mathcal{K}(1, n)$ of all Kauffman states of $L(1, n)$ that sends (J, I) to

$$s(J,I) = (\underbrace{1,\ldots,1}_{j_1}, \underbrace{-1,\ldots,-1}_{i_1-j_1}, \underbrace{1,\ldots,1}_{j_2-i_1}, \underbrace{-1,\ldots,-1}_{i_2-j_2}, \ldots, \underbrace{1,\ldots,1}_{j_{t+1}-i_t}, \underbrace{-1,\ldots,-1}_{n-j_{t+1}}).$$

Define a family of roof states

$$\mathcal{R}_n = \{(C_{s(J,I)})_I \mid (J, I) \in \mathcal{D}_n\}$$

and a map ψ_n from $\mathcal{R}_n$ to the set of all subsets of $\{0, 1, \ldots, n\}$ by

$$\psi_n(R) = \mathcal{J}(R).$$

Let $\mathcal{U}_n = \psi_n(\mathcal{R}_n)$, i.e., $\mathcal{U}_n$ is the set of $J = \{j_1, j_2, \ldots, j_{t+1}\}$ such that $0 \leq j_k \leq n$ and $j_{k+1} - j_k > 1$ for all k.

LEMMA 3.8. *ψ_n is a bijection between $\mathcal{R}_n$ and $\mathcal{U}_n$.*

PROOF. The surjectivity of ψ_n is clear, so it suffices to show that ψ_n is injective. Suppose that $\psi_n(R) = \psi_n(R')$ for some $R = (C_{s(J,I)})_I \in \mathcal{R}_n$ and $R' = (C_{s(J',I')})_{I'} \in \mathcal{R}_n$. Notice that, for a fixed $j \neq 0, n$, an arc $e_j \in R$ if and only if the j-th entry of $s(J, I)$ is 1 and the $(j + 1)$-th entry of $s(J, I)$ is -1. Analogously, $e_0 \in R$ if and only if the first entry of $s(J, I)$ is -1, and $e_n \in R$ if and only if the last entry of $s(J, I)$ is 1. Thus, $\mathcal{J}(R) = J$. The same argument holds for R', so $\mathcal{J}(R') = J'$. Therefore,

$$J = \mathcal{J}(R) = \psi_n(R) = \psi_n(R') = \mathcal{J}(R') = J'$$

and consequently $R = (C_{s(J,I)})_I = (C_{s(J',I')})_{I'} = R'$. □

Let C be a crossingless connection with $n_t(C) = n$ and $\mathrm{ht}(C) \geq 1$. Given $J \in \mathcal{U}_n$, define C^J as the unique crossingless connection C' such that

$$C = \psi_n^{-1}(J) *_v C'$$

provided that C' exists, and set $C^J = K_0$ otherwise. Thus, geometrically C^J is obtained from C by removing all e_j's for $j \in J$ together with their ends and then moving left- and right-boundary points along the sides of C to its top (if necessary), so that $\mathrm{ht}(C^J) = \mathrm{ht}(C) - 1$. For example, R^J for the roof state R given in Figure 3.7(b) and $J = \{4, 6\}$ is shown in Figure 3.7(c). We also let $C^J_I = (C^J)_I$ to make notations shorter.

LEMMA 3.9. *For $C \in \mathrm{Cat}(m,n)$ with $m \geq 1$, let*

$$\mathcal{S}(C) = \{(J,I) \in \mathcal{D}_n \mid C_I^J \neq K_0\}.$$

Then

$$C(A) = \sum_{(J,I)\in\mathcal{S}(C)} A^{-n+2(\|J\|-\|I\|)}\, C_I^J(A), \tag{3.5}$$

where $\|K\|$ stands for the sum of (distinct) elements of K.

PROOF. Expanding the first row of $L(m,n)$ yields

$$L(m,n) = \sum_{s\in\mathcal{K}(1,n)} A^{p(s)-n(s)} \cdot C_s *_v L(m-1,n).$$

Since there is a bijection between $s \in \mathcal{K}(1,n)$ and $(J,I) \in \mathcal{D}_n$ and $\mathcal{D}_n \subset \mathcal{U}_n \times \mathcal{L}_n$,

$$L(m,n) = \sum_{(J,I)\in\mathcal{D}_n} A^{-n+2(\|J\|-\|I\|)} \cdot \psi_n^{-1}(J) *_v \varphi_n^{-1}(I) *_v L(m-1,n).$$

Notice that diagrams $\psi_n^{-1}(J) *_v \varphi_n^{-1}(I) *_v L(m-1,n)$ and $\psi_n^{-1}(J) *_v L(m-1,n-2|I|) *_v \varphi_n^{-1}(I)$ are regularly isotopic, hence

$$L(m,n) = \sum_{(J,I)\in\mathcal{D}_n} A^{-n+2(\|J\|-\|I\|)} \cdot \psi_n^{-1}(J) *_v L(m-1,n-2|I|) *_v \varphi_n^{-1}(I).$$

Furthermore, using (2.1) for $L(m-1,n-2|I|)$ yields

$$L(m,n) = \sum_{(J,I)\in\mathcal{D}_n} \sum_{C'\in\mathrm{Cat}(m-1,n-2|I|)} A^{-n+2(\|J\|-\|I\|)}\, C'(A) \cdot \psi_n^{-1}(J) *_v C' *_v \varphi_n^{-1}(I).$$

Notice that, if for $(J,I) \in \mathcal{D}_n$ and $C' \in \mathrm{Cat}(m-1,n-2|I|)$ we let $C = \psi_n^{-1}(J) *_v C' *_v \varphi_n^{-1}(I)$, then $C \in \mathrm{Cat}(m,n)$ and $(J,I) \in \mathcal{S}(C)$. Conversely, for $C \in \mathrm{Cat}(m,n)$ and $(J,I) \in \mathcal{S}(C)$, if we let $C' = C_I^J$, then $(J,I) \in \mathcal{D}_n$ and $C' \in \mathrm{Cat}(m-1,n-2|I|)$. Therefore

$$\begin{aligned} L(m,n) &= \sum_{C\in\mathrm{Cat}(m,n)} \sum_{(J,I)\in\mathcal{S}(C)} A^{-n+2(\|J\|-\|I\|)}\, C_I^J(A) \cdot \psi_n^{-1}(J) *_v C_I^J *_v \varphi_n^{-1}(I) \\ &= \sum_{C\in\mathrm{Cat}(m,n)} \Big(\sum_{(J,I)\in\mathcal{S}(C)} A^{-n+2(\|J\|-\|I\|)}\, C_I^J(A) \Big) C. \end{aligned}$$

It follows that $C(A)$ is given by (3.5). □

COROLLARY 3.10. *Coefficients of $C(A)$ are non-negative integers for any Catalan state C.*

PROOF. The statement follows from (3.5) by induction on $\mathrm{ht}(C)$. □

Given a roof state R with $n_t(R) = n$ and $\mathrm{ht}(R) \geq 1$, define

$$\mathcal{H}(R) = \{(J,I) \in \mathcal{D}_n \mid J \subseteq \mathcal{J}(R)\}.$$

PROPOSITION 3.11. *Let $(R,\tilde{I}) \in \mathcal{W}$ with $n_t(R) = n$ and $\mathrm{ht}(R) \geq 1$. Then $(R,\tilde{I}) \trianglelefteq (R^J, I \oplus \tilde{I})$ for all $(J,I) \in \mathcal{H}(R)$ and*

$$\Theta_A(R,\tilde{I};\cdot) = \sum_{(J,I)\in\mathcal{H}(R)} A^{-n+2(\|J\|-\|I\|)}\, \Theta_A(R^J, I \oplus \tilde{I};\cdot). \tag{3.6}$$

We will refer to the formula (3.6) as the *first-row expansion* of $(R,\tilde{I})$.

PROOF. By Proposition 3.7(i), it suffices to show that the above statement holds for $\tilde{I} = \emptyset$.

Given $(J, I) \in \mathcal{H}(R)$, clearly $I \ominus \emptyset = I \in \mathcal{L}_n$ and $n_b(R^J) = n_b(R)$. Thus, conditions (a) and (c) in definition of $\trianglelefteq$ are satisfied. Moreover, since $(J, I) \in \mathcal{D}_n$, we see that $|J| = |I| + 1$. Therefore,

$$n_t(R^J) + 2|I| = n_t(R) + 2 - 2|J| + 2|I| = n_t(R),$$

so condition (b) also holds. Hence, $(R, \emptyset) \trianglelefteq (R^J, I)$ for all $(J, I) \in \mathcal{H}(R)$.

Assume that F is a floor state such that $C = R *_v F$ is a Catalan state. Since $C_I^J = R^J *_v F_I$ and $[[R^J *_v F_I]]_A = 0$ if $(J, I) \in \mathcal{H}(R) \setminus \mathcal{S}(C)$, where $\mathcal{S}(C)$ is defined in the statement of Lemma 3.9, by (3.5)

$$[[R *_v F]]_A = C(A) = \sum_{(J,I)\in\mathcal{H}(R)} A^{-n+2(\|J\|-\|I\|)} [[R^J *_v F_I]]_A.$$

If $R *_v F$ is not a Catalan state, as one may verify, $R^J *_v F_I$ is also not a Catalan state for any $(J, I) \in \mathcal{H}(R)$, hence

$$[[R *_v F]]_A = 0 = \sum_{(J,I)\in\mathcal{H}(R)} A^{-n+2(\|J\|-\|I\|)} [[R^J *_v F_I]]_A.$$

Consequently,

$$\Theta_A(R, \emptyset; F) = \sum_{(J,I)\in\mathcal{H}(R)} A^{-n+2(\|J\|-\|I\|)} \Theta_A(R^J, I; F)$$

for any floor state F. □

REMARK 3.12. Let $\mathcal{P} = \{(R^J, I \oplus \tilde{I}) \mid (J, I) \in \mathcal{H}(R)\}$, then (3.6) can be written as

$$\Theta_A(R, \tilde{I}; \cdot) = \sum_{(R',I')\in\mathcal{P}} Z_{R',I'}(A)\, \Theta_A(R', I'; \cdot), \tag{3.7}$$

where

$$Z_{R',I'}(A) = \sum_{(J,I)\in\mathcal{A}_{R',I'}} A^{-n+2(\|J\|-\|I\|)}$$

and $\mathcal{A}_{R',I'} = \{(J, I) \in \mathcal{H}(R) \mid (R^J, I \oplus \tilde{I}) = (R', I')\}$. We will also be referring to (3.7) as the first-row expansion of $(R, \tilde{I})$.

THEOREM 3.13. *For any Catalan state C, $C(A) \neq 0$ if and only if C is realizable.*

PROOF. Let $C \in \text{Cat}(m, n)$. If $C(A) \neq 0$ then by (2.2) the set $\mathcal{K}(C)$ is nonempty, so C is realizable. We show that if C is realizable then $C(A) \neq 0$. By Theorem 2.3, it suffices to show that, if C satisfies conditions $\#(C \cap l_i^h) \leq n$ for $i = 1, 2, \ldots, m-1$ and $\#(C \cap l_j^v) \leq m$ for $j = 1, 2, \ldots, n-1$, then $C(A) \neq 0$.

We prove the above statement by induction on m. Indeed, when $m = 0$ then clearly $C(A) = 1$. In the case $m = 1$, $\#(C \cap l_j^v) = 1$ for $j = 1, 2, \ldots, n-1$, by Remark 2.5, C is a horizontal product of Catalan states $C_i \in \text{Cat}(1, 1)$ and therefore $C(A) = \prod_{i=1}^n C_i(A) = A^k$ for some k. Thus $C(A) \neq 0$ in both cases.

Let $m \geq 2$, then using Remark 2.5 we may assume that $\#(C \cap l_j^v) \leq m-2$ for $j = 1, 2, \ldots, n-1$. If $|\mathcal{J}(C)| = 0$ then $\#(C \cap l_1^h) = n+2$ contradicting our assumption, so $|\mathcal{J}(C)| \geq 1$. Assume that $j' \in \mathcal{J}(C)$ and let $C' = C_I^J$ for $(J, I) = (\{j'\}, \emptyset)$. Since C' and C intersect horizontal lines the same number of times and $\#(C' \cap l_j^v) \leq \#(C \cap l_j^v) + 1 \leq m-1$ for $j = 1, 2, \ldots, n-1$, by the induction hypothesis, $C'(A) \neq 0$. According to Lemma 3.9,

$$C(A) = \sum_{(J,I) \in \mathcal{S}(C)} A^{-n+2(\|J\|-\|I\|)} C_I^J(A)$$

and, by Corollary 3.10, coefficients of $C_I^J(A)$ are non-negative. Since $C'(A) \neq 0$ appears in the above sum, it follows that $C(A) \neq 0$. □

Recall, for a Catalan state C with no bottom returns, $Q_q(T(C), v_0, \alpha)$ denotes its plucking polynomial of the plane rooted tree.

PROPOSITION 3.14. *A Catalan state C with no bottom returns is realizable if and only if $Q_q(T(C), v_0, \alpha) \neq 0$.*

PROOF. If C is realizable then $C(A) \neq 0$ by Theorem 3.13 and consequently, $Q_q(T(C), v_0, \alpha) \neq 0$ by Theorem 2.7. It remains to show that if $C \in \mathrm{Cat}(m, n)$ with no bottom returns is non-realizable then $Q_q(T(C), v_0, \alpha) = 0$. Using Lemma 2.2 of [**9**], for C with no bottom returns it must be $\#(C \cap l_j^v) \leq m$ for all $j = 1, 2, \ldots, n-1$. Since C is not realizable, there is $1 \leq i \leq m-1$ such that $\#(C \cap l_i^h) > n$ by Theorem 2.3, and in particular, $m \geq 2$.

Our proof is by induction on $m \geq 2$. For $m = 2$, since $\#(C \cap l_1^h) > n$, it must be $\#(C \cap l_1^h) = n+2$. Consequently, $T(C)$ has no leaves v with $\alpha(v) = 1$, so $Q_q(T(C), v_0, \alpha) = 0$ by Definition 2.6. For $m > 2$, if $\#(C \cap l_1^h) > n$ then $Q_q(T(C), v_0, \alpha) = 0$ by the same argument as in case $m = 2$. Assume that $\#(C \cap l_i^h) > n$ for some $i > 1$. By Definition 2.6,

$$Q_q(T(C), v_0, \alpha) = \sum_{v \in L_1(T(C))} q^{r(T(C), v_0, v)} Q_q(T(C) - v, v_0, \alpha_v).$$

For each $v \in L_1(T(C))$, let $C_v = C^{\{j_v\}} \in \mathrm{Cat}(m-1, n)$, where j_v is the index of e_j that corresponds to v. Clearly, C_v is not realizable since $\#(C_v \cap l_{i-1}^h) = \#(C \cap l_i^h) > n$. Moreover, $T(C_v) = T(C) - v$ and α_v is its delay, so

$$Q_q(T(C) - v, v_0, \alpha_v) = Q_q(T(C_v), v_0, \alpha_v) = 0$$

by induction hypothesis. Consequently, $Q_q(T(C), v_0, \alpha) = 0$. □

REMARK 3.15. From the result above, the formula for $C(A)$ given in Theorem 2.7 can be used for all Catalan states C with no bottom returns if we set $\beta(C) = 0$ when C is non-realizable and define $Q^*(T(C), v_0, \alpha) = 0$ whenever $Q(T(C), v_0, \alpha) = 0$.

4. Θ_A-state expansion

In this section we show that coefficient of any Catalan state C with a given roof state R (i.e., $C = R *_v F$ for some floor state F) can be found using coefficients of Catalan states with no top returns. For this purpose, we introduce a Θ_A-state expansion for (R, I) (see Definition 4.1) that expresses $\Theta_A(R, I; \cdot)$ as a linear combination (over the field of rational functions) of $\Theta_A(R', I'; \cdot)$'s, where R' is a middle state. The coefficient of C can be found by evaluating Θ_A-state expansion for $(R, \emptyset)$ at F. Since each $R' *_v F_{I'}$ is either a Catalan state with no top returns or K_0, $\Theta_A(R', I'; F) = [[R' *_v F_{I'}]]_A$ can be found by either applying Theorem 2.7 or simply $\Theta_A(R', I'; F) = 0$. As we will prove, Θ_A-state expansion exists for every (R, I) (Theorem 4.7) and the proof of Theorem 4.7 yields an algorithm (Algorithm 4.1) for finding a Θ_A-state expansion for any pair (R, I). This gives us an efficient method for finding coefficients of Catalan states with any given roof state.

DEFINITION 4.1. A Θ_A*-state expansion* for $(R, I) \in \mathcal{W}$ is a relation given by

$$\Theta_A(R, I; \cdot) = \sum_{(R',I') \in \mathcal{P}'} Q_{R',I'}(A)\, \Theta_A(R', I'; \cdot), \tag{4.1}$$

where $\mathcal{P}'$ is a finite subset of $\mathcal{W}$ that satisfies conditions

(i) R' in each pair $(R', I') \in \mathcal{P}'$ is a middle state and
(ii) $\mathcal{P}' \trianglerighteq (R, I)$,

and $0 \neq Q_{R',I'}(A) \in \mathbb{Q}(A)$ for every $(R', I') \in \mathcal{P}'$.

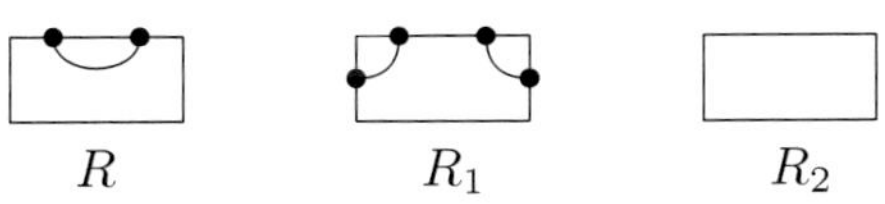

FIGURE 4.1. Roof states R, R_1, and R_2

EXAMPLE 4.2. A Θ_A-state expansion for $(R, \emptyset)$, where R is shown in Figure 4.1, is given by

$$\Theta_A(R, \emptyset; \cdot) = \frac{1}{A^{-2} + A^2}\, \Theta_A(R_1, \emptyset; \cdot) - \frac{1}{A^{-2} + A^2}\, \Theta_A(R_2, \{1\}; \cdot),$$

where R_1 and R_2 are shown in Figure 4.1. The equation above can be obtained using the first-row expansion (3.6) for $(R_1, \emptyset)$. Moreover, for any realizable $C \in \mathrm{Cat}(m, 2)$ with no top and bottom returns, by Theorem 2.4,

$$\Theta_A(R, \emptyset; \cdot) = \frac{1}{(A^{-2} + A^2)C(A)}\, \Theta_A(C *_v R_1, \emptyset; \cdot) - \frac{1}{A^{-2} + A^2}\, \Theta_A(R_2, \{1\}; \cdot). \tag{4.2}$$

As one may verify, (4.2) is a Θ_A-state expansion for $(R, \emptyset)$. Thus, in fact there are infinitely many different Θ_A-state expansions for $(R, \emptyset)$.

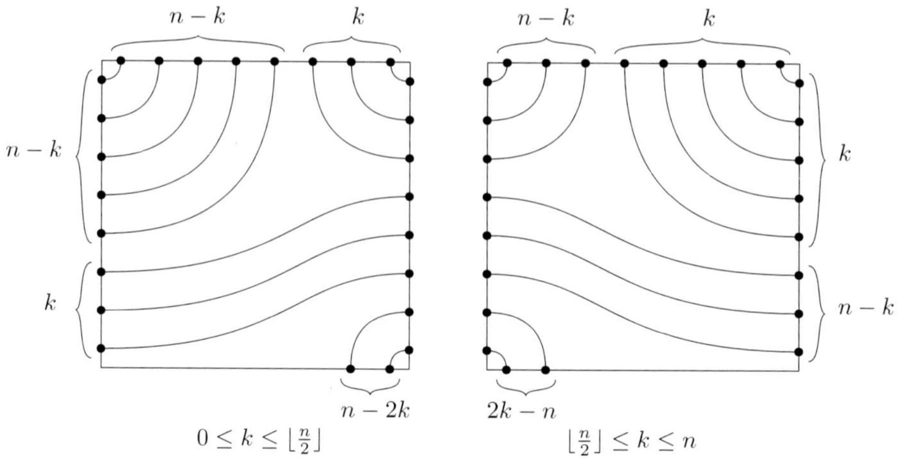

FIGURE 4.2. Middle states $R_{n,k}$

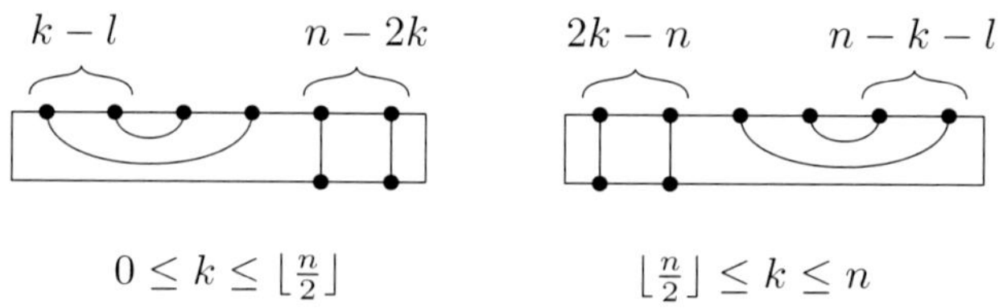

FIGURE 4.3. Top states $R'_{n,k,l}$

LEMMA 4.3. *Let $R_{n,k}$ and $R'_{n,k,l}$, $0 \le k \le n$ and $0 \le l \le \min\{k, n-k\}$, be as in Figure 4.2 and Figure 4.3, respectively. Then there are Laurent polynomials $Z_{R,I}(A)$ such that*

$$\Theta_A(R_{n,k}, \emptyset; \cdot) = \sum_{(R,I)\in\mathcal{P}} Z_{R,I}(A)\, \Theta_A(R, I; \cdot), \tag{4.3}$$

where $\mathcal{P} = \{(R'_{n,k,|I|}, I) \mid I \in \mathcal{L}_n,\ |I| \le \min\{k, n-k\}\} \trianglerighteq (R_{n,k}, \emptyset)$.

PROOF. For $n = 0$ the statement is clearly true, so we assume that $n = \mathrm{ht}(R_{n,k}) \ge 1$. Using the first-row expansion (3.7) for $(R_{n,k}, \emptyset)$ we see that

$$\Theta_A(R_{n,k}, \emptyset; \cdot) = \sum_{(R',I')\in\mathcal{P}_1} Z_{R',I'}(A)\, \Theta_A(R', I'; \cdot), \tag{4.4}$$

where $\mathcal{P}_1 = \{((R_{n,k})^J, I) \mid (J, I) \in \mathcal{H}(R_{n,k})\}$ and $Z_{R',I'}(A)$'s are Laurent polynomials. We see that, for each $(R', I') \in \mathcal{P}_1$, $\mathrm{ht}(R') = n-1$. If $n \ge 2$, we may apply (3.7) to each $(R', I') \in \mathcal{P}_1$ and therefore $\Theta_A(R', I'; \cdot)$ can be written as a sum over a set $\mathcal{P}_{R',I'}$. Notice that, since $\mathcal{P}_{R',I'} \trianglerighteq (R', I')$ and $\mathcal{P}_1 \trianglerighteq (R_{n,k}, \emptyset)$ by Proposition 3.11, it follows by Lemma 3.6 that $\mathcal{P}_{R',I'} \trianglerighteq (R_{n,k}, \emptyset)$ for every $(R', I') \in \mathcal{P}_1$. After substituting for $\Theta_A(R', I'; \cdot)$'s into (4.4) and collecting like terms, we see that

$$\Theta_A(R_{n,k}, \emptyset; \cdot) = \sum_{(R'',I'')\in\mathcal{P}_2} Z_{R'',I''}(A)\, \Theta_A(R'', I''; \cdot),$$

where $\mathcal{P}_2 = \bigcup_{(R',I')\in\mathcal{P}_1} \mathcal{P}_{R',I'}$. Clearly, $\mathcal{P}_2 \trianglerighteq (R_{n,k},\emptyset)$, $\mathrm{ht}(R'') = 0$ for each $(R'',I'') \in \mathcal{P}_2$, and each $Z_{R'',I''}(A)$ is a Laurent polynomial. A finite number of iterations of the above procedure results in

$$\Theta_A(R_{n,k},\emptyset;\cdot) = \sum_{(R,I)\in\mathcal{P}_n} Z_{R,I}(A)\,\Theta_A(R,I;\cdot), \tag{4.5}$$

where $\mathcal{P}_n \trianglerighteq (R_{n,k},\emptyset)$, $\mathrm{ht}(R) = 0$ for each $(R,I) \in \mathcal{P}_n$, and $Z_{R,I}(A)$'s are Laurent polynomials.

We first show that $\mathcal{P}_n \subseteq \mathcal{P}$, i.e., we prove that for each $(R,I) \in \mathcal{P}_n$, $R = R'_{n,k,|I|}$, $I \in \mathcal{L}_n$, and $|I| \le \min\{k, n-k\}$. Since $\mathcal{P}_n \trianglerighteq (R_{n,k},\emptyset)$, we see that $I \in \mathcal{L}_n$ for all $(R,I) \in \mathcal{P}_n$. Moreover, if $R = R'_{n,k,|I|}$ then $|I| \le \min\{k, n-k\}$. Hence, to complete our argument, it suffices to show that $R = R'_{n,k,|I|}$.

Let $(\tilde{R},\tilde{I}) \in \mathcal{P}_i$, $1 \le i \le n-1$. Since $R_{n,k}$ has neither left nor right returns, $\tilde{R}$ obtained from $R_{n,k}$ by applying first-row expansions has also neither left nor right returns and consequently $|\mathcal{J}(\tilde{R})| \le |\mathcal{J}(R_{n,k})| = 2$. Moreover, if $|\mathcal{J}(\tilde{R})| = 0$ then each arc of $\tilde{R}$ must have one of its ends on the bottom side of $\tilde{R}$. Since $\mathrm{ht}(\tilde{R}) = n - i > 0$, $\tilde{R}$ has two arcs $\tilde{c}$ and $\tilde{c}'$ with ends y_1 and y'_1, respectively. However, since $\tilde{R}$ is obtained by the first-row expansions of $(R_{n,k},\emptyset)$, arcs $\tilde{c}$ and $\tilde{c}'$ should correspond to arcs joining bottom-boundary points with left- and right-boundary points of $R_{n,k}$, respectively. Since $R_{n,k}$ does not have two arcs with such property, $|\mathcal{J}(\tilde{R})| = 0$ is not possible. It follows that $|\mathcal{J}(\tilde{R})| = 1$ or 2 for each $(\tilde{R},\tilde{I}) \in \mathcal{P}_i$. In the former case, let $\mathcal{J}(\tilde{R}) = \{j\}$ and then by (3.6),

$$\Theta_A(\tilde{R},\tilde{I};\cdot) = A^{-n_t(\tilde{R})+2j}\,\Theta_A(\tilde{R}',\tilde{I}';\cdot),$$

where $\tilde{R}' = \tilde{R}^{\{j\}}$, $n_t(\tilde{R}') = n_t(\tilde{R})$, $\mathcal{J}(\tilde{R}') = \{j\}$ or $\emptyset$, and $\tilde{I}' = \tilde{I}$.

In the latter case, let $\mathcal{J}(\tilde{R}) = \{j_1 < j_2\}$ and then by (3.6),

$$\begin{aligned}\Theta_A(\tilde{R},\tilde{I};\cdot) &= A^{-n_t(\tilde{R})+2j_1}\,\Theta_A(\tilde{R}'_1,\tilde{I}'_1;\cdot) + A^{-n_t(\tilde{R})+2j_2}\,\Theta_A(\tilde{R}'_2,\tilde{I}'_2;\cdot)\\ &\quad + \sum_{j_1<a<j_2} A^{-n_t(\tilde{R})+2(j_1+j_2-a)}\,\Theta_A(\tilde{R}''_a,\tilde{I}''_a;\cdot),\end{aligned}$$

where in the above sum

- $\tilde{R}'_1 = \tilde{R}^{\{j_1\}}$, $n_t(\tilde{R}'_1) = n_t(\tilde{R})$, $\mathcal{J}(\tilde{R}'_1) = \{j_1, j_2-1\}$ or $\{j_2-1\}$, and $\tilde{I}'_1 = \tilde{I}$;
- $\tilde{R}'_2 = \tilde{R}^{\{j_2\}}$, $n_t(\tilde{R}'_2) = n_t(\tilde{R})$, $\mathcal{J}(\tilde{R}'_2) = \{j_1+1, j_2\}$ or $\{j_1+1\}$, and $\tilde{I}'_2 = \tilde{I}$;
- $\tilde{R}''_a = \tilde{R}^{\{j_1,j_2\}}$, $n_t(\tilde{R}''_a) = n_t(\tilde{R}) - 2$, $\mathcal{J}(\tilde{R}''_a) = \{j_1, j_2-2\}$, $\{j_1\}$, $\{j_2-2\}$, or $\emptyset$, and $\tilde{I}''_a = \{a\} \oplus \tilde{I}$;

as shown in Figure 4.4.

Since $(R,I) \in \mathcal{P}_n$ is obtained by the first-row expansion of some $(\tilde{R},\tilde{I}) \in \mathcal{P}_{n-1}$, either $|\mathcal{J}(R)| = 0$, 1, or 2. We show that $|\mathcal{J}(R)| = 2$ is not possible. Indeed, if $\mathcal{J}(R) = \{j_1 < j_2\}$, then (R,I) is a result of n first-row expansions of pairs $(\tilde{R},\tilde{I})$ with $|\mathcal{J}(\tilde{R})| = 2$. As we argued above, in such a case each $(\tilde{R},\tilde{I}) = (\tilde{R}'_1,\tilde{I}'_1)$, $(\tilde{R}'_2,\tilde{I}'_2)$, or $(\tilde{R}''_a,\tilde{I}''_a)$ for some a. Let c_1 be the number of times $(\tilde{R},\tilde{I}) = (\tilde{R}'_1,\tilde{I}'_1)$, c_2 be the number of times $(\tilde{R},\tilde{I}) = (\tilde{R}'_2,\tilde{I}'_2)$, and c_3 be the number of times $(\tilde{R},\tilde{I}) = (\tilde{R}''_a,\tilde{I}''_a)$ for some a. Clearly, $c_1 + c_2 + c_3 = n$ and, as one can see by analyzing these three cases, $|I| = c_3$, $j_1 = c_2$, and $j_2 = n - c_1 - 2c_3$. Therefore,

$$1 \le j_2 - j_1 = -|I| \le 0,$$

a contradiction.

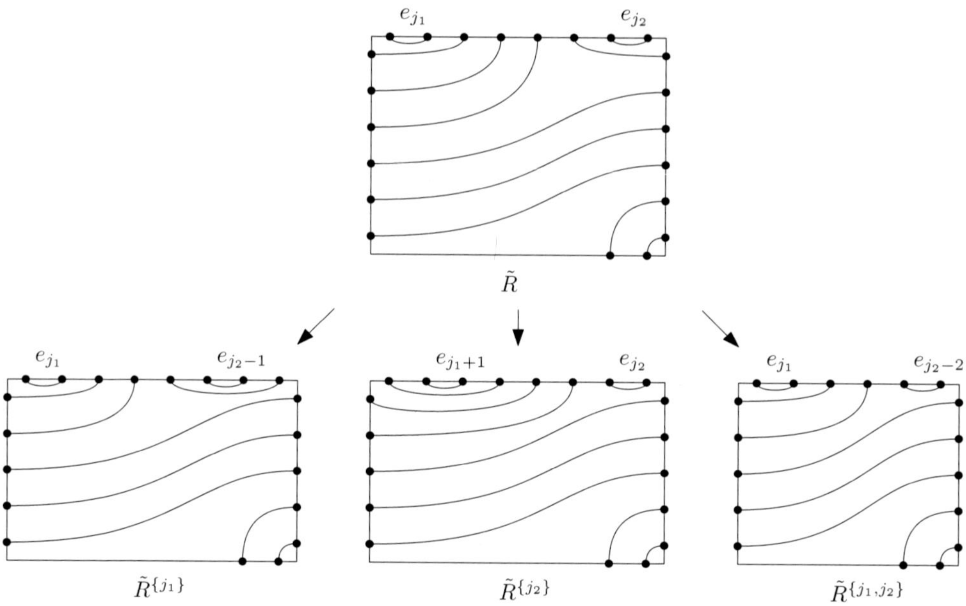

FIGURE 4.4. Case $\mathcal{J}(\tilde{R}) = \{j_1, j_2\}$

Therefore, $|\mathcal{J}(R)| \leq 1$ and we see that $\mathrm{ht}(R) = 0$, $n_b(R) = n_b(R_{n,k}) = |n-2k|$, and all top returns of R are to the left (respectively right) of all arcs with one end on the bottom if $0 \leq k \leq \lfloor \frac{n}{2} \rfloor$ (respectively $\lfloor \frac{n}{2} \rfloor \leq k \leq n$). It follows that $R = R'_{n,k,l}$ for some l and, since $n_t(R) = n - 2|I|$ and $n_t(R'_{n,k,l}) = n - 2l$, it must be $l = |I|$.

We now show that $\mathcal{P} \subseteq \mathcal{P}_n$. Let $(R'_{n,k,l}, I) \in \mathcal{P}$, where $I = \{i_1, i_2, \ldots, i_l\} \in \mathcal{L}_n$ and $l = |I| \leq \min\{k, n-k\}$, and let $(R_0, I_0) = (R_{n,k}, \emptyset)$. It suffices to find a sequence of pairs $(R_{j+1}, I_{j+1}) \in \mathcal{P}_{j+1}$, $0 \leq j \leq n-1$, such that $(R_n, I_n) = (R'_{n,k,l}, I)$ and $(R_{j+1}, I_{j+1}) = ((R_j)^{J'_j}, I'_j \oplus I_j)$ for some $(J'_j, I'_j) \in \mathcal{H}(R_j)$.

If $0 \leq k \leq \lfloor \frac{n}{2} \rfloor$, let

$$(J'_j, I'_j) = \begin{cases} (\{0, n-2j\}, \{i_{l-j}\}), & \text{for } 0 \leq j \leq l-1, \\ (\{n-2l\}, \emptyset), & \text{for } l \leq j \leq k-1, \\ (\{k-l\}, \emptyset), & \text{for } k \leq j \leq n-1. \end{cases}$$

We argue that $(J'_j, I'_j) \in \mathcal{H}(R_j)$. For $0 \leq j \leq l-1$, since R_j is as shown in Figure 4.5 and $\{i_{l-j}\} \preceq \{i_1, \ldots, i_{l-j-1}\} \oplus \{i_{l-j}\} = \{i_1, \ldots, i_{l-j}\} = I \ominus \{i_{l+1-j}, \ldots, i_l\} \in L_{n-2j}$ by Proposition 3.2(i) and Proposition 3.5(i), it follows that $\mathcal{J}(R_j) = \{0, n-2j\}$ and $0 < i_{l-j} < n-2j$. Therefore, $(\{0, n-2j\}, \{i_{l-j}\}) \in \mathcal{D}_{n-2j}$ and $\{0, n-2j\} \subseteq \mathcal{J}(R_j)$, i.e., $(J'_j, I'_j) = (\{0, n-2j\}, \{i_{l-j}\}) \in \mathcal{H}(R_j)$ for all $0 \leq j < l$.

After first l steps R_l is as shown in Figure 4.5 with $j = l$ and, as one may verify, R_j is the roof state shown in Figure 4.6 for $l < j < k$. Therefore, $\mathcal{J}(R_j) = \{j-l, n-2l\}$ and consequently $(J'_j, I'_j) = (\{n-2l\}, \emptyset) \in \mathcal{H}(R_j)$ for all $l \leq j < k$.

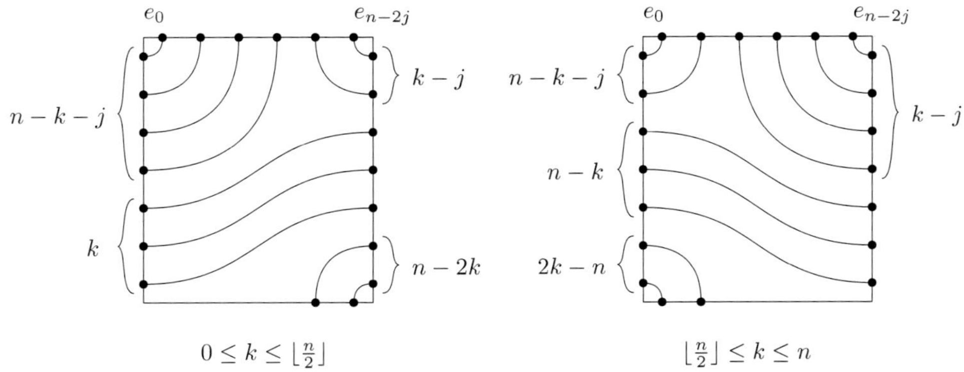

FIGURE 4.5. Roof states R_j for $0 \leq j \leq l$

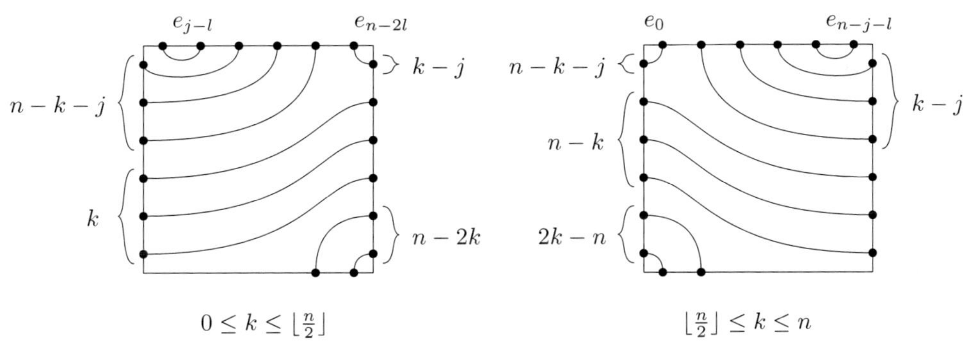

FIGURE 4.6. Roof states R_j for $l < j < \min\{k, n-k\}$

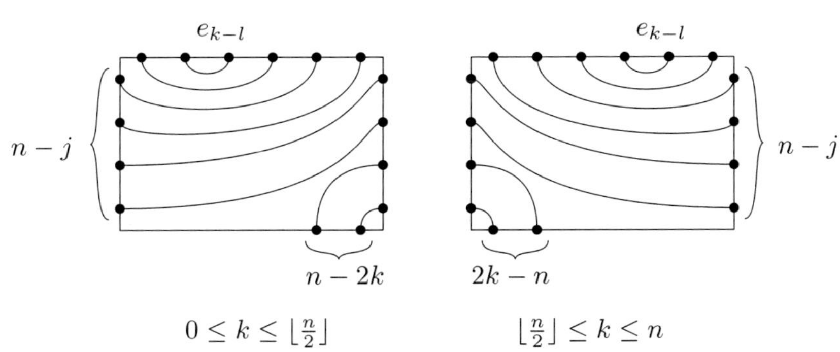

FIGURE 4.7. Roof states R_j for $\min\{k, n-k\} \leq j \leq n$

Finally, we see that R_j is the roof state shown in Figure 4.7 for $k \leq j < n$. Therefore, $\mathcal{J}(R_j) = \{k-l\}$ and consequently $(J'_j, I'_j) = (\{k-l\}, \emptyset) \in \mathcal{H}(R_j)$ for all $k \leq j \leq n-1$. Clearly $R_n = R'_{n,k,l}$ is as shown in Figure 4.7 with $j = n$ and

$$I_n = \underbrace{\emptyset \oplus \cdots \oplus \emptyset}_{n-l} \oplus \{i_1\} \oplus \{i_2\} \oplus \cdots \oplus \{i_l\} = I$$

by Proposition 3.2(i).

If $\lfloor \frac{n}{2} \rfloor \leq k \leq n$, let

$$(J'_j, I'_j) = \begin{cases} (\{0, n-2j\}, \{i_{l-j}\}), & \text{for } 0 \leq j \leq l-1, \\ (\{0\}, \emptyset), & \text{for } l \leq j \leq n-k-1, \\ (\{k-l\}, \emptyset), & \text{for } n-k \leq j \leq n-1. \end{cases}$$

Using analogous arguments as in the previous case with diagrams on the right of Figures 4.5–4.7, we can show that $(J'_j, I'_j) \in \mathcal{H}(R_j)$ for all $0 \leq j \leq n-1$, $R_n = R'_{n,k,l}$, and $I_n = I$.

Since as it was shown above $\mathcal{P}_n = \mathcal{P}$ in (4.5), the proof is completed. □

LEMMA 4.4. *Coefficients $Z_{R,I}(A)$ in Lemma 4.3 are given by*

$$Z_{R,I}(A) = Z_{R'_{n,k,|I|}, I}(A) = A^{(n-2k)(\min\{k,n-k\}-|I|)} \, [[R''_{n,k,|I|} *_v \varphi_n^{-1}(I)]]_A,$$

where $R''_{n,k,l}$ are shown in Figure 4.8. In particular, $Z_{R,I}(A) \neq 0$.

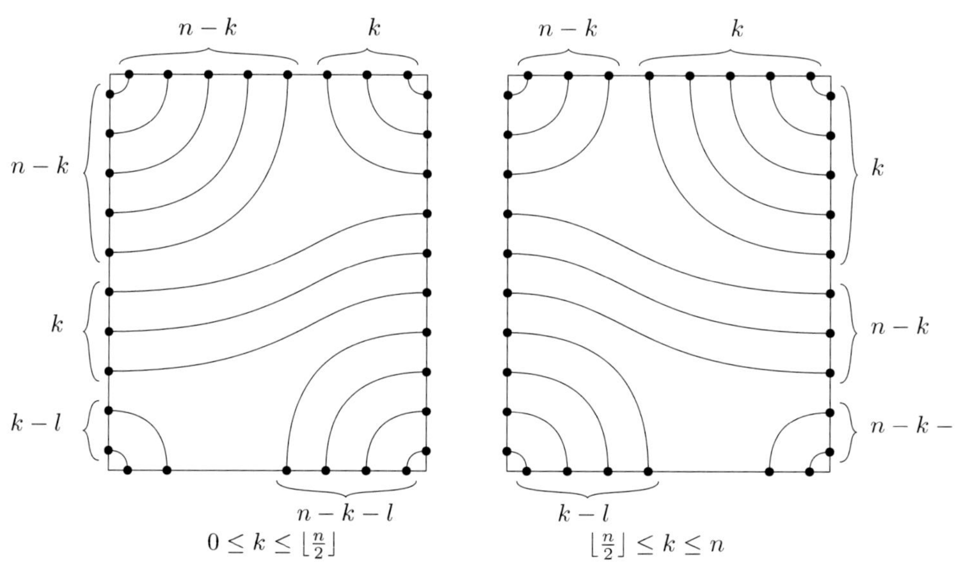

FIGURE 4.8. Middle states $R''_{n,k,l}$

PROOF. Given $(R'_{n,k,|I|}, I) \in \mathcal{P}$, let $F'_{n,k,I} = M_{n-2|I|,k'-|I|} *_v \varphi_n^{-1}(I)$, where $M_{n,k}$ is a middle state shown in Figure 4.9(a) and $k' = \min\{k, n-k\}$. For each $I' \in \mathcal{L}_n$ with $I' \prec I$, define Catalan state $C = R'_{n,k,|I'|} *_v (F'_{n,k,I})_{I'}$. Since

$$\#(C \cap l^v_{k-|I'|}) \geq k' - |I'| > k' - |I| = \mathrm{ht}(C)$$

(see Figure 4.9(b) for case $0 \leq k \leq \lfloor \frac{n}{2} \rfloor$), by Theorem 2.3, C is not realizable and consequently

$$\Theta_A(R'_{n,k,|I'|}, I'; F'_{n,k,I}) = C(A) = 0.$$

For each $I' \in \mathcal{L}_n$ with $I \prec I'$, we see that $(F'_{n,k,I})_{I'} = K_0$, so

$$\Theta_A(R'_{n,k,|I'|}, I'; F'_{n,k,I}) = 0.$$

Therefore, formula (4.3) evaluated at the floor state $F'_{n,k,I}$ reduces into

$$\Theta_A(R_{n,k}, \emptyset; F'_{n,k,I}) = Z_{R,I}(A)\, \Theta_A(R'_{n,k,|I|}, I; F'_{n,k,I}).$$

Since

$$\Theta_A(R_{n,k}, \emptyset; F'_{n,k,I}) = [[R_{n,k} *_v M_{n-2|I|,k'-|I|} *_v \varphi_n^{-1}(I)]]_A = [[R''_{n,k,|I|} *_v \varphi_n^{-1}(I)]]_A$$

and, as one may verify,

$$\Theta_A(R'_{n,k,|I|}, I; F'_{n,k,I}) = [[R'_{n,k,|I|} *_v M_{n-2|I|,k'-|I|}]]_A = A^{-(n-2k)(k'-|I|)},$$

it follows that $Z_{R,I}(A)$ is as we claimed.

Since the Catalan state $C' = R''_{n,k,|I|} *_v \varphi_n^{-1}(I)$ has no returns on three sides, by Lemma 2.2 of [**9**], C' is realizable. Consequently, by Theorem 3.13, $C'(A) \neq 0$, i.e., $Z_{R,I}(A) \neq 0$. □

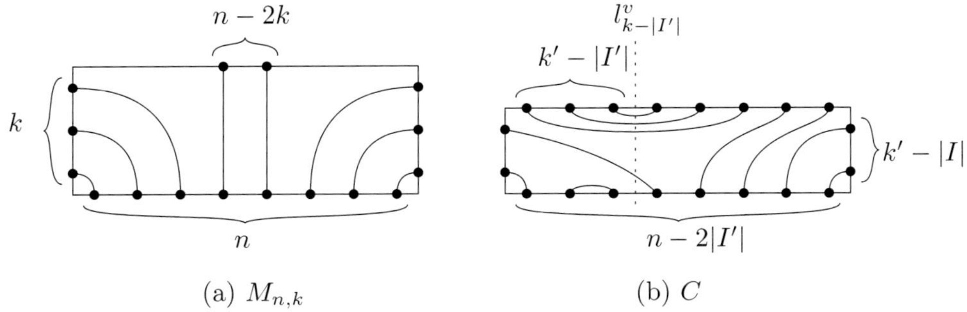

FIGURE 4.9. Middle state $M_{n,k}$ and Catalan state C

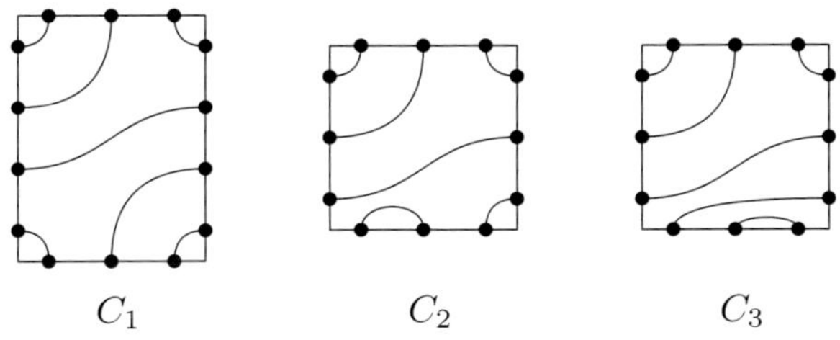

FIGURE 4.10. Catalan states C_1, C_2, and C_3

EXAMPLE 4.5. We find (4.3) for $n = 3$ and $k = 1, 2$. When $n = 3$ and $k = 1$, since $\mathcal{L}_3 = \{\emptyset, \{1\}, \{2\}\}$ and $\min\{k, n-k\} = 1$,

$$\begin{aligned}\Theta_A(R_{3,1}, \emptyset; \cdot) = Z_{R'_{3,1,0}, \emptyset}(A)\, \Theta_A(R'_{3,1,0}, \emptyset; \cdot) + Z_{R'_{3,1,1}, \{1\}}(A)\, \Theta_A(R'_{3,1,1}, \{1\}; \cdot) \\ + Z_{R'_{3,1,1}, \{2\}}(A)\, \Theta_A(R'_{3,1,1}, \{2\}; \cdot),\end{aligned}$$

where

$$Z_{R'_{3,1,0}, \emptyset}(A) = A^{(3-2)(1-0)}\, C_1(A) = A^{-7} + A^{-3} + A,$$

$$Z_{R'_{3,1,1}, \{1\}}(A) = A^{(3-2)(1-1)}\, C_2(A) = A^{-5} + A^{-1},$$

$$Z_{R'_{3,1,1}, \{2\}}(A) = A^{(3-2)(1-1)}\, C_3(A) = A^{-3},$$

and coefficients of Catalan states C_1, C_2, and C_3 shown in Figure 4.10 are computed using Theorem 2.7.

Analogously, equation (4.3) for $n = 3$ and $k = 2$ is

$$\begin{aligned}\Theta_A(R_{3,2}, \emptyset; \cdot) = Z_{R'_{3,2,0}, \emptyset}(A)\, \Theta_A(R'_{3,2,0}, \emptyset; \cdot) + Z_{R'_{3,2,1}, \{1\}}(A)\, \Theta_A(R'_{3,2,1}, \{1\}; \cdot) \\ + Z_{R'_{3,2,1}, \{2\}}(A)\, \Theta_A(R'_{3,2,1}, \{2\}; \cdot),\end{aligned}$$

where

$$Z_{R'_{3,2,0}, \emptyset}(A) = A^{(3-4)(1-0)}\, \overline{C_1}(A) = A^{-1}\, C_1(A^{-1}) = A^{-1} + A^3 + A^7,$$

$$Z_{R'_{3,2,1}, \{1\}}(A) = A^{(3-4)(1-1)}\, \overline{C_3}(A) = C_3(A^{-1}) = A^3,$$

and

$$Z_{R'_{3,2,1}, \{2\}}(A) = A^{(3-4)(1-1)}\, \overline{C_2}(A) = C_2(A^{-1}) = A + A^5.$$

Formulas for $\overline{C_1}(A)$, $\overline{C_2}(A)$, and $\overline{C_3}(A)$ above are obtained from $C_1(A)$, $C_2(A)$, and $C_3(A)$ by applying Remark 2.2(i).

LEMMA 4.6. *Given $(R, I) \in \mathcal{W}$, assume that there is a finite subset $\mathcal{P}'$ of $\mathcal{W}$ with the following properties:*

(i) $\mathcal{P}' \trianglerighteq (R, I)$;
(ii) *(R', I') has a Θ_A-state expansion for every $(R', I') \in \mathcal{P}'$; and*
(iii) *there are rational functions $Q_{R',I'}(A)$, $(R', I') \in \mathcal{P}'$, such that*

$$\Theta_A(R, I; \cdot) = \sum_{(R',I') \in \mathcal{P}'} Q_{R',I'}(A)\, \Theta_A(R', I'; \cdot).$$

Then (R, I) also has a Θ_A-state expansion.

PROOF. For each $(R', I') \in \mathcal{P}'$, let

$$\Theta_A(R', I'; \cdot) = \sum_{(\tilde{R}', \tilde{I}') \in \tilde{\mathcal{P}}'_{R',I'}} Q'_{(R',I'),(\tilde{R}',\tilde{I}')}(A)\, \Theta_A(\tilde{R}', \tilde{I}'; \cdot)$$

be its Θ_A-state expansion. Then clearly $\tilde{R}'$ in each pair

$$(\tilde{R}', \tilde{I}') \in \tilde{\mathcal{S}}' = \bigcup_{(R',I') \in \mathcal{P}'} \tilde{\mathcal{P}}'_{R',I'}$$

is a middle state. Moreover, since $\tilde{\mathcal{P}}'_{R',I'} \trianglerighteq (R',I')$ and $\mathcal{P}' \trianglerighteq (R,I)$, by Lemma 3.6, $\tilde{\mathcal{P}}'_{R',I'} \trianglerighteq (R,I)$ for every $(R',I') \in \mathcal{P}'$ and consequently, $\tilde{\mathcal{S}}' \trianglerighteq (R,I)$. Furthermore, for each $(\tilde{R}',\tilde{I}') \in \tilde{\mathcal{S}}'$ define

$$\tilde{Q}'_{(R',I'),(\tilde{R}',\tilde{I}')}(A) = \begin{cases} Q'_{(R',I'),(\tilde{R}',\tilde{I}')}(A), & \text{if } (\tilde{R}',\tilde{I}') \in \tilde{\mathcal{P}}'_{R',I'}, \\ 0, & \text{otherwise,} \end{cases}$$

then

$$\begin{aligned} \Theta_A(R,I;\cdot) &= \sum_{(R',I')\in\mathcal{P}'} \sum_{(\tilde{R}',\tilde{I}')\in\tilde{\mathcal{P}}'_{R',I'}} Q_{R',I'}(A)\, Q'_{(R',I'),(\tilde{R}',\tilde{I}')}(A)\, \Theta_A(\tilde{R}',\tilde{I}';\cdot) \\ &= \sum_{(R',I')\in\mathcal{P}'} \sum_{(\tilde{R}',\tilde{I}')\in\tilde{\mathcal{S}}'} Q_{R',I'}(A)\, \tilde{Q}'_{(R',I'),(\tilde{R}',\tilde{I}')}(A)\, \Theta_A(\tilde{R}',\tilde{I}';\cdot) \\ &= \sum_{(\tilde{R}',\tilde{I}')\in\tilde{\mathcal{S}}'} \tilde{Q}_{\tilde{R}',\tilde{I}'}(A)\, \Theta_A(\tilde{R}',\tilde{I}';\cdot), \end{aligned}$$

where

$$\tilde{Q}_{\tilde{R}',\tilde{I}'}(A) = \sum_{(R',I')\in\mathcal{P}'} Q_{R',I'}(A)\, \tilde{Q}'_{(R',I'),(\tilde{R}',\tilde{I}')}(A)$$

are rational functions. Finally, if we let

$$\mathcal{S}' = \{(\tilde{R}',\tilde{I}') \in \tilde{\mathcal{S}}' \mid \tilde{Q}_{\tilde{R}',\tilde{I}'}(A) \neq 0\},$$

then

$$\Theta_A(R,I;\cdot) = \sum_{(R',I')\in\mathcal{S}'} \tilde{Q}_{R',I'}(A)\, \Theta_A(R',I';\cdot)$$

is a Θ_A-state expansion for (R,I). □

THEOREM 4.7. *Every $(R,I) \in \mathcal{W}$ has a Θ_A-state expansion.*

PROOF. Our proof is by induction on $n = n_t(R)$. For $n = 0, 1$ or when R is a middle state, the statement is obvious since we can take $\mathcal{P}' = \{(R,I)\}$ and $Q_{R,I}(A) = 1$ in (4.1).

Assume that (4.1) exists for all roof states with $n_t(R) < n$ and arbitrary $I \in \mathrm{Fin}(\mathbb{N})$. We notice that, if there exists a Θ_A-state expansion for (R,I) then Proposition 3.7(ii) yields a Θ_A-state expansion for $(R *_v M, I)$ for any middle state M with $n_t(M) = n_b(R)$. Thus, it suffices to consider cases when R is a top state.

For a top state R with $n_t(R) = n$, define

$$q(R) = \begin{cases} \min\{j, n-j\} - u + n - v, & \text{if } \mathcal{J}(R) \neq \emptyset, \\ 0, & \text{otherwise,} \end{cases}$$

where $j = j(R) = \min \mathcal{J}(R)$, $u = u(R) = |\{i \in \varphi_n(\overline{R}^*) \mid i \leq j\}|$, and $v = v(R) = \min(\mathcal{J}(R) \cup \{n\} \setminus \{j\})$ (see Figure 4.11). Notice that $u \leq \min\{j, n-j\}$ and $v \leq n$, so $q(R) \geq 0$. Moreover, $q(R) = 0$ if and only if $R = L(0,n)$ or $R = R'_{n,j,0}$, where $R'_{n,j,0}$ is shown in Figure 4.3. Clearly, $R = L(0,n)$ is a middle state, so $\mathcal{J}(R) = \emptyset$ and thus $q(R) = 0$. For $R = R'_{n,j,0}$, we see that $u = \min\{j, n-j\}$ and $v = n$, so $q(R) = 0$ as well. Conversely, if $q(R) = 0$ and R is not a middle state, then $0 \leq \min\{j, n-j\} - u = v - n \leq 0$. Thus, it must be $\min\{j, n-j\} = u$ and $v = n$. Since $v = n$ if and only if $\mathcal{J}(R) = \{j\}$, and $\min\{j, n-j\} = u$ if and only if arcs joining top-boundary points with bottom-boundary points of R are all either to the

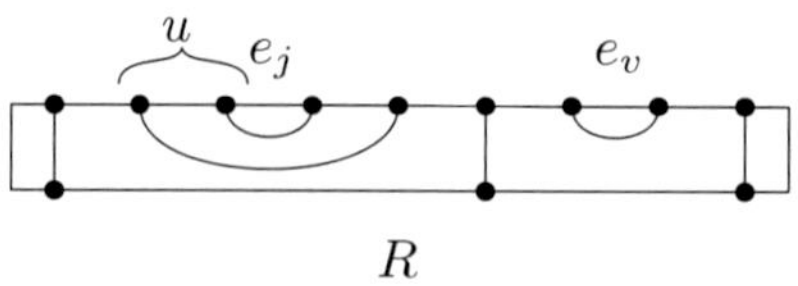

FIGURE 4.11. $j = j(R)$, $u = u(R)$, and $v = v(R)$ for top state R with $\mathcal{J}(R) \neq \emptyset$

left or to the right of the u parallel copies of e_j, it follows that $R = R'_{n,j,0}$. If R is a top state which is also a middle state, then clearly $R = L(0, n)$.

We use induction on $q = q(R)$ to show that for any top state R with $n_t(R) = n$ and arbitrary $I \in \mathrm{Fin}(\mathbb{N})$, there is a Θ_A-state expansion for (R, I).

If $q = 0$, then as we argued above $R = L(0, n)$ or $R = R'_{n,j,0}$ for some j. If $R = L(0, n)$, then R is a middle state, so (R, I) has a Θ_A-state expansion. For $R = R'_{n,j,0}$, we see that by Lemma 4.3 and Proposition 3.7(i),

$$\begin{aligned}\Theta_A(R_{n,j}, I; \cdot) &= Z_{R'_{n,j,0},\emptyset}(A)\, \Theta_A(R, \emptyset \oplus I; \cdot) \\ &\quad + \sum_{(R',I') \in \mathcal{P}'} Z_{R',I'}(A)\, \Theta_A(R', I' \oplus I; \cdot),\end{aligned} \tag{4.6}$$

where

$$\mathcal{P}' = \{(R'_{n,j,|J|}, J) \mid J \in \mathcal{L}_n,\ 0 < |J| \leq \min\{j, n-j\}\} \trianglerighteq (R_{n,j}, \emptyset)$$

and coefficients $Z_{R',I'}(A)$ are given in Lemma 4.4. Let

$$\tilde{\mathcal{P}}' = \{(R_{n,j}, I)\} \cup \{(R', I' \oplus I) \mid (R', I') \in \mathcal{P}'\},$$

then by Proposition 3.7(i), $\tilde{\mathcal{P}}' \trianglerighteq (R_{n,j}, I)$. Since $n_t(R) = n_t(R_{n,j})$ and $n_b(R) = n_b(R_{n,j})$, it follows that $\tilde{\mathcal{P}}' \trianglerighteq (R, I)$. Moreover, each $(\tilde{R}', \tilde{I}') \in \tilde{\mathcal{P}}'$ has a Θ_A-state expansion since either $\tilde{R}' = R_{n,j}$ is a middle state or $n_t(\tilde{R}') < n$ in which case induction can be applied to $(\tilde{R}', \tilde{I}')$. Furthermore, after solving (4.6) for $\Theta_A(R, I; \cdot)$,

$$\Theta_A(R, I; \cdot) = \frac{1}{Z_{R'_{n,j,0},\emptyset}(A)} \Big(\Theta_A(R_{n,j}, I; \cdot) - \sum_{(R',I') \in \mathcal{P}'} Z_{R',I'}(A)\, \Theta_A(R', I' \oplus I; \cdot) \Big).$$

Therefore, $\tilde{\mathcal{P}}'$ satisfies all assumptions of Lemma 4.6, consequently (R, I) has a Θ_A-state expansion.

Given $q > 0$, assume that for any top state R with $n_t(R) = n$ and $q(R) < q$, and arbitrary $I \in \mathrm{Fin}(\mathbb{N})$, there is a Θ_A-state expansion for (R, I). For a top state R with $n_t(R) = n$ and $q(R) = q$, let $\tilde{R} = \psi_n^{-1}(\{j\}) *_v R$ and let

$$\mathcal{P}' = \{(\tilde{R}^{\tilde{J}}, \tilde{I} \oplus I) \mid (\tilde{J}, \tilde{I}) \in \mathcal{H}(\tilde{R}) \setminus (\{j\}, \emptyset)\}.$$

We show that every $(R',I') \in \mathcal{P}'$ has a Θ_A-state expansion. If $(R',I') \in \mathcal{P}'$ with $n_t(R') = n$ then $R' = \tilde{R}^{J'}$ for some $J' = \{k\}$, where $k \in \mathcal{J}(\tilde{R}) \setminus \{j\}$, i.e., $k \geq v+1$. Since $j(R') = j+1$, $u(R') = u+1$, and $v(R') \geq v+1$, it follows that $q(R') < q$. Therefore, for each pair $(R',I') \in \mathcal{P}'$ either $n_t(R') < n$ or $n_t(R') = n$ and $q(R') < q$, hence either by induction on n or on q, each (R',I') has a Θ_A-state expansion. In particular, we proved that $(R,I) \notin \mathcal{P}'$ and since $(R,I) = (\tilde{R}^{\{j\}}, \emptyset \oplus I)$, by (3.7), it follows that

$$\Theta_A(\tilde{R}, I; \cdot) = A^{-n+2j}\, \Theta_A(R, I; \cdot) + \sum_{(R',I') \in \mathcal{P}'} Z_{R',I'}(A)\, \Theta_A(R', I'; \cdot). \tag{4.7}$$

Define $\delta : \mathbb{Z} \to \{0,1\}$ by $\delta(x) = 1$ if $x > 0$ and $\delta(x) = 0$ otherwise. We show that $(\tilde{R}, I)$ has a Θ_A-state expansion. Let $\tilde{R} = \tilde{R}' *_v \tilde{M}'$, where $\tilde{R}'$ is a top state and $\tilde{M}'$ is a middle state. Then clearly $j(\tilde{R}') = j$, $u(\tilde{R}') = u + \delta(\min\{j, n-j\} - u)$, and $v(\tilde{R}') = v + \delta(n - v)$. One checks that $u(\tilde{R}') + v(\tilde{R}') > u + v$ and consequently $q(\tilde{R}') < q$. By induction $(\tilde{R}', I)$ has a Θ_A-state expansion, so Proposition 3.7(ii) implies that there is a Θ_A-state expansion for $(\tilde{R}, I)$.

Let $\tilde{\mathcal{P}}' = \{(\tilde{R}, I)\} \cup \mathcal{P}'$. Since $\mathcal{P}' \trianglerighteq (\tilde{R}, I)$ by Proposition 3.11, $n_t(R) = n_t(\tilde{R})$, and $n_b(R) = n_b(\tilde{R})$, it follows that $\tilde{\mathcal{P}}' \trianglerighteq (R, I)$. Moreover, as we showed above, each $(\tilde{R}', \tilde{I}') \in \tilde{\mathcal{P}}'$ has a Θ_A-state expansion and, using (4.7),

$$\Theta_A(R, I; \cdot) = A^{n-2j}\Big(\Theta_A(\tilde{R}, I; \cdot) - \sum_{(R',I') \in \mathcal{P}'} Z_{R',I'}(A)\, \Theta_A(R', I'; \cdot)\Big).$$

Therefore, $\tilde{\mathcal{P}}'$ satisfies all assumptions of Lemma 4.6 and consequently (R, I) has a Θ_A-state expansion. □

The proof of Theorem 4.7 yields an algorithm for finding Θ_A-state expansion for (R, I).

REMARK 4.8. In line 3 of Algorithm 4.1, we check if a roof state R satisfies the *horizontal-line conditions*, i.e., $\#(R \cap l_i^h) \leq n_t(R)$ for all $i = 1, 2, \ldots, \mathrm{ht}(R)$. This allows us to deal with trivial cases more efficiently. Indeed, by Theorem 2.3, if R does not satisfy this condition, then $\Theta_A(R, I; F) = 0$ for any floor state F. In particular, if $n_b(R) > n_t(R)$ then $\Theta_A(R, I; \cdot) \equiv 0$.

EXAMPLE 4.9. Let R be the roof state in Figure 4.12. We apply Algorithm 4.1 to find a Θ_A-state expansion for $(R, \emptyset)$ first and then use it to compute $C(A)$ for the Catalan state $C = R *_v F$, where C and F are shown in Figure 4.12. Let $M = (M_{2,1})^*$ be a π-rotation of the middle state $M_{2,1}$ shown in Figure 4.9(a). Using (3.6) for $(R'_{4,1,0} *_v M, \emptyset)$, where top state $R'_{n,k,l}$ is shown in Figure 4.3, we see that

$$\begin{aligned}\Theta_A(R, \emptyset; \cdot) = {} & A^2\, \Theta_A(R'_{4,1,0} *_v M, \emptyset; \cdot) - A^6\, \Theta_A(R'_{4,2,0}, \emptyset; \cdot) - A^4\, \Theta_A(R'_{2,1,0}, \{2\}; \cdot) \\ & - A^2\, \Theta_A(R'_{2,1,0}, \{3\}; \cdot).\end{aligned}$$

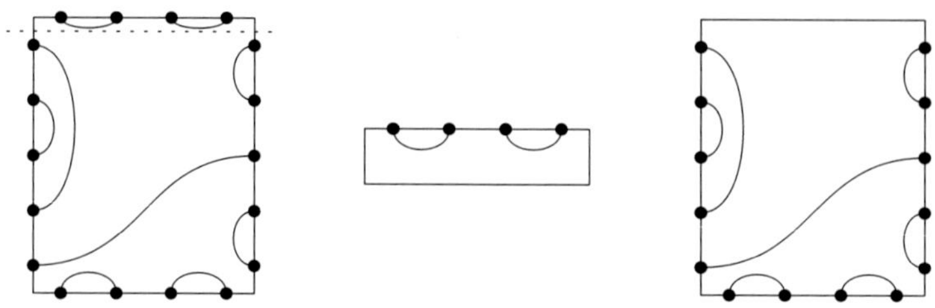

$C = R *_v F$ $\quad$ R $\quad$ F

FIGURE 4.12. Catalan state C, roof state R, and floor state F

Algorithm 4.1 Θ_A-state expansion for $(R, I) \in \mathcal{W}$

procedure THETA(R, I)
$\quad n \leftarrow n_t(R)$
$\quad$ **if not** $\#(R \cap l_i^h) \leq n$ for all i **then**
$\quad\quad$ **return** 0
$\quad$ **else if** R is a middle state **then**
$\quad\quad$ **return** $\Theta_A(R, I; \cdot)$
$\quad$ **else if** $\mathrm{ht}(R) > 0$ **then**
$\quad\quad (R, M) \leftarrow R/l_0^h$
$\quad\quad \sum_i Q_i(A)\, \Theta_A(R_i, I_i; \cdot) \leftarrow$ THETA(R, I)
$\quad\quad$ **return** $\sum_i Q_i(A)\, \Theta_A(R_i *_v M, I_i; \cdot)$
$\quad j \leftarrow \min \mathcal{J}(R)$
$\quad$ **if** $R = R'_{n,j,0}$ **then**
$\quad\quad$ ExpSum $\leftarrow \Theta_A(R_{n,j}, I; \cdot)$
$\quad\quad$ **for** $I' \in \mathcal{L}_n$ with $0 < |I'| \leq \min\{j, n-j\}$ **do**
$\quad\quad\quad Z(A) \leftarrow Z_{R'_{n,j,|I'|}, I'}(A)$ from Lemma 4.4
$\quad\quad\quad$ ExpSum $\leftarrow$ ExpSum $- Z(A) \cdot$ THETA$(R'_{n,j,|I'|}, I' \oplus I)$
$\quad\quad Z(A) \leftarrow Z_{R'_{n,j,0}, \emptyset}(A)$ from Lemma 4.4
$\quad\quad$ **return** $(Z(A))^{-1} \cdot$ ExpSum
$\quad$ **else**
$\quad\quad R \leftarrow \psi_n^{-1}(\{j\}) *_v R$
$\quad\quad$ ExpSum $\leftarrow$ THETA(R, I)
$\quad\quad$ **for** $(J', I') \in \mathcal{H}(R) \setminus \{(\{j\}, \emptyset)\}$ **do**
$\quad\quad\quad$ ExpSum $\leftarrow$ ExpSum $- A^{-n+2\|J'\|-2\|I'\|} \cdot$ THETA$(R^{J'}, I' \oplus I)$
$\quad\quad$ **return** $A^{n-2j} \cdot$ ExpSum

For a positive integer n, let $[n]_q = 1 + q + \cdots + q^{n-1}$ and let $[n] = [n]_{q=A^4}$. By Lemma 4.3, Lemma 4.4, and Proposition 3.7(ii),

$$\begin{aligned}
&\Theta_A(R'_{4,1,0} *_v M, \emptyset; \cdot) \\
&\quad = \frac{A^{14}}{[4]}\, \Theta_A(R_{4,1} *_v M, \emptyset; \cdot) - \frac{A^2[3]}{[4]}\, \Theta_A(R'_{4,1,1} *_v M, \{1\}; \cdot) \\
&\quad\quad - \frac{A^4[2]}{[4]}\, \Theta_A(R'_{4,1,1} *_v M, \{2\}; \cdot) - \frac{A^6}{[4]}\, \Theta_A(R'_{4,1,1} *_v M, \{3\}; \cdot)
\end{aligned}$$

and

$$\begin{aligned}\Theta_A(R'_{4,2,0},\emptyset;\cdot) &= \frac{A^8[2]}{[4][3]}\,\Theta_A(R_{4,2},\emptyset;\cdot) - \frac{A^4[2]}{[4]}\,\Theta_A(R'_{4,2,1},\{1\};\cdot)\\ &\quad - \frac{A^2[2]^2}{[4]}\,\Theta_A(R'_{4,2,1},\{2\};\cdot) - \frac{A^4[2]}{[4]}\,\Theta_A(R'_{4,2,1},\{3\};\cdot)\\ &\quad - \frac{A^8[2]}{[4][3]}\,\Theta_A(R'_{4,2,2},\{1,2\};\cdot) - \frac{A^6[2]^2}{[4][3]}\,\Theta_A(R'_{4,2,2},\{1,3\};\cdot),\end{aligned}$$

where roof states $R_{n,k}$ are shown in Figure 4.2. Moreover, by Lemma 4.3, Lemma 4.4, and Proposition 3.7(i),

$$\Theta_A(R'_{2,1,0},I;\cdot) = \frac{A^2}{[2]}\,\Theta_A(R_{2,1},I;\cdot) - \frac{A^2}{[2]}\,\Theta_A(R'_{2,1,1},\{1\}\oplus I;\cdot)$$

for $I = \{1\},\{2\},\{3\}$.

Since $R'_{4,1,1} *_v M = M$, $R'_{4,2,1} = R'_{2,1,0}$, and $R'_{4,2,2} = R'_{2,1,1} = R_{0,0}$, it follows that

$$\begin{aligned}\Theta_A(R,\emptyset;\cdot) &= \frac{A^{16}}{[4]}\,\Theta_A(R_{4,1} *_v M,\emptyset;\cdot) - \frac{A^4[3]}{[4]}\,\Theta_A(M,\{1\};\cdot)\\ &\quad - \frac{A^6[2]}{[4]}\,\Theta_A(M,\{2\};\cdot) - \frac{A^8}{[4]}\,\Theta_A(M,\{3\};\cdot) - \frac{A^{14}[2]}{[4][3]}\,\Theta_A(R_{4,2},\emptyset;\cdot)\\ &\quad + \frac{A^{12}}{[4]}\,\Theta_A(R_{2,1},\{1\};\cdot) + \frac{A^6(A^4-1)}{[4]}\,\Theta_A(R_{2,1},\{2\};\cdot)\\ &\quad - \frac{A^4}{[4]}\,\Theta_A(R_{2,1},\{3\};\cdot) + \frac{A^6}{[3][2]}\,\Theta_A(R_{0,0},\{1,2\};\cdot)\\ &\quad + \frac{A^4}{[3]}\,\Theta_A(R_{0,0},\{1,3\};\cdot).\end{aligned}$$

We see that $F_{\{2\}} = F_{\{1,2\}} = K_0$ and

$$\Theta_A(R_{4,1} *_v M,\emptyset;F) = A^{-24}\,[2]^3[3]^2, \quad \Theta_A(R_{4,2},\emptyset;F) = A^{-14}\,[2]^2[3]^2,$$

$$\Theta_A(R_{2,1},\{1\};F) = \Theta_A(M,\{1\};F) = \Theta_A(R_{2,1},\{3\};F) = \Theta_A(M,\{3\};F) = A^{-4}\,[2],$$

by Theorem 2.7 and $\Theta_A(R_{0,0},\{1,3\};F) = 0$ by Theorem 2.3, hence

$$\begin{aligned}C(A) = \Theta_A(R,\emptyset;F) &= \frac{A^{-8}\,[2]^2(1+3A^4+2A^8+3A^{12}+A^{16})}{[4]}\\ &= A^{-8}(1+A^4)(1+3A^4+A^8).\end{aligned}$$

Given $1 \le j \le n-1$, we let $1 \le u \le \min\{j, n-j\}$ and denote by

$$\mathcal{C}_{n,u} = \{(I_1, I_2, \ldots, I_t) \mid \emptyset = I_1 \prec I_2 \prec \cdots \prec I_t \in \mathcal{L}_n,\ |I_t| \le u,\ t \ge 1\}.$$

For $c = (I_1, \ldots, I_t) \in \mathcal{C}_{n,u}$ let $I_c = I_t$ and define

$$\pi_{n,j,c}(A) = \frac{1}{Z_{n-2|I_c|,j-|I_c|,\emptyset}(A)} \prod_{i=1}^{t-1} \Big(- \frac{Z_{n-2|I_i|,j-|I_i|,I_{i+1}\ominus I_i}(A)}{Z_{n-2|I_i|,j-|I_i|,\emptyset}(A)} \Big), \tag{4.8}$$

where $Z_{n,j,I}(A) = Z_{R'_{n,j,|I|},I}(A)$ is as in Lemma 4.4.

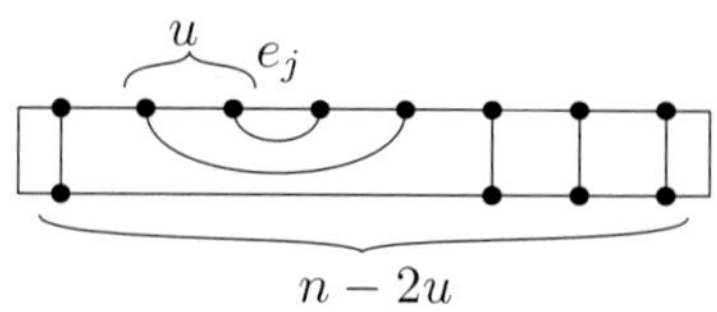

FIGURE 4.13. Top state $T_{n,j,u}$

COROLLARY 4.10. *Let $T_{n,j,u}$ be the top state shown in Figure 4.13 with $u \geq 1$. Then*

$$\begin{aligned}(4.9)\qquad &\Theta_A(T_{n,j,u}, \emptyset; \cdot) \\ &= A^{(n-2j)d} \sum_{c \in \mathcal{C}_{n,u}} \pi_{n,j,c}(A)\, \Theta_A(R_{n-2|I_c|,j-|I_c|} *_v M_{n-2u,d}, I_c; \cdot),\end{aligned}$$

where $d = \min\{j, n-j\} - u$, $R_{n,k}$ is shown in Figure 4.2, $M_{n,k}$ is shown in Figure 4.9(a), and $\pi_{n,j,c}(A)$ is given by (4.8).

PROOF. Let $\tilde{R} = R'_{n,j,0} *_v M_{n-2u,d}$, where $R'_{n,j,0}$ is as in Figure 4.3 and denote by $Z_{n,j,I}(A)$ the Laurent polynomial $Z_{R'_{n,j,|I|},I}(A)$ given in Lemma 4.4. Using Lemma 4.3 and Proposition 3.7(ii),

$$\begin{aligned}(4.10)\qquad \Theta_A(\tilde{R}, \emptyset; \cdot) &= \frac{1}{Z_{n,j,\emptyset}(A)} \Theta_A(R_{n,j} *_v M_{n-2u,d}, \emptyset; \cdot) \\ &\quad - \frac{1}{Z_{n,j,\emptyset}(A)} \sum_{(R',I') \in \mathcal{P}'} Z_{n,j,I'}(A)\, \Theta_A(R' *_v M_{n-2u,d}, I'; \cdot)\end{aligned}$$

where $\mathcal{P}' = \{(R'_{n,j,|I'|}, I') \mid I' \in \mathcal{L}_n,\ 0 < |I'| \leq \min\{j, n-j\}\}$.

Since $R'_{n,j,|I'|} = R'_{n-2|I'|,j-|I'|,0}$, by Lemma 4.3 and Proposition 3.7(i) and (ii), each $\Theta_A(R' *_v M_{n-2u,d}, I'; \cdot)$ on the right-hand side of (4.10) can be written as

$$\begin{aligned}(4.11)\quad &\Theta_A(R' *_v M_{n-2u,d}, I'; \cdot) \\ &= \frac{1}{Z_{n-2|I'|,j-|I'|,\emptyset}(A)} \Big(\Theta_A(R_{n-2|I'|,j-|I'|} *_v M_{n-2u,d}, \emptyset \oplus I'; \cdot) \\ &\quad - \sum_{(R'',I'') \in \mathcal{P}''_{R',I'}} Z_{n-2|I'|,j-|I'|,I''}(A)\, \Theta_A(R'' *_v M_{n-2u,d}, I'' \oplus I'; \cdot) \Big),\end{aligned}$$

where

$$\mathcal{P}''_{R',I'} = \{(R'_{n-2|I'|,j-|I'|,|I''|}, I'') \mid I'' \in \mathcal{L}_{n-2|I'|},\ 0 < |I''| \leq \min\{j, n-j\} - |I'|\}.$$

Substituting (4.11) into (4.10) yields

$$\begin{aligned}
&\Theta_A(\tilde{R},\emptyset;\cdot)\\
&= \frac{1}{Z_{n,j,\emptyset}(A)}\,\Theta_A(R_{n,j} *_v M_{n-2u,d},\emptyset;\cdot)\\
&\quad - \sum_{(R',I')\in\mathcal{P}'} \frac{Z_{n,j,I'}(A)}{Z_{n,j,\emptyset}(A)}\,\frac{1}{Z_{n-2|I'|,j-|I'|,\emptyset}(A)}\,\Theta_A(R_{n-2|I'|,j-|I'|} *_v M_{n-2u,d}, I';\cdot)\\
&\quad + \sum_{(R',I')\in\mathcal{P}'}\ \sum_{(R'',I'')\in\mathcal{P}''_{R',I'}} \frac{Z_{n,j,I'}(A)}{Z_{n,j,\emptyset}(A)}\,\frac{Z_{n-2|I'|,j-|I'|,I''}(A)}{Z_{n-2|I'|,j-|I'|,\emptyset}(A)}\\
&\qquad\qquad \cdot\,\Theta_A(R'' *_v M_{n-2u,d}, I''\oplus I';\cdot).
\end{aligned}$$

After recursively applying Lemma 4.3 and Proposition 3.7(i) and (ii), we see that

$$\Theta_A(\tilde{R},\emptyset;\cdot) = \sum_{c\in\mathcal{C}_{n,u+d}} \pi_{n,j,c}(A)\,\Theta_A(R_{n-2|I_c|,j-|I_c|} *_v M_{n-2u,d}, I_c;\cdot). \tag{4.12}$$

Applying d times the first-row expansion (3.6) to $(\tilde{R},\emptyset)$ gives

$$\Theta_A(\tilde{R},\emptyset;\cdot) = A^{(-n+2j)d}\,\Theta_A(T_{n,j,u},\emptyset;\cdot). \tag{4.13}$$

Since all terms $\Theta_A(R_{n-2|I_c|,j-|I_c|} *_v M_{n-2u,d}, I_c;\cdot)$ with $|I_c|>u$ in the sum are zero by Remark 4.8, we see that (4.9) follows from (4.12) and (4.13). □

Note that (4.9) after collecting like-terms can be used to obtain Θ_A-state expansions for the family of top states $T_{n,j,u}$. Consequently, by Proposition 3.7(ii), we can obtain Θ_A-state expansions for infinitely many roof states $R = T_{n,j,u} *_v M$, where M is a middle state with $n_t(M) = n-2u$.

We say that a coefficient $C(A)$ of a Catalan state C is *unimodal* if there is an integer k and a unimodal polynomial[8] $p(x)$ such that $C(A) = A^k p(A^4)$. Clearly, coefficients of Catalan states of $L(m,n)$ for $n=1,2$ are unimodal and, using Proposition 5.4 of **[11]**, it can also easily be seen that coefficients of Catalan states of $L(m,3)$ are unimodal. Furthermore, preliminary results that we obtained using Θ_A-state expansion for Catalan states of $L(m,4)$ indicate that quite likely coefficients for these states are unimodal as well. However, this is not the case for coefficients of Catalan states of $L(m,5)$ as the following example shows.

EXAMPLE 4.11. Let C be the Catalan state in Figure 4.14, then $C = R'_{5,1,0} *_v F$ for the top state $R'_{n,k,l}$ in Figure 4.3 and the floor state F shown in Figure 4.14. Since $F_I = K_0$ for all $I\in\mathcal{L}_5\setminus\{\emptyset,\{4\}\}$, by Lemma 4.3, one shows that

$$\begin{aligned}
C(A) = \Theta_A(R'_{5,1,0},\emptyset;F) &= \frac{1}{Z_{R'_{5,1,0},\emptyset}(A)}\,\Theta_A(R_{5,1},\emptyset;F)\\
&\quad - \frac{Z_{R'_{5,1,1},\{4\}}(A)}{Z_{R'_{5,1,0},\emptyset}(A)}\,\Theta_A(R'_{5,1,1},\{4\};F),
\end{aligned}$$

[8]A polynomial $p(x) = \sum_{i=0}^{n} a_i x^i$ is unimodal if its coefficients $a_0, a_1, \ldots, a_n$ form a unimodal sequence.

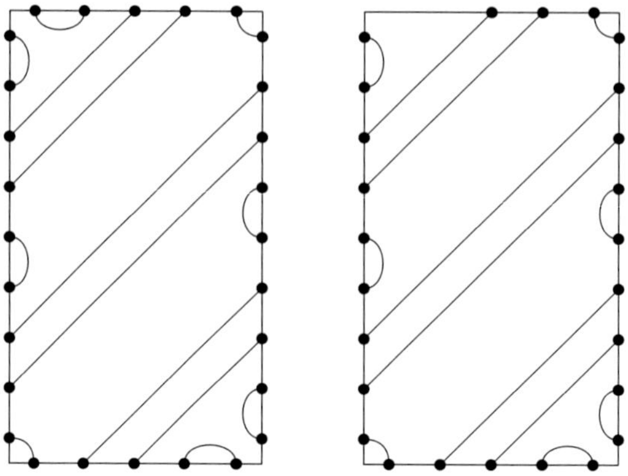

FIGURE 4.14. Catalan state C and floor state F

where $Z_{R'_{5,1,0},\emptyset}(A) = A^{-23}(1 + A^4 + A^8 + A^{12} + A^{16})$ and $Z_{R'_{5,1,1},\{4\}}(A) = A^{-15}$ are given in Lemma 4.4. Using Theorem 2.7 one verifies that

$$\Theta_A(R_{5,1}, \emptyset; F) = A^{-24}(A^{-12} + A^{-8} + A^{-4} + 2 + A^4)^3$$

and

$$\Theta_A(R'_{5,1,1}, \{4\}; F) = A^{-9}.$$

Therefore,

$$C(A) = A^{-37}(1 + 2A^4 + 3A^8 + 7A^{12} + 8A^{16} + 7A^{20} + 9A^{24} + 5A^{28} + A^{32}).$$

Acknowledgment

The authors would like to thank Professor Józef H. Przytycki for all valuable comments and suggestions.

References

[1] R. P. Bakshi, S. Mukherjee, J. H. Przytycki, M. Silvero, and X. Wang, *On multiplying curves in the Kauffman bracket skein algebra of the thickened four-holed sphere*, J. Knot Theory Ramifications **30** (2021), no. 14, Paper No. 2141001, 29, DOI 10.1142/S0218216521410017. MR4407082

[2] F. Bonahon and H. Wong, *Representations of the Kauffman bracket skein algebra I: invariants and miraculous cancellations*, Invent. Math. **204** (2016), no. 1, 195–243, DOI 10.1007/s00222-015-0611-y. MR3480556

[3] D. Bullock and J. H. Przytycki, *Multiplicative structure of Kauffman bracket skein module quantizations*, Proc. Amer. Math. Soc. **128** (2000), no. 3, 923–931, DOI 10.1090/S0002-9939-99-05043-1. MR1625701

[4] Z. Cheng, S. Mukherjee, J. H. Przytycki, X. Wang, and S. Y. Yang, *Realization of plucking polynomials*, J. Knot Theory Ramifications **26** (2017), no. 2, 1740016, 9, DOI 10.1142/S0218216517400168. MR3604498

[5] Z. Cheng, S. Mukherjee, J. H. Przytycki, X. Wang, and S. Y. Yang, *Strict unimodality of plucking polynomials of rooted trees*, J. Knot Theory Ramifications **27** (2018), no. 7, 1841009, 19, DOI 10.1142/S0218216518410092. MR3835751

[6] Z. Cheng, S. Mukherjee, J. H. Przytycki, X. Wang, and S. Y. Yang, *Rooted trees with the same plucking polynomial*, Osaka J. Math. **56** (2019), no. 3, 661–674. MR3982046

[7] J.-P. Cho, *The Kauffman bracket skein algebra of the punctured torus*, PhD dissertation, 2012.

[8] J.-P. Cho and R. Gelca, *Representations of the Kauffman bracket skein algebra of the punctured torus*, Fund. Math. **225** (2014), no. 1, 45–56, DOI 10.4064/fm225-1-3. MR3205564
[9] M. K. Dabkowski, C. Li, and J. H. Przytycki, *Catalan states of lattice crossing*, Topology Appl. **182** (2015), 1–15, DOI 10.1016/j.topol.2014.11.015. MR3305606
[10] M. K. Dabkowski and A. D. Merrill, *Coefficients of Catalan states without nesting and maximal number of returns*, J. Knot Theory Ramifications **30** (2021), no. 12, Paper No. 2141007, 24, DOI 10.1142/S0218216521410078. MR4394064
[11] M. K. Dabkowski and J. H. Przytycki, *Catalan states of lattice crossing: an application of plucking polynomial*, Topology Appl. **254** (2019), 12–28, DOI 10.1016/j.topol.2018.12.003. MR3894207
[12] M. K. Dabkowski and M. Rakotomalala, *B-type Catalan states of lattice crossing*, J. Knot Theory Ramifications **32** (2023), no. 8, Paper No. 2340005, 28, DOI 10.1142/S0218216523400059. MR4641890
[13] C. Frohman and R. Gelca, *Skein modules and the noncommutative torus*, Trans. Amer. Math. Soc. **352** (2000), no. 10, 4877–4888, DOI 10.1090/S0002-9947-00-02512-5. MR1675190
[14] C. Frohman and J. Kania-Bartoszynska, *The structure of the Kauffman bracket skein algebra at roots of unity*, Math. Z. **289** (2018), no. 3-4, 889–920, DOI 10.1007/s00209-017-1980-2. MR3830231
[15] C. Frohman, J. Kania-Bartoszynska, and T. Lê, *Unicity for representations of the Kauffman bracket skein algebra*, Invent. Math. **215** (2019), no. 2, 609–650, DOI 10.1007/s00222-018-0833-x. MR3910071
[16] R. Gelca, *A product-to-sum formula for the quantum group of* $SL(2,\mathbb{C})$, Preprint, `arXiv:math/0201122`, 2002.
[17] R. Gelca and A. Uribe, *The Weyl quantization and the quantum group quantization of the moduli space of flat* SU(2)*-connections on the torus are the same*, Comm. Math. Phys. **233** (2003), no. 3, 493–512, DOI 10.1007/s00220-002-0759-3. MR1962120
[18] T. T. Q. Lê, *Triangular decomposition of skein algebras*, Quantum Topol. **9** (2018), no. 3, 591–632.
[19] C. Li, *Temperley-Lieb algebra approach to Catalan states of generalized crossing and lattice crossing*, J. Knot Theory Ramifications **25** (2016), no. 5, 1650027, 27, DOI 10.1142/S0218216516500279. MR3488314
[20] M. McLendon, *Traces on the skein algebra of the torus*, Topology Appl. **154** (2007), no. 18, 3140–3144, DOI 10.1016/j.topol.2007.08.005. MR2364642
[21] J. H. Przytycki, *Skein modules of* 3*-manifolds*, Bull. Polish Acad. Sci. Math. **39** (1991), no. 1-2, 91–100. MR1194712
[22] J. H. Przytycki, *Fundamentals of Kauffman bracket skein modules*, Kobe J. Math. **16** (1999), no. 1, 45–66. MR1723531
[23] J. H. Przytycki, *The Kauffman bracket skein algebra of a surface times the interval has no zero divisors*, Proceedings of the 46th Japan Topology Symposium at Hokkaido University, July 26-29 (1999), 52–61.
[24] J. H. Przytycki, *Knots and graphs: two centuries of interaction*, Knot theory and its applications, Contemp. Math., vol. 670, Amer. Math. Soc., Providence, RI, 2016, pp. 171–257, DOI 10.1090/conm/670/13449. MR3568528
[25] J. H. Przytycki, *q-polynomial invariant of rooted trees*, Arnold Math. J. **2** (2016), no. 4, 449–461, DOI 10.1007/s40598-016-0053-7. MR3564883
[26] J. H. Przytycki and A. S. Sikora, *On skein algebras and* $\mathrm{Sl}_2(\mathbf{C})$*-character varieties*, Topology **39** (2000), no. 1, 115–148, DOI 10.1016/S0040-9383(98)00062-7. MR1710996
[27] J. H. Przytycki and A. S. Sikora, *Skein algebras of surfaces*, Trans. Amer. Math. Soc. **371** (2019), no. 2, 1309–1332, DOI 10.1090/tran/7298. MR3885180
[28] V. G. Turaev, *The Conway and Kauffman modules of a solid torus* (Russian, with English summary), Zap. Nauchn. Sem. Leningrad. Otdel. Mat. Inst. Steklov. (LOMI) **167** (1988), no. Issled. Topol. 6, 79–89, 190, DOI 10.1007/BF01099241; English transl., J. Soviet Math. **52** (1990), no. 1, 2799–2805. MR964255

Department of Mathematical Sciences, The University of Texas at Dallas, Richardson, TX 75080

Email address: mdab@utdallas.edu

Department of Mathematical Sciences, The University of Texas at Dallas, Richardson, TX 75080

Email address: cheyu.wu@utdallas.edu

Contemporary Mathematics
Volume **824**, 2025
https://doi.org/10.1090/conm/824/16457

Structure of the chromatic polynomial

Radmila Sazdanovic and Daniel Scofield

ABSTRACT. Recently, big data techniques such as machine learning and topological data analysis have made their way to theoretical mathematics. Motivated by the recent work with polynomial invariants for knots, we use manifold learning and topological data analysis techniques to explore the structure and properties of the point cloud consisting of the chromatic polynomials of graphs up to 10 crossings. Although chromatic, as well as the Tutte polynomial fail to distinguish graphs, according to a conjecture by Bollobás, Pebody and Riordan they approximate the space of random graphs. In this work we compare structures in the chromatic data revealed using filtered PCA and Ball Mapper techniques, and relate them with a range of numerical invariants for graphs.

1. Introduction

Graph invariants are properties of graphs that are invariant under graph isomorphism. Their definition implies they can be used to distinguish graphs: if values of the invariant are different the graphs are non-isomorphic, but it is possible to have non-isomorphic graphs with the same invariant. Invariants are abundant in graph theory, as well as in other areas of mathematics such as low-dimensional topology. They may be numerical, polynomials, or more involved algebraic constructions such as homology theories. Examples of such invariants for graphs include numbers of vertices, edges, and cycles, connectivity, vertex degrees, and chromatic number, the chromatic, Tutte and magnitude polynomials, the Stanley chromatic function [**35**, **51**, **54**, **56**], and the corresponding homology theories categorifying these polynomial invariants [**21**, **24**, **25**, **49**].

Recent years have brought significant advances of artificial intelligence and machine learning methods in graph theory, which have proven challenging partially due to the heterogeneity of graphs and the wide range of their applications. In this paper we take a different approach motivated by the conjecture of Bollobás, Pebody and Riordan stating that almost every pair of independently chosen random graphs are distinguished by their Tutte and chromatic polynomials. Hence,

---

2020 *Mathematics Subject Classification.* Primary 05C31, 55N31.
The first author was partially supported by NSF grant DMS-1854705.

instead of studying graphs directly, we examine the data derived from their chromatic polynomials. This idea of using invariants instead of objects has successfully been implemented in knot theory [**16**, **26**, **28**, **36**, **43**]. More precisely, we use ideas introduced in the work of Dłotko, Gurnari, Hajij, Levitt, and the first author, applying manifold learning and topological data analysis, and compare the results of these two approaches [**36**, **43**].

In this paper, Principal Component Analysis (PCA), a widely used dimension reduction technique, filtered PCA introduced in [**36**], and TDA's Ball Mapper algorithm are applied to chromatic polynomial data for graphs with a fixed number of vertices. The dimensionality of chromatic data for graphs of all orders up to 10 appears to be one, with the first principal component influenced by all nontrivial chromatic coefficients. The corresponding Ball Mapper graph (BMGraph) has remarkable and intriguing resemblance with the PCA structure. Though the evidence is only empirical, the first principal component (PC1) corresponds to the elongated, linear structure of the Ball Mapper graph, and the second principal component (PC2) corresponds, informally, to its "width".

The structure of the chromatic data is further explored with respect to various graph invariants using Ball Mapper as a tool for visualizing functions between high dimensional spaces. The linear structure of the Ball Mapper, analogous to the first Principal component, is determined by the number of edges in a graph, which is consistent with the prominence of edges E in coefficient formulas and the PCA observations. When colored by the corresponding values of PC2, the linear structure of the BMGraph corresponding to PC1 splits into concurrent/parallel strands determined by increasing values of PC2. Similarly, for a fixed number of edges and vertices, the collection of clusters in the Ball Mapper graph corresponds to a slice of points in the 2-dimensional PCA projection. Clusters on both ends of each slice in the PCA projection are graphs whose chromatic coefficients are extremal in the Rodriguez/Satyanarayana poset [**46**].

Our analysis facilitates the study of numerical graph invariants through dimensionality reduction and clustering. Recent work on graph compression summarized in Section 2.3 suggests a correlation between the chromatic polynomial and various measures of graph irregularity. While these measures are incomparable, our two-dimension projection of chromatic data, combined with Ball Mapper, reveals clusters of graphs which tend to maximize or minimize these invariants among all graphs with the same numbers of vertices and edges, see Section 4.4.

For example, on one end of the spectrum defined by our second principal component PC2, we find clusters of graphs with smallest chromatic coefficients and maximum irregularity scores, which tend to have more triangles, more cliques (complete subgraphs), and higher differences between vertex degrees; e.g., a complete graph with one pendant edge. The other end of the spectrum features graphs with largest chromatic coefficients and minimum irregularity scores; which tend to have a more regular structure, greater numbers of large cycles, and generally avoid cliques; e.g., Turan graphs as typical examples.

This work leaves a number of natural questions unanswered and raises more. Some of the questions to be addressed in the upcoming work are the following.

QUESTION 1.1. One of the main requirements in this approach is that the results are stable with respect to the filtration of our data, in this case by the number of vertices. It is quite remarkable that both dimensionality and Ball Mapper structure were stable, in addition to being consistent across methods. While filtered PCA [**36**] provides a reliable way of dimensionality reduction for data where finding a representative sample is a challenge, the next natural step is to apply these methods to a set of random graphs.

QUESTION 1.2. Big data techniques, such as machine learning and topological data analysis have only recently been applied to data arising from theoretical mathematics where they could provide valuable insights and conjectures. This was achieved in knot theory [**16**] where the relation between knot invariants was discovered by AI and later proven [**15**]. In this paper we provide insights into the missing theoretical characterization of chromatically maximal graphs, or the features corresponding to graph irregularity, which is quite elusive as the most prominent irregularity measures seem to be mutually incomparable [**2**]. The scope and types of results in graph or knot theory that can be obtained using these approaches is an interesting open question.

QUESTION 1.3. Methods used in this paper can be readily extended to other graph polynomials. The Tutte polynomial is particularly interesting as it specializes both to the chromatic and the Jones polynomial for knots. Determining the dimensionality of Tutte data is an exciting open problem since it should be at least 3, since that is the dimensionality of the Jones polynomial data [**36**]. Furthermore, Relational Ball Mapper or Mapper-on-Ball Mapper [**43**] constructions can be used to explore relations between the Tutte and chromatic, or Tutte and Jones polynomial.

To summarize, this approach provides a new way of vectorizing graphs, and opens up another area of theoretical mathematics to big data techniques. In addition to immediate extensions of this work to other polynomials such as Tutte there are exciting but less clear goals to determining statistical nature of polynomial and other graph invariants and relations between them, as well as potential applications to other sciences.

2. Background

In this section we provide necessary background in graph theory, focusing on the chromatic polynomial and irregularity measures.

Let $G = (V, E)$ be a graph, with vertex set V and edge multiset E. If G has an edge between vertices $v_i, v_j \in V$, we write the corresponding element in E as $\{v_i, v_j\}$.

DEFINITION 2.1. The order of a graph G is the number of vertices $n = |V|$. A loop in graph G is an edge of the form $\{x, x\}$ for some $x \in V$. A multiple edge is an element that occurs multiple times in the multiset E. We say that G is a simple graph if it has no loops or multiple edges.

DEFINITION 2.2. Let v be a vertex of a simple graph G. The degree of vertex v, denoted $\deg(v)$, is the number of edges incident to v.

DEFINITION 2.3. A path in graph G is a finite sequence of vertices $v_1, v_2, \ldots, v_n$ such that $\{v_i, v_{i+1}\}$ is an edge of G for $i = 1, 2, \ldots, n-1$ and all v_i are distinct.

DEFINITION 2.4. A graph $G = (V, E)$ is connected if any two vertices in G are joined by a path along edges of G. A bridge in a connected graph G is an edge $e \in E$ such that the graph $(V, E - \{e\})$ is disconnected. A pendant edge is an edge incident to a vertex of degree one.

DEFINITION 2.5. The girth of graph G is the length of the shortest cycle in G.

DEFINITION 2.6. Let $H = (V_H, E_H)$ be a subgraph of graph $G = (V, E)$. We say H is an induced subgraph if for every $\{v_i, v_j\} \in E$ with $v_i, v_j \in V_H$, the edge $\{v_i, v_j\}$ is in E_H.

DEFINITION 2.7. The complete graph K_n is the graph which consists of n vertices and $\binom{n}{2}$ edges connecting each distinct pair of vertices. A clique of size m is a subset of vertices of G whose induced subgraph is isomorphic to K_m. The clique number $\omega(G)$ is the maximum size over all cliques in G.

2.1. The chromatic polynomial.

DEFINITION 2.8 ([**20, 45, 56**]). Let $G = (V, E)$ be a graph. A mapping $f : V \to \{1, 2, \ldots, \lambda\}$ is called a λ-coloring of G if for any pair of vertices $x, y \in V$ such that $\{x, y\} \in E$, $f(x) \neq f(y)$. Two λ-colorings f, g are distinct if $f(x) \neq g(x)$ for some vertex $x \in V$. The chromatic polynomial of the graph G, denoted $P_G(\lambda)$, is equal to the number of distinct λ-colorings of G.

The degree of the chromatic polynomial is equal to the order of G, and thus we write the polynomial with coefficients c_{n-i} as follows:

$$P_G(\lambda) = c_n\lambda^n + c_{n-1}\lambda^{n-1} + c_{n-2}\lambda^{n-2} + \ldots + c_1\lambda + c_0.$$

Since no graph has a 0-coloring, the roots of $P_G(\lambda)$ include zero, which forces the free term to be zero $c_0 = 0$ for all graphs.

The chromatic polynomial of a disjoint union of graphs is the product of chromatic polynomials of its components. Additionally, adding multiple edges does not change the chromatic polynomial, while the existence of a loop trivializes it. Therefore, in this paper we consider only simple connected graphs.

THEOREM 2.9 ([**38**]). *If G is a graph with m edges, girth $g > 2$, and n_g cycles of length g, then the first g coefficients of the chromatic polynomial are:*

$$c_{n-i} = \begin{cases} (-1)^i \dbinom{m}{i} & 0 \leq i < g-1, \\ (-1)^{g-1}\left(\dbinom{m}{g-1} - n_g\right) & i = g-1. \end{cases}$$

For a detailed summary of graph invariants which are determined by the chromatic polynomial see [**40**].

PROPOSITION 2.10 ([**40**]). *For a connected graph G, the chromatic polynomial $P_G(\lambda)$ determines the numbers of vertices, edges, triangles, and blocks in G, as well as the girth g and the number of cycles of length g.*

It was long conjectured [**45**] that the coefficients of $P_G(\lambda)$ are unimodal in the following sense: if $h_i = |c_{n-i}|$, there exists some j with $2 \leq j \leq n-1$ such that

$$h_n \leq h_{n-1} \leq \ldots \leq h_{j-1} \leq h_j \geq h_{j+1} \geq \ldots \geq h_0.$$

This result has since been proven [**27**] by showing that the stronger property of log-concavity holds for this sequence: $h_{i-1}h_{i+1} \leq h_i^2$ for all $0 < i < n$.

DEFINITION 2.11. A graph is chromatically unique if no non-isomorphic graph has the same chromatic polynomial.

Cycle graphs, θ-graphs [**10**] and graphs which consist of wheel graphs with all but three or four adjacent spokes removed [**11**] are families of chromatically unique graphs.

There are also infinitely many graphs which are not chromatically unique; i.e., non-isomorphic graphs which share a chromatic polynomial. Examples include trees with n vertices, whose chromatic polynomials all have the form $\lambda(\lambda-1)^{n-1}$ regardless of structure [**45**, Theorem 13].

For any connected graph G, the first six coefficients of the chromatic polynomial have known formulas computed by counting subgraphs of G.

THEOREM 2.12 ([**6**, **22**]). *The first four coefficients of the chromatic polynomial $P_G(\lambda)$ are given by the following formulas: $c_n = 1$, $c_{n-1} = -m$, $c_{n-2} = \binom{m}{2} - t_1$, and $c_{n-3} = -\binom{m}{3} + (m-2)t_1 + t_2 - 2t_3$, where m is the number of edges in G, t_1 is the number of triangles, t_2 is the number of induced 4-cycles (pure squares with no diagonal present), and t_3 is the number of complete subgraphs on 4 vertices.*

*The 5th and 6th coefficients are given by the following formulas, where the t_is count induced subgraphs of G on 5 and 6 vertices specified in [**6**]. For example, t_4 is the number of induced 5-cycles (pure pentagons) and t_8 is the number of induced graphs isomorphic to K_5.*

$$\begin{aligned}
c_{n-4} = {} & \binom{m}{4} - \binom{m-2}{2}t_1 + \binom{t_1}{2} - (m-3)t_2 - (2m-9)t_3 - t_4 + t_5 + 2t_6 \\
& + 3t_7 - 6t_8 \\
c_{n-5} = {} & -\binom{m}{5} + \binom{m-2}{3}t_1 - (m-4)\binom{t_1}{2} + \binom{m-3}{2}t_2 - (t_2 - 2t_3)t_1 + t_4 \\
& - (m^2 - 10m + 30)t_3 - (m-3)t_5 - 2(m-5)t_6 - 3(m-6)t_7 + 6(m-8)t_8 \\
& + t_9 - t_{10} - 2t_{11} - 2t_{12} - t_{13} + t_{14} - t_{15} - 3t_{16} - 4t_{17} - 4t_{18} + 2t_{19} - 4t_{20} \\
& - t_{21} + 4t_{22} + 3t_{23} + 4t_{24} + 5t_{25} + 4t_{26} + 6t_{27} + 8t_{28} + 16t_{29} + 12t_{30} - 24t_{31}
\end{aligned}$$

Note that Whitney's original paper introducing the chromatic polynomial gives an interpretation of each c_{n-i} in terms of spanning subgraphs of G [**56**]. Gian-Carlo Rota showed that each coefficient can be expressed as a Möbius function on a poset whose elements are subsets of E, and that the sum of the coefficients of $P_G(\lambda)$ is zero [**47**]. Moreover, Meredith showed that the size of every coefficient is bounded by $|c_{n-i}| \leq \binom{m}{i}$ [**38**]. These upper bounds are achieved for the first g coefficients of the chromatic polynomial when all cycles of G have length at least g.

In [**31**] the author defines a Tutte polynomial poset and characterizes the minimal graphs in this poset. With a similar motivation, we define a chromatic polynomial poset as follows.

DEFINITION 2.13. Let $G(n, m)$ denote the set of simple connected graphs with n vertices and m edges. The chromatic polynomial poset is defined on $G(n, m)$ by $H \leq G$ if and only if $P_G(\lambda) - P_H(\lambda) = r(x)$ for some polynomial $r(x)$ with non-negative coefficients. In other words, if we denote chromatic polynomial coefficients by c_i, then $H \leq G$ if and only if $|c_i(H)| \leq |c_i(G)|$ for all $1 \leq i \leq n$.

Rodriguez and Satyanarayana [**46**] give a complete characterization of the graphs which are minimal in the chromatic polynomial poset.

DEFINITION 2.14 ([**20**]). Let $J(n, m)$ be the family of graphs $G \in G(n, m)$ such that all blocks of G are complete graphs, including possibly triangles K_3 and double edges K_2. Let $L(n, m)$ be the family of graphs $G \in G(n, m)$ such that one block B of G has clique number $\omega(B) \geq |V(B)| - 1$ and all other blocks are K_2s.

Note that B is a complete graph or contains a complete subgraph containing all but one of the vertices of B, while all other blocks are bridges or pendant edges.

PROPOSITION 2.15 ([**46**]). *Any pair of graphs $G, H \in L(n, m) \cup J(n, m)$, have the same chromatic polynomials $P_G(\lambda) = P_H(\lambda)$.*

THEOREM 2.16 ([**46**]). *Any graph $H \in L(n, m) \cup J(n, m)$ is a minimal element of the poset $G(n, m)$.*

In Section 4.4, we describe extremal graphs in our data set using the following definition based on the chromatic polynomial poset.

DEFINITION 2.17. A chromatically minimal (resp., chromatically maximal) graph with n vertices and m edges is one which is a minimal (resp. maximal) element of the poset $G(n, m)$.

2.2. Irregularity measures for graphs. A graph G is k-regular if $\deg(v) = k$ for each vertex $v \in V(G)$. Numerous invariants have been proposed to describe how far a given graph is from being regular, i.e., how irregular it is. A generally accepted criterion for an irregularity measure is that it should return a value of zero if and only if a graph is regular. The difference between the maximum and minimum vertex degrees in a graph is one example of an irregularity measure. Two other commonly cited measures are spectral irregularity and variance irregularity of a graph.

DEFINITION 2.18 (Spectral irregularity [**55**]). Let $\lambda_1(G)$ be the largest eigenvalue of the adjacency matrix of graph G and let $\overline{d}(G)$ be the average degree of the vertices of G. The spectral irregularity measure of G is $\epsilon(G) = \lambda_1(G) - \overline{d}(G)$, which is zero if G is regular and positive otherwise.

DEFINITION 2.19 (Variance irregularity [**5**]). Suppose G has order n and the vertices are labeled, with $d_i = deg(v_i)$. The variance irregularity measure of G is

$$\sigma(G) = \frac{1}{n}\sum_{i=1}^{n} d_i^2 - \frac{1}{n^2}\left(\sum_{i=1}^{n} d_i\right)^2.$$

For other proposed irregularity measures see [**1**, **4**, **39**] or a survey [**17**, **41**] that also talks about their comparative strengths.

The following classes of graphs are the most irregular under spectral and variance irregularity.

DEFINITION 2.20 ([**3**, **41**]). Let G be a connected graph with n vertices and m edges. Let d, t be the unique integer values with $2 \leq d$ and $0 \leq t < d$ such that $m = \binom{d}{2} + t$. Then G is the quasi-complete graph $QC(n, m)$ if it is the unique graph such that:

(1) G has a clique C of size $n - 1$;
(2) The single vertex v' not included in C is adjacent to exactly t of the other vertices.

DEFINITION 2.21 ([**3**, **41**]). Let G be a connected graph with n vertices and m edges. Let d, t be the unique integer values with $2 \leq d$ and $0 \leq t < d$ such that $m = \binom{n}{2} - \binom{d}{2} - t$. Then G is the quasi-star graph $QS(n, m)$ if it is the unique graph that:

(1) contains a set S of $n - d - 1$ vertices, each of which is adjacent to all other vertices of G;
(2) contains one vertex adjacent to exactly $n - t$ vertices of S;
(3) the only edges of G are the ones already specified.

PROPOSITION 2.22 (Proposition 1, [**5**]). *Among all graphs in $G(n, m)$, the maximum spectral irregularity is achieved by quasi-complete $QC(n, m)$, and the maximum variance irregularity is achieved by $QC(n, m)$ for $m > \frac{1}{2}\binom{n}{2} + \frac{n}{2}$ and by quasi-star fraphs $QS(n, m)$ such that $m < \frac{1}{2}\binom{n}{2} - \frac{n}{2}$.*

On the other hand, we have graphs which are as close to regular as possible, such as Turan graphs.

DEFINITION 2.23. A Turan graph $T(n, r)$ is the graph formed by partitioning a set of n vertices into r groups, with sizes as equal as possible, and connecting two vertices by an edge if and only if the vertices belong to separate groups. If we let $t, p \in \mathbb{Z}$ such that $n = tr + p$, each vertex of $T(n, r)$ has degree either $n - t$ or $n - t - 1$.

Turan graphs are a family of extremal graphs whose degree sequences are constant or nearly constant and they are characterized by the following theorem.

THEOREM 2.24 (Turan's Theorem, [**53**]). *Turan graph on n vertices with r groups $T(n, r)$ has the maximal number of edges among all order n graphs that contain no $(r + 1)$-clique.*

2.3. Threshold graphs and compression. Compression of a graph is an operation first introduced in [**32**] and known by a variety of names, including the Kelmans transformation [**13**] and swing surgery [**48**]. See the introduction of [**31**] for details.

DEFINITION 2.25. The neighborhood of vertex v in graph G, denoted $N_G(v)$, is the set of all vertices which are connected to v by an edge. We say that vertex v dominates vertex u if $N_G(u) \subset N_G[v]$, where $N_G[v] = N_G(v) \cup \{v\}$.

DEFINITION 2.26. A graph G is threshold if for all vertices $u, v \in V(G)$, either u dominates v or v dominates u.

For other characterizations of threshold graphs, see [**13, 33**].

DEFINITION 2.27. Let $G = (V, E)$ be a graph and let $u, v \in V$. The compression of G from vertex u to vertex v is an operation defined as follows: for each vertex $x \in N_G(u) - N_G(v) - \{v\}$ we delete all edges of the form $\{x, u\}$ from G and replace them with corresponding edges of the form $\{x, v\}$. The new graph produced by the compression of G from u to vertex v is denoted $G_{u \to v}$.

Compression preserves the number of vertices and edges of a graph, and the result is independent of the order of in which vertices u and v are used, up to isomorphism (i.e., $G_{u\to v}$ and $G_{v \to u}$ are isomorphic graphs). Compression of a graph monotonically increases a number of graph invariants, such as the spectral radius and the number of independent sets of order k, while decreasing other invariants such as the number of spanning trees and the vertex connectivity [**13, 30, 31**].

Any connected graph can be transformed into a connected threshold graph by repeated applications of graph compression [**7, 48**]. Thus, threshold graphs are often extremal with respect to these types of invariants.

3. Methods and data

3.1. Principal component analysis. Principal component analysis (PCA) is a common technique for exploring multivariate data. The algorithm finds a set of directions within the feature space which explain most of the variance in the data. The number of these directions is often far smaller than the number of original features, and so PCA can be used as a method of dimensionality reduction when visualizing and interpreting a high-dimensional data set.

Let $X \in \mathbb{R}^{n \times p}$ be a matrix representing a data set of n individuals, each with p features. PCA takes the sample covariance matrix S for the data set and finds an orthonormal eigensystem $\{(\lambda_j, a_j)\}_{j=1}^{p}$ for S with eigenvalues $\lambda_1 \geq \lambda_2 \geq \ldots \geq \lambda_p$. The eigenvector a_1 is the unit vector in $\mathbb{R}^p$ such that the entries of Xa_1 have maximal variance (up to an overall sign change for a_1). The second eigenvector a_2 maximizes variance subject to the constraint that a_2 is orthogonal to a_1, and all successive vectors a_j maximize variance among directions orthogonal to $\{a_1, \ldots, a_{j-1}\}$.

DEFINITION 3.1 ([**29**]). Given a data set $X \subset \mathbb{R}^{n \times p}$ and PCA eigensystem $\{(\lambda_j, a_j)\}_{j=1}^{p}$, the jth principal component of X is the product Xa_j. For the jth principal component of X, the p elements of the vector a_j are called the PC loadings for the jth principal component, and the n elements of Xa_j are called the PC scores.

Note that if a_j is an eigenvector of S with eigenvalue λ_j, then so is its opposite $-a_j$. Thus the signs of PC loadings and PC scores are arbitrary and may be reversed depending on the implementation of PCA in a given instance.

DEFINITION 3.2 ([**29**]). The explained variance associated with the jth principal component of the data set X is the eigenvalue λ_j, and the normalized explained variance $\overline{\lambda_j}$ is defined by $\overline{\lambda_j} = \frac{1}{\sum_{i=1}^{p} \lambda_i} \lambda_j$.

The proportion of variance explained by the first j principal components of X will be denoted $S_j = \sum_i^j \overline{\lambda_i}$.

The eigenvector a_1 is a unit vector that determines a line in the feature space $\mathbb{R}^p$. This line indicates the axis in $\mathbb{R}^p$ along which the data set represented by X has the greatest variation. Similarly, a_2 gives the direction orthogonal to a_1 for which the data set has the second-greatest variation, and so on for $2 < j \leq p$.

Suppose that $x \in \mathbb{R}^p$ is a column of X representing one individual in the data set. The PC score for the jth principal component represents the length of the projection of x onto the line corresponding to eigenvector a_j. The implementation of PCA in the Python `scikit-learn` package [**44**] centers the data by moving the mean to the origin. Thus a PC score of zero in the jth component would indicate that x lies at the center of the distribution with respect to the jth direction of variation. A positive or negative score indicates that x lies further from the center on either side along this direction.

3.2. Ball Mapper. Ball Mapper [**19, 42**] is a recent Mapper-type algorithm for visualizing and analyzing the local and global structure of a data set. Like the widely-used Mapper algorithm [**50**], Ball Mapper takes a data set in $\mathbb{R}^n$ as an input and returns a graph which describes the proximity of points within the data. While Mapper is highly dependent on a choice of filter function for the data set and a choice of clustering method, Ball Mapper's approach to obtaining the cover is simpler and requires only one input parameter, a radius ϵ.

Given a set of points $X \subset \mathbb{R}^d$ and $\epsilon > 0$, see Figure 1(A), Ball Mapper finds a collection $C \subset X$, called an ϵ–net Figure 1(B), such that no point in X is located further than distance ϵ from an element of C. This collection determines an overlapping cover $B(X, \epsilon) = \{B_1, B_2, \ldots, B_i, \ldots\}$: a set of d-dimensional balls with radius ϵ that contain all of X, Figure 1(C). The algorithm then constructs the nerve of $B(X, \epsilon)$, called Ball Mapper graph (BMGraph) Figure 1(D). In a Ball Mapper graph, vertices v_i correspond to the balls B_i in the cover, and two vertices v_i, v_j are connected by an edge if and only if B_i and B_j contain shared points in X.

The Ball Mapper parameter ϵ determines the scale of the features highlighted in the output. By varying ϵ, we create a linear sequence of images which highlight structures at successive scales, inspired by the filtration in persistent homology (but in this case we do not have the stability result). Additionally, we can choose to filter our data according to a range of values of a given criterion and obtain an output that can be displayed as a sequence of graphs, see Figure 4.

Results in this paper are obtained using the R implementation of Ball Mapper [**18**]. This version of Ball Mapper supports additional visualization features; e.g., the disks corresponding to vertices of the Ball Mapper graph are scaled according to the number of points they contain. Thus larger disks in the visualization correspond to larger clusters of points within radius ϵ of each other. Another useful feature is

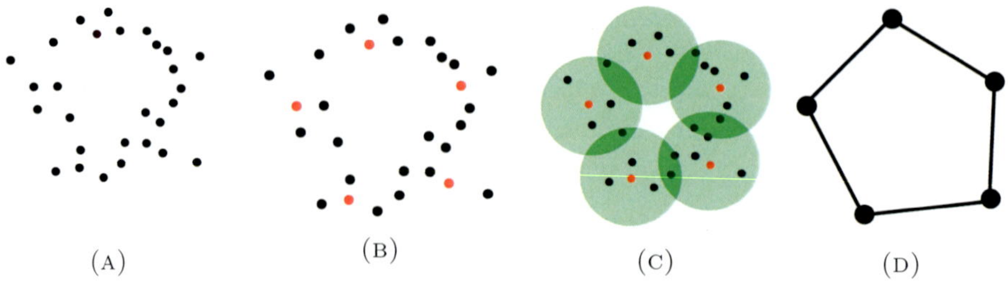

FIGURE 1. Ball Mapper construction: Illustration. The input point cloud is shown in (A), ϵ–net (B), Ball Mapper cover (C), corresponding Ball Mapper graph (D).

coloring the Ball Mapper graph using values of some function $f : X \to \mathbb{R}$ defined on the input data. The color of each Ball Mapper disc is the average value of the functions f on the points of X in that cluster. For convenience, we include the colored vertical bar on the right of each image which describes a uniform scale illustrating the distribution of colors from the minimum to maximum values of f within the data set.

3.3. Graph data. The focus of this paper is on the chromatic polynomials and our data contains chromatic polynomials for all simple connected graphs up to ten vertices for a total of 11,716,571 graphs. Our database contains many additional properties including numbers of vertices and edges, spectral irregularity, variance irregularity, numbers of subgraphs including cycles and K_4s, maximal and minimal vertex degrees, and identifications of all graphs which belong to classes including trees, Turan graphs, or bipartite graphs.

Chromatic polynomials are computed using the Sage `GraphGenerators` class implemented in Python 3 [**23, 52**]. The point cloud consists of the set of all chromatic polynomial coefficients obtained in the following way. Given graph G with chromatic polynomial $P_G(\lambda) = c_n\lambda^n + c_{n-1}\lambda^{n-1} + \ldots + c_1\lambda$, we represent $P_G(\lambda)$ by its coefficient vector $Q(G) = (|c_n|, |c_{n-1}|, \ldots, |c_1|, 0, \ldots, 0)$. For simplicity and computational efficiency, we take absolute values of each coefficient, since the sign is given by $(-1)^k$ for each coefficient c_{n-k} and thus remains constant within each feature of our data set. Our data set consists of coefficient vectors with at most 10 non-zero entries, so we standardize length by adding zero entries for graphs with less than 10 vertices. Table 1 contains sample vectors for the triangle C_3, the complete graph K_5, and the cycle C_{10} with 10 edges.

TABLE 1. Chromatic point cloud examples: chromatic coefficient vectors (absolute values) for the triangle, complete graph on 5 vertices, the cycle graph with all ten non-zero entries.

G	Chromatic polynomial $\mathbf{P_G}(\lambda)$	Vector $\mathbf{Q}(G)$
C_3	$\lambda^3 - 3\lambda^2 + 2\lambda$	$(1, 3, 2, 0, 0, 0, 0, 0, 0, 0)$
K_5	$\lambda^5 - 10\lambda^4 + 35\lambda^3 - 50\lambda^2 + 24\lambda$	$(1, 10, 35, 50, 24, 0, 0, 0, 0, 0)$
C_{10}	$\lambda^{10} - 10\lambda^9 + 45\lambda^8 - 120\lambda^7 + 210\lambda^6 - 252\lambda^5 + 210\lambda^4 - 120\lambda^3 + 45\lambda^2 - 9\lambda$	$(1, 10, 45, 120, 210, 252, 210, 120, 45, 9)$

In the sections that follow, we visualize data sets corresponding to graphs with a fixed number of vertices. Within each data set, the varying sizes of chromatic coefficients make it difficult to see important features during visualization. We apply a log transformation $f(x) = \ln(1 + x)$ to the data set derived from graphs with n vertices and then separately normalize each feature to the range $[0, 1]$ with min-max scaling over this particular data set.

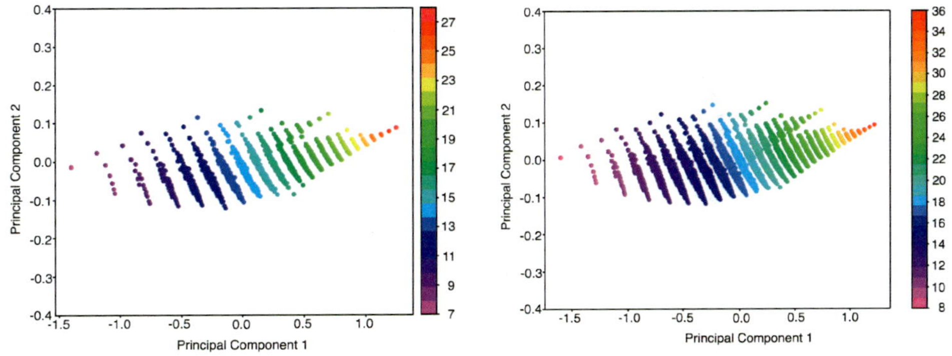

(A) (B)

FIGURE 2. A PCA projection of the chromatic polynomial data to 2 dimensions determined by the first two principal components and colored by number of edges for graphs with eight (A) and nine vertices (B).

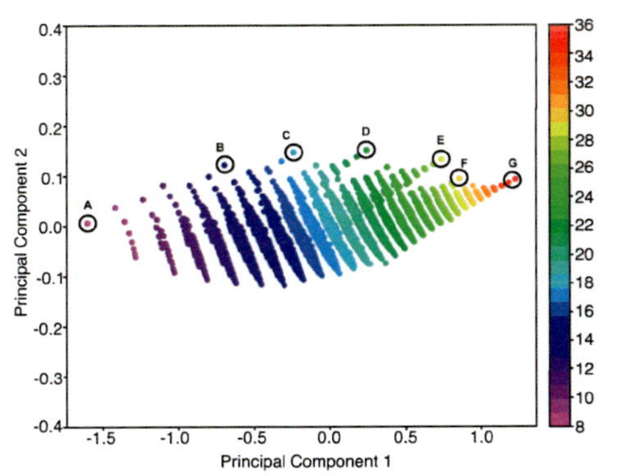

FIGURE 3. The PCA projection from Figure 2(B) with clusters containing some well known graphs circled and labeled. In addition to multiple non-isomorphic graphs the following clusters contain A: all 9-vertex trees, B: complete graph K_5 with 4 pendant edges, C: complete graph K_6 with 3 pendant edges, D: complete graph K_7 with 2 pendant edges, E: complete graph K_8 with 1 pendant edge, F: complete graph K_8 with triangle attached along 1 edge, G: complete graph K_9, and K_9 with 1 edge removed.

4. Chromatic polynomial data: Structure and dimensionality

The main results of this paper rely on introducing big data analysis to graph theory in a novel way inspired by applications to knot theory [**36, 43**]. In these approaches a graph is represented by a unique vector, consisting of coefficients of the chromatic polynomial. In this paper we focus on filtered PCA and Ball Mapper, but this graph representation lends itself to any of the big data techniques, including machine learning and other artificial intelligence tools.

Input data consists of chromatic polynomials for graphs up to ten crossings as described in Section 3.3. This point cloud naturally embeds in $\mathbb{R}^9$ and we are using Euclidean distance as our metric for this initial analysis.

Note that Theorem 2.12 provides an interpretation of the first few chromatic coefficients in terms of number of vertices and edges, as well as the number of special subgraphs such as cycles, complete graphs, and many others. Moreover, Bollobás, Pebody and Riordan conjecture that almost every pair of independently chosen random graphs are distinguished by their Tutte and chromatic polynomials.

Conjecture 4.1 ([**8**]). *Let $G(n, 1/2)$ be the space of random graphs on a fixed set of n vertices in which each pair of vertices is joined independently with probability $1/2$. Let $T(G), p(G)$ denote the Tutte polynomial and chromatic polynomial of G respectively. Then almost every graph $G \in G(n, 1/2)$ is such that $T(G') = T(G)$ implies G' is isomorphic to G. In addition, almost every graph $G \in G(n, 1/2)$ is such that $p(G') = p(G)$ implies G' is isomorphic to G.*

While the conjecture remains open, it has been shown that the probability that two independently chosen random graphs have the same Tutte polynomial is of order $O(1/\log n)$ [**37**].

4.1. Dimension of the chromatic data.

Inspired by the big data approach to the Jones polynomial, a polynomial invariant for knots [**36**], we apply Principal Component Analysis (PCA) to the chromatic data for graphs with increasing number of vertices, up to ten maximum. The *dimension* of the Jones polynomial is defined to be the smallest value **d** for which the normalized explained variance of the first **d** principal components sums to more than 95% across all considered orders [**36**].

According to this definition and PCA computations, chromatic polynomial data is essentially one-dimensional, and this dimensionality is stable with respect to increasing the order of a graphs, see Table 2. This table presents a snapshot of PCA computations for chromatic polynomial data for graphs of orders $n = 8, 9, 10$ and the first three principal components for each data set. In particular the last two rows in Table 2 contain the normalized (NEV) and cumulative explained variance (CEV) for $n = 8, 9, 10$ indicating that almost all variation is explained by the first principal component. More precisely, the cumulative explained variance of PC1, S_1, is greater than 0.99 for all orders up to 10. In Section 4.3 we revisit this question since the structure of the Ball Mapper graph is linear, see comparison with the PCA in Figure 6.

TABLE 2. The normalized explained variance $\overline{\lambda_j}(n)$ (NEV) and cumulative normalized explained variance $S_j(n)$ (CEV) for the first three principal components of the chromatic polynomial data of graphs with orders 8, 9, and 10 respectively.

	$n = 8$			$n = 9$			$n = 10$		
	PC1	PC2	PC3	PC1	PC2	PC3	PC1	PC2	PC3
NEV	0.99188	0.00792	0.00018	0.99178	0.00806	0.00015	0.99189	0.00796	0.00015
CEV	0.99188	0.99980	0.99998	0.99178	0.99984	0.99999	0.99189	0.99985	0.99999

4.2. Graph-theoretic interpretation of the principal components. To get an insight into the graph-theoretic relevance of the principal components and the dimension reduction of our data from 10 to one (or two), we provide visualisations of the first two principal directions and further insights obtained using Ball Mapper; see Figure 6.

Focusing on the two most significant principal components on chromatic polynomials of graphs with fixed order, we visualize the structure of each data set by plotting each graph as a coordinate pair (x_1, x_2) pair where x_i represents the ith PC score of that graph, see Figure 2. This figure illustrates the persistent structure found for graphs with eight and nine vertices and the stability of the dimension reduction with respect to the number of vertices of a graph. The first principal component corresponds to the horizontal axis in Figure 2, which appears to be roughly correlated with an increase in the number of edges. Graphs with a fixed number of edges are found in clusters separated by thin regions of empty space in this projection. Figure 3 identifies the location of certain well-known graphs, such as trees and complete graphs, at the boundary of the projection. This boundary also contains maximum coefficient graphs; see Figure 12.

The data set of chromatic polynomials for graphs of 9 vertices has 8 dimensions, corresponding to the eight coefficients which are not trivially equal to 0 or 1. The loadings for PC1 and PC2 in Table 3 are the entries of the eigenvectors that give the directions of greatest variation in this eight-dimensional data set. We observe that all loadings for PC1 are positive, so that movement along this axis corresponds to an overall increase or decrease in coefficient size. This agrees with our previous observation that PC1 is correlated with the number of edges in a graph. For PC2, the loadings for the last four coefficients have the opposite sign from the first four, indicating that an increase in PC2 score correlates with an decrease in relative size of the last few coefficients.

TABLE 3. PC loadings for two most significant principal components, PC1 and PC2, for the chromatic polynomial data for all graphs of order 9, with chromatic polynomial determined by eight coefficients $P_G(\lambda) = \lambda^9 + c_8\lambda^8 + \ldots + c_1\lambda$

	c_8	c_7	c_6	c_5	c_4	c_3	c_2	c_1
PC1	0.34694	0.34683	0.34761	0.34876	0.35005	0.35172	0.35579	0.379545
PC2	0.54568	0.39243	0.24013	0.08614	−0.06850	−0.22145	−0.37150	−0.53983

PC loadings provide a way to think about principal components in terms of the coefficients of the chromatic polynomial, and in turn, based on the formulas in Theorem 2.12, features of a graph. We hypothesize that PC1 measures the average size of the coefficient vector as a whole, and that PC2 detects the relative difference between the first and last coefficients.

4.3. Chromatic data: Ball Mapper structure. In this section we apply Ball Mapper to the chromatic point cloud consisting of coefficient vectors of all graphs up to $n = 10$ vertices.

First, we provide the evidence for the stability of the Ball Mapper approach to the chromatic polynomial data with respect to the order of a graph. In Figures 4 and 5 we provide evidence for a persistent structure across data sets of graphs with order equal to eight, nine, and ten respectively. Graphs of each given order are clustered in long branch-like features, with neighboring branches of adjacent orders

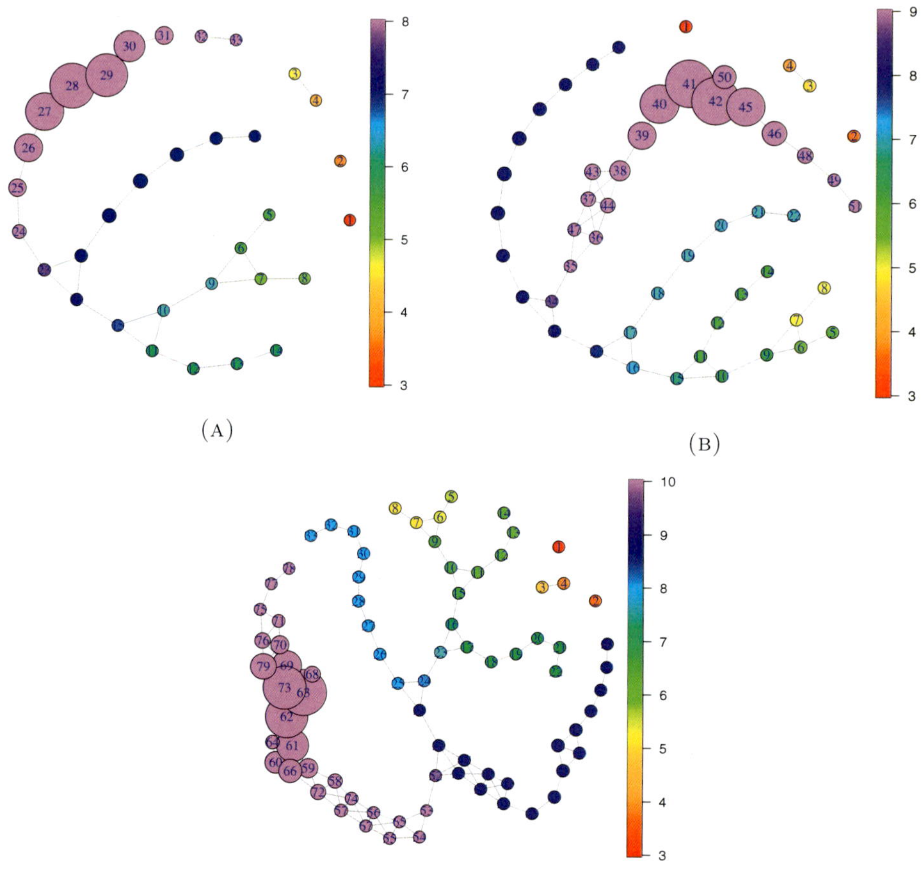

(A) (B) (C)

FIGURE 4. Stability of BMGraphs for chromatic data with respect to the number of vertices with graphs of order at most 8 (A), nine (B) and ten (C) with the Ball Mapper parameter $\epsilon = 0.1$, colored by the number of vertices.

across all three BMGraphs. These branch-like features are also reflected in the two-dimensional PCA plot for all graphs up to 10 vertices combined, see Figure 5. This structure invites and justifies exploring the data corresponding to graphs of a given order.

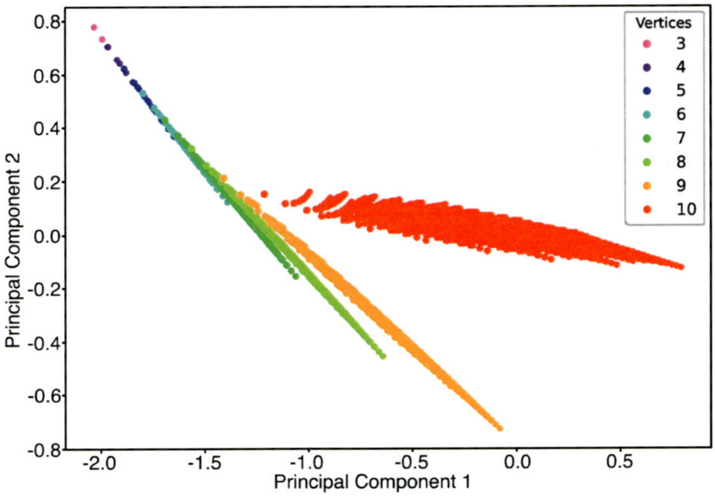

FIGURE 5. A two-dimensional PCA projection of the chromatic polynomial data for all graphs with 3-10 vertices. The projection is based on the first two principal components for normalized data, colored by number of vertices.

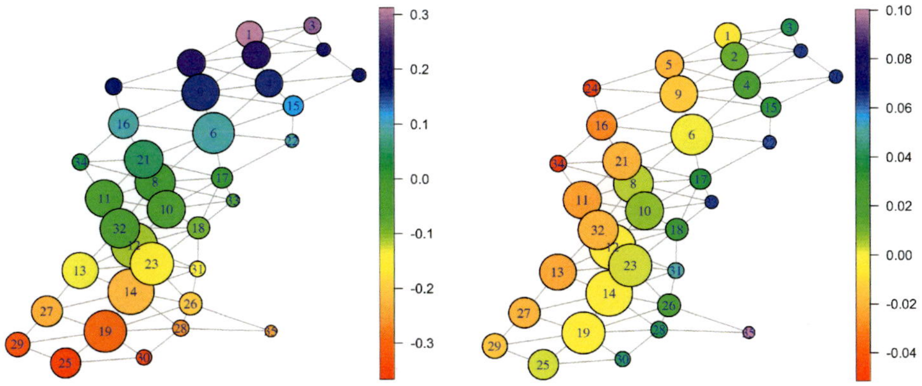

(A) (B)

FIGURE 6. Clusters 9-12 in BMGraph of the Chromatic polynomial data shown in Figure 8 at radius $\epsilon = 0.06$ colored by the average PC1 score (A), and the average PC2 score (B) of graphs contained in each cluster.

Figure 7 and Figure 8 provide a detailed structure of branches consisting only of graphs with exactly 8 and 9 vertices respectively. Both Ball Mapper graphs have linear structure consistent with the PCA analysis in Section 4.1. To confirm the informal correspondence between the 1-dimensionality of the data obtained by the PCA approach and the "linear" structure of the BMGraph, we color each cluster in the BMGraph by the average PC scores of the graphs it contains, Figure 6. The remarkable agreement between the two methods suggests that the first principal direction corresponds to the linear feature of the Ball Mapper structure.

The linear direction within BMGraphs corresponds to the number of edges in a graph, see Figures 7(B) and 8(B), as well as the L^2 norm of the chromatic coefficient vectors, see Figure 7(C), 8(C). This result can be explained by the fact that the number of edges E contributes to the largest terms of each formula in Theorem 2.12.

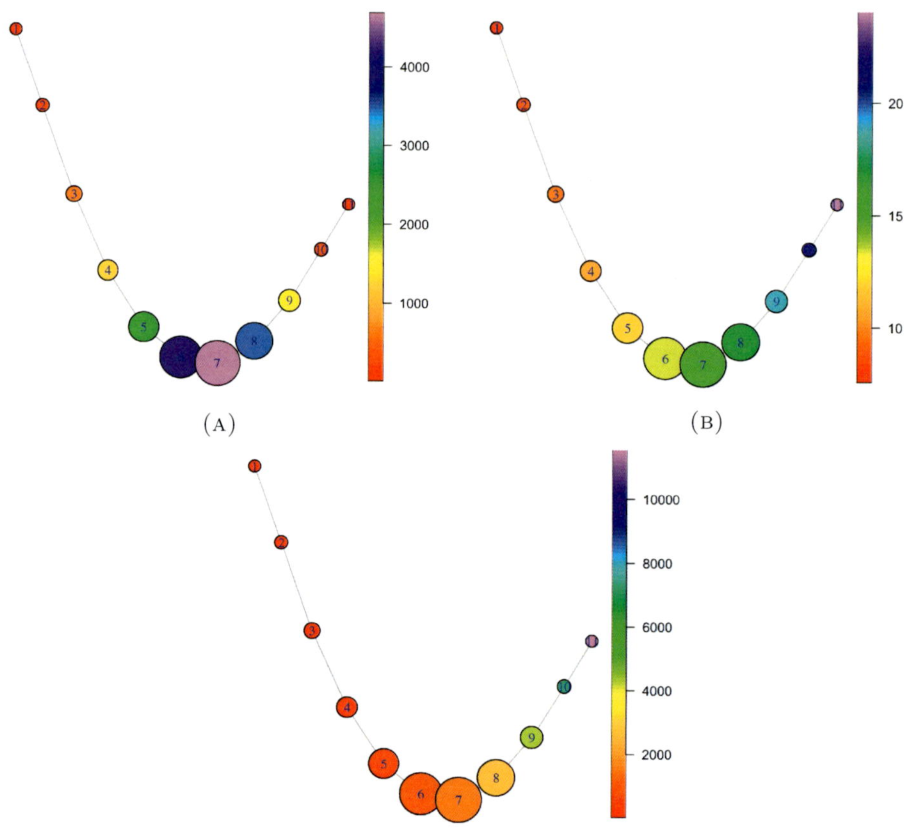

FIGURE 7. BMGraph of the chromatic polynomial data of all graphs with exactly 8 vertices and the Ball Mapper parameter $\epsilon = 0.22$ colored by cluster size (A), number of edges (B), and the norm of the chromatic polynomial (C).

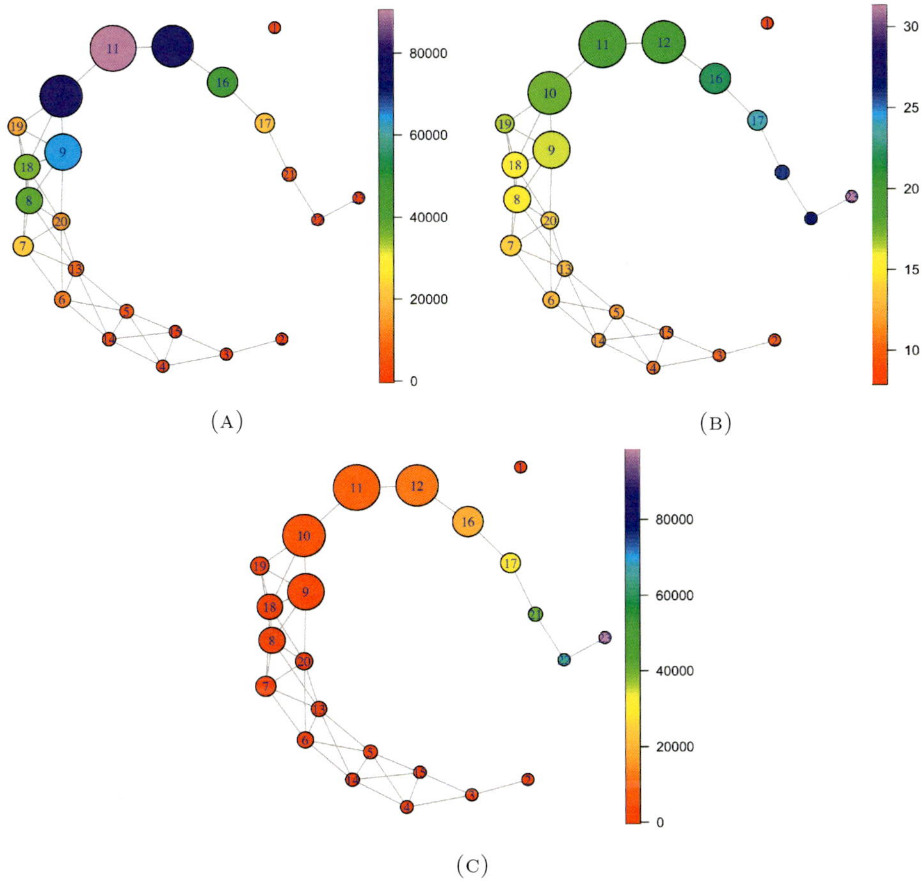

FIGURE 8. BMGraph of the chromatic polynomial data of all graphs with exactly 9 vertices and the Ball Mapper parameter $\epsilon = 0.15$ colored by cluster size (A), number of edges (B), and the norm of the chromatic polynomial (C).

Based on the BMGraph in Figure 8(A) and associated data, four largest clusters contain more than 50,000 out of total of 261079 points/graphs: cluster 9 (64442), cluster 10 (80017), cluster 11 (90054), and cluster 12 (78878). Therefore, we analyze the collection of these 4 clusters using a smaller value of radius ϵ, see BMGraphs on Figure 9. As expected, the major PC1 direction of this particular embedding on the BMGraph is correlated with the number of edges Figure 9(A), and the PC2 direction is roughly consistent with the L^2 norm of the chromatic coefficients, Figure 9(B). The average number of triangles in a graph contained in any given cluster, as well as the average degree distance seem to be varying in the PC2 direction, complementary to the main linear structure, from north-west to south-east in Figures 9(C-D).

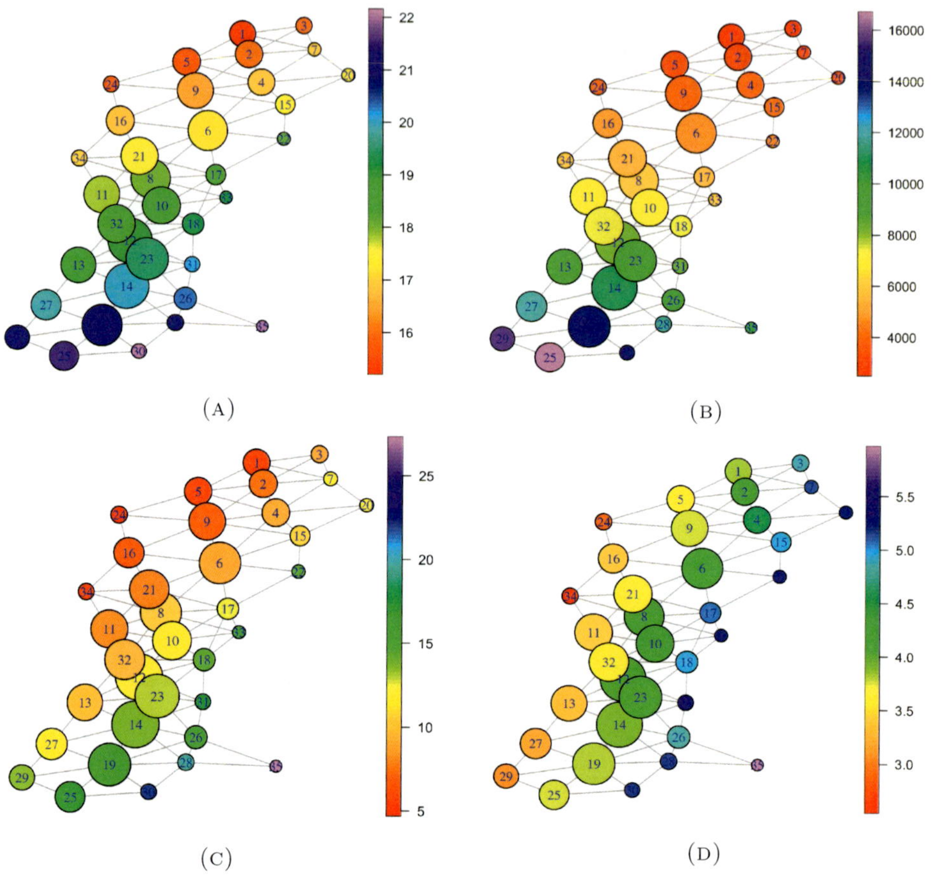

FIGURE 9. Clusters 9-12 in BMGraph of the chromatic polynomial data shown in Figure 8 at radius $\epsilon = 0.06$ colored by: number of edges (A) and by L^2 norm of chromatic coefficients (B), average number of triangles in a graph (C), average difference between the maximal and minimal degree (D).

4.4. Extremal chromatic polynomials and irregularity. Section 2.2 provides a few details on regular graphs and irregularity measures, but is by no means comprehensive. Note that there is a lot of ambiguity when it comes to irregularity measures. It was recently shown that no two of the following irregularity measures are mutually comparable: spectral irregularity, variance irregularity, Albertson index, and total irregularity [**2**].

Section 2.3 offers a brief survey of results on graph compression, which uniformly increases or decreases values of a number of graph invariants related to the chromatic and Tutte polynomials.

PROPOSITION 4.2. *Suppose that G' is obtained from G by compression. Then*

(1) *The coefficients of the chromatic polynomial of G' are smaller than those of G: $|c_i(G')| \leq |c_i(G)|$ for all $1 \leq i \leq n$.*
(2) *Graph G' has greater variance irregularity than G: $\sigma(G') \geq \sigma(G)$.*
(3) *Graph G' has greater spectral irregularity than G: $\epsilon(G') \geq \epsilon(G)$.*

PROOF. Part (1) is a restatement of Theorem 6.3 of [**13**]. Part (2) derived from Lemma 2.3 of [**14**] with Definition 2.19 in mind. Part (3) follows from Theorem 2.1 of [**12**] with respect to Definition 2.18. □

As a consequence, if we consider graphs with a fixed number of vertices and edges, then part (1) claims that threshold graphs and their co-chromatic graphs are among the chromatically minimal and part (2) says that threshold graphs maximize variance irregularity.

The rest of this section provides visualizations of these observations and a large-scale comparison between the spectral and variance irregularity measures, highlighting the relation between threshold graphs (the extremal graphs with respect to compression) and the coefficients of the chromatic polynomial. For example, the graphs with extremal PCA2 scores and minimal coefficient graphs, as in Figure 12(C), are either threshold or co-chromatic with threshold graphs, see Figure 10.

The BMGraph 11(A) is colored by spectral irregularity and (B) by variance irregularity emphasizing that, in these particular embeddings of BMGraphs, clusters on the left- and right-hand sides contain graphs with extreme irregularity measures, with the average irregularity monotonically increasing from left to right. The same phenomenon appears in Figure 9(D) when the coloring function is the difference between maximal and minimal vertex degrees.

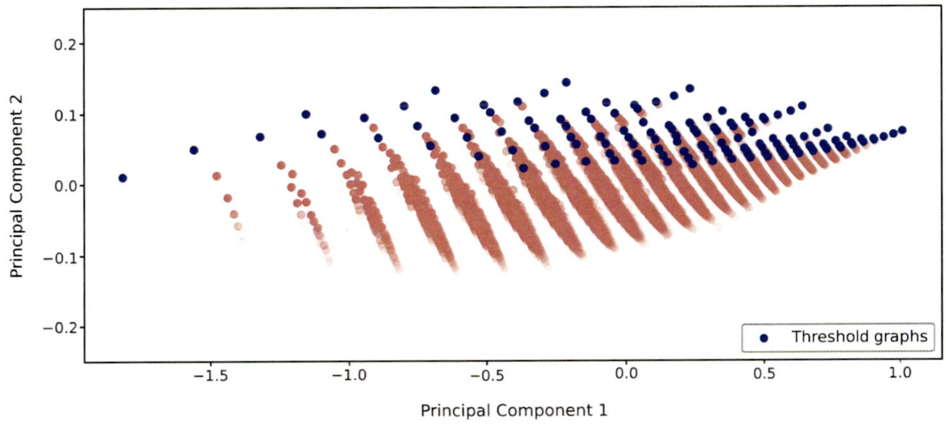

FIGURE 10. A PCA projection of the chromatic polynomial data for all graphs of order 9 into 2-dimensions determined by the two most significant principal directions, PC1 and PC2, colored by presence of threshold graphs denoted by blue color.

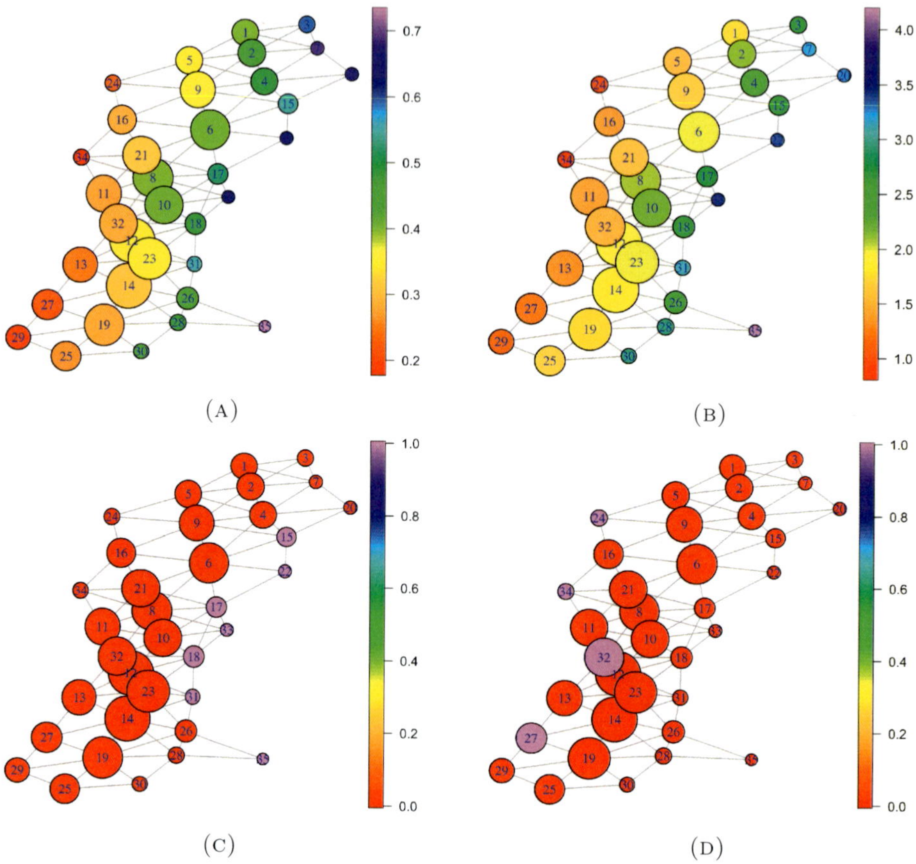

FIGURE 11. Clusters 9-12 in GMGraph of the Chromatic polynomial data shown in Figure 8 at radius $\epsilon = 0.06$ colored by average measure of spectral irregularity in each cluster (A), variance irregularity (B), presence of chromatically minimal (purple color) (C) and presence of chromatically maximal graphs (D).

These observations based on Ball Mapper graphs are matched by visualizations of the chromatic data in the 2-dimensional PCA projection, see Figure 12. In both projections, higher PC2 score corresponds to higher values of graph irregularity measures within the subset consisting of graphs with equal number of edges. However, the overall distributions of spectral Figures 12(A), and variance irregularity Figures 12(B) are different, with the variance regularity being lower on the middle left section.

The location of chromatically minimal and maximal graphs of order 9 (see Definition 2.17) in the BMGraph and 2-dimensional PCA projection is shown on Figures 11(C), 11(D) and 12(C), 12(D), respectively. Both methods show that chromatically minimal graphs are found in regions of highest irregularity. This is expected since the families of minimal graphs characterized by Definition 2.14

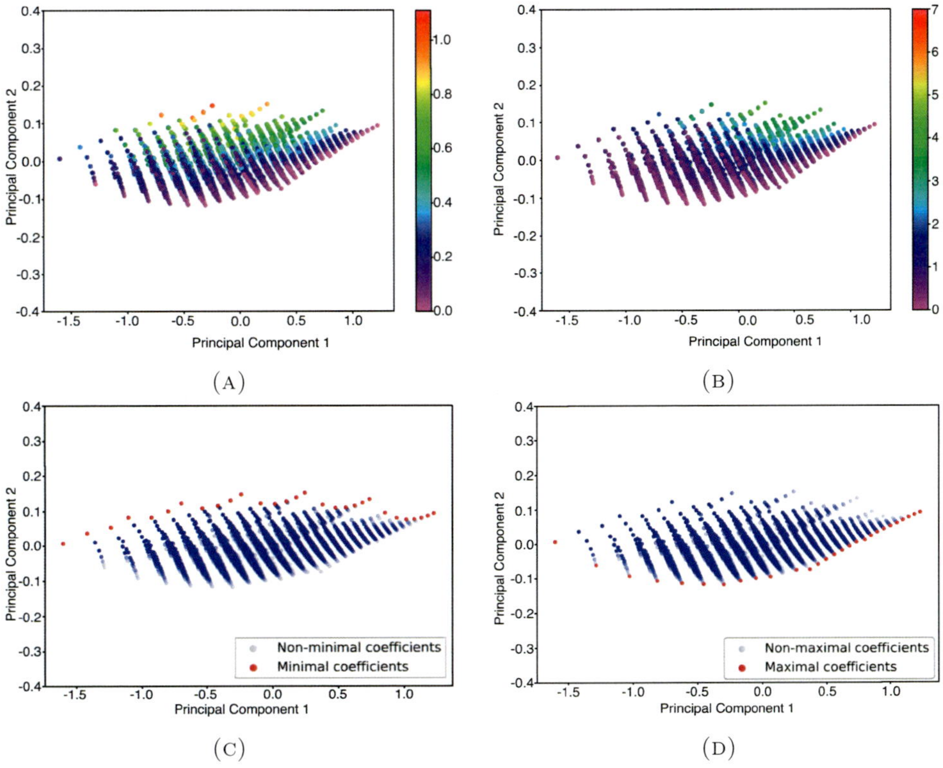

FIGURE 12. A PCA projection of the chromatic polynomial data for all graphs of order 9 into 2-dimensions determined by two most significant principal directions, PC1 and PC2, colored by spectral (A) and variance irregularity measure (B), the presence of minimum-coefficient (C) and maximum-coefficient graphs (D) denoted by red color.

include members of the high-irregularity connected quasi-complete graphs and connected quasi-stars (Section 2.2). For example, it follows from Definition 2.20 that the connected quasi-complete graph $QC(n, m)$ is an element of the family $L(n, m)$.

While, to the best of our knowledge, there exists no similar characterization for chromatically maximal graphs, their location among the most regular graphs is supported by results from extremal graph theory. Lazebnik and others prove partial characterizations for graphs which maximize evaluations of the chromatic polynomial $P_G(\lambda)$ at given values of λ [**34**]. These classes of graphs include Turan graphs (Definition 2.23) and are generally similar to Turan graphs in structure. Turan graphs are extremal graphs which are chromatically unique [**9**] and close to regular, see Section 2.2. The location of all Turan graphs of order 9 within the 2-dimensional PCA projection, among other chromatically maximal graphs, is highlighted in red in 12(D).

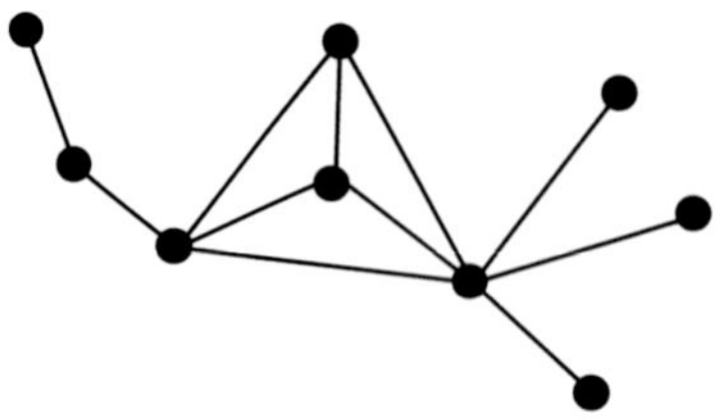

(A)

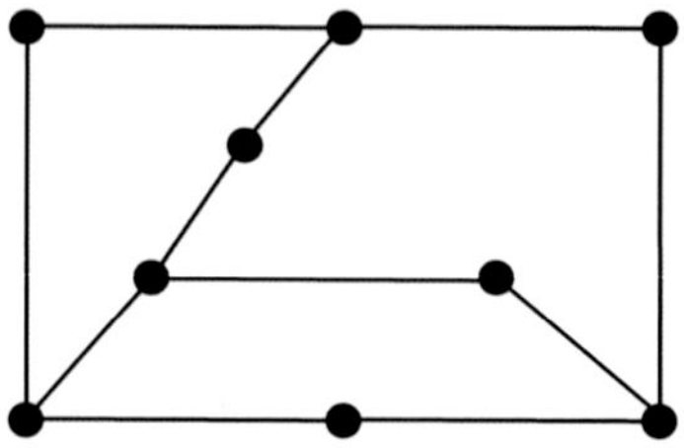

(B)

FIGURE 13. Extremal graphs in $G(9, 11)$ with coefficient vectors equal to: (1, 11, 51, 131, 205, 201, 121, 41, 6) (A), and (1, 11, 55, 165, 328, 446, 406, 224, 56) (B).

Computations for graphs up to nine vertices imply that:

- The minimal and maximal elements of the chromatic polynomial poset $G(n, m)$ achieve the extremal PC2 scores among all graphs with n vertices and m edges when the principal components are computed over the data set of all n-vertex graphs.
- The Turan graph $T(n, m)$ is maximal in the poset $G(n, m)$.

Further investigation of chromatically maximal graphs in our data suggests that there is a relationship between the "tail" of the polynomial (the last few coefficients) and graph irregularity. For example, consider the graphs in Figure 13 with $n = 9$ vertices and $E = 11$ edges. The graph on the left is one of three co-chromatic non-isomorphic graphs which are minimal in $G(9, 11)$, with higher irregularity measures and significantly smaller tail than the graph in Figure 13(B), which is maximal in $G(9, 11)$.

Acknowledgments

The authors extend their heartfelt gratitude to Pawel Dłotko and Davide Gurnari for many helpful conversations and support with Ball Mapper.

References

[1] H. Abdo, S. Brandt, and D. Dimitrov, *The total irregularity of a graph*, Discrete Math. Theor. Comput. Sci. **16** (2014), no. 1, 201–206. MR3214266

[2] H. Abdo, D. Dimitrov, and I. Gutman, *Graph irregularity and its measures*, Appl. Math. Comput. **357** (2019), 317–324, DOI 10.1016/j.amc.2019.04.013. MR3938885

[3] R. Ahlswede and G. O. H. Katona, *Graphs with maximal number of adjacent pairs of edges*, Acta Math. Acad. Sci. Hungar. **32** (1978), no. 1-2, 97–120, DOI 10.1007/BF01902206. MR505076

[4] M. O. Albertson, *The irregularity of a graph*, Ars Combin. **46** (1997), 219–225. MR1470801

[5] F. K. Bell, *A note on the irregularity of graphs*, Linear Algebra Appl. **161** (1992), 45–54, DOI 10.1016/0024-3795(92)90004-T. MR1142729

[6] H. Bielak, *A new method for counting chromatic coefficients*, Ann. Univ. Mariae Curie-Skłodowska Sect. AI Inform. **3** (2005), 179–189. MR2580273

[7] Z. R. Bogdanowicz, *Spanning trees in undirected simple graphs*, Stevens Institute of Technology, 1985.

[8] B. Bollobás, L. Pebody, and O. Riordan, *Contraction-deletion invariants for graphs*, J. Combin. Theory Ser. B **80** (2000), no. 2, 320–345, DOI 10.1006/jctb.2000.1988. MR1794697

[9] C.-Y. Chao and G. A. Novacky Jr., *On maximally saturated graphs*, Discrete Math. **41** (1982), no. 2, 139–143, DOI 10.1016/0012-365X(82)90200-X. MR676873

[10] C. Y. Chao and E. G. Whitehead Jr., *On chromatic equivalence of graphs*, Theory and applications of graphs (Proc. Internat. Conf., Western Mich. Univ., Kalamazoo, Mich., 1976), Lecture Notes in Math., Vol. 642, Springer, Berlin-New York, 1978, pp. 121–131. MR505733

[11] C. Y. Chao and E. G. Whitehead Jr., *Chromatically unique graphs*, Discrete Math. **27** (1979), no. 2, 171–177, DOI 10.1016/0012-365X(79)90107-9. MR537472

[12] P. Csikvári, *On a conjecture of V. Nikiforov*, Discrete Math. **309** (2009), no. 13, 4522–4526, DOI 10.1016/j.disc.2009.02.013. MR2519191

[13] P. Csikvári, *Applications of the Kelmans transformation: extremality of the threshold graphs*, The Electronic Journal of Combinatorics (2011), P182–P182.

[14] J. Cutler and A. J. Radcliffe, *Extremal graphs for homomorphisms*, J. Graph Theory **67** (2011), no. 4, 261–284, DOI 10.1002/jgt.20530. MR2839366

[15] A. Davies, A. Juhász, M. Lackenby, and N. Tomasev, *The signature and cusp geometry of hyperbolic knots*, Preprint, `arXiv:2111.15323`, 2021.

[16] A. Davies, P. Veličković, L. Buesing, S. Blackwell, D. Zheng, N. Tomašev, R. Tanburn, P. Battaglia, C. Blundell, A. Juhász, et al., *Advancing mathematics by guiding human intuition with AI*, Nature **600** (2021), no. 7887, 70–74.

[17] D. Dimitrov and R. Škrekovski, *Comparing the irregularity and the total irregularity of graphs*, Ars Math. Contemp. **9** (2015), no. 1, 45–50, DOI 10.26493/1855-3974.341.bab. MR3377090

[18] P. Dłotko, *Ballmapper: Create a ball mapper graph of the input data*, R package version 0.1.0, 2019.

[19] P. Dłotko, *Ball mapper: A shape summary for topological data analysis*, Preprint, `arXiv:1901.07410`, 2019.

[20] F. M. Dong, K. M. Koh, and K. L. Teo, *Chromatic polynomials and chromaticity of graphs*, World Scientific Publishing Co. Pte. Ltd., Hackensack, NJ, 2005, DOI 10.1142/9789812569462. MR2159409

[21] M. Eastwood and S. Huggett, *Euler characteristics and chromatic polynomials*, European J. Combin. **28** (2007), no. 6, 1553–1560, DOI 10.1016/j.ejc.2006.09.005. MR2339484

[22] E. J. Farrell, *On chromatic coefficients*, Discrete Math. **29** (1980), no. 3, 257–264, DOI 10.1016/0012-365X(80)90154-5. MR560769

[23] Conda forge community, *The conda-forge Project: Community-based Software Distribution Built on the conda Package Format and Ecosystem*, July 2015.

[24] L. Helme-Guizon and Y. Rong, *A categorification for the chromatic polynomial*, Algebr. Geom. Topol. **5** (2005), 1365–1388, DOI 10.2140/agt.2005.5.1365. MR2171813

[25] R. Hepworth and S. Willerton, *Categorifying the magnitude of a graph*, Homology Homotopy Appl. **19** (2017), no. 2, 31–60, DOI 10.4310/HHA.2017.v19.n2.a3. MR3683605

[26] M. C. Hughes, *A neural network approach to predicting and computing knot invariants*, J. Knot Theory Ramifications **29** (2020), no. 3, 2050005, 20, DOI 10.1142/S0218216520500054. MR4101599

[27] J. Huh, *Milnor numbers of projective hypersurfaces and the chromatic polynomial of graphs*, J. Amer. Math. Soc. **25** (2012), no. 3, 907–927, DOI 10.1090/S0894-0347-2012-00731-0. MR2904577

[28] V. Jejjala, A. Kar, and O. Parrikar, *Deep learning the hyperbolic volume of a knot*, Phys. Lett. B **799** (2019), 135033, 7, DOI 10.1016/j.physletb.2019.135033. MR4031161

[29] I. T. Jolliffe and J. Cadima, *Principal component analysis: a review and recent developments*, Philos. Trans. Roy. Soc. A **374** (2016), no. 2065, 20150202, 16, DOI 10.1098/rsta.2015.0202. MR3479904

[30] N. Kahl, *Graph vulnerability parameters, compression, and threshold graphs*, Discrete Appl. Math. **292** (2021), 108–116, DOI 10.1016/j.dam.2020.12.025. MR4203522

[31] N. Kahl, *Extremal graphs for the Tutte polynomial*, J. Combin. Theory Ser. B **152** (2022), 121–152, DOI 10.1016/j.jctb.2021.09.004. MR4321647

[32] A. K. Kelmans, *On graphs with randomly deleted edges*, Acta Math. Acad. Sci. Hungar. **37** (1981), no. 1-3, 77–88, DOI 10.1007/BF01904874. MR616881

[33] L. Keough and A. J. Radcliffe, *Graphs with the fewest matchings*, Combinatorica **36** (2016), no. 6, 703–723, DOI 10.1007/s00493-014-3188-y. MR3597588

[34] F. Lazebnik, *The maximum number of colorings of graphs of given order and size: a survey*, Discrete Math. **342** (2019), no. 10, 2783–2791, DOI 10.1016/j.disc.2018.08.010. MR3990664
[35] T. Leinster, *The magnitude of metric spaces*, Doc. Math. **18** (2013), 857–905. MR3084566
[36] J. S. F. Levitt, M. Hajij, and R. Sazdanovic, *Big data approaches to knot theory: understanding the structure of the Jones polynomial*, J. Knot Theory Ramifications **31** (2022), no. 13, Paper No. 2250095, 20, DOI 10.1142/s021821652250095x. MR4522846
[37] M. Loebl, J. Matoušek, and O. Pangrác, *Triangles in random graphs*, Discrete Math. **289** (2004), no. 1-3, 181–185, DOI 10.1016/j.disc.2004.08.008. MR2106042
[38] G. H. J. Meredith, *Coefficients of chromatic polynomials*, J. Combinatorial Theory Ser. B **13** (1972), 14–17, DOI 10.1016/0095-8956(72)90003-2. MR309781
[39] V. Nikiforov, *Eigenvalues and degree deviation in graphs*, Linear Algebra Appl. **414** (2006), no. 1, 347–360, DOI 10.1016/j.laa.2005.10.011. MR2209250
[40] M. Noy, *Graphs determined by polynomial invariants*, Theoret. Comput. Sci. **307** (2003), no. 2, 365–384, DOI 10.1016/S0304-3975(03)00225-1. Random generation of combinatorial objects and bijective combinatorics. MR2022584
[41] J. A. de Oliveira, C. S. Oliveira, C. Justel, and N. M. M. de Abreu, *Measures of irregularity of graphs*, Pesquisa Operacional **33** (2013), no. 3, 383–398.
[42] D. Paweł, *Ball mapper: a shape summary for topological data analysis*, `arXiv:1901.07410`, 2019.
[43] D. Paweł, D. Gurnari, and R. Sazdanovic, *Knot invariants and their relations: a topological perspective*, Preprint, `arXiv:2109.00831`, 2021.
[44] F. Pedregosa, G. Varoquaux, A. Gramfort, and et al., *Scikit-learn: machine learning in Python*, J. Mach. Learn. Res. **12** (2011), 2825–2830. MR2854348
[45] R. C. Read, *An introduction to chromatic polynomials*, J. Combinatorial Theory **4** (1968), 52–71. MR224505
[46] J. Rodriguez and A. Satyanarayana, *Chromatic polynomials with least coefficients*, Discrete Math. **172** (1997), no. 1-3, 115–119, DOI 10.1016/S0012-365X(96)00274-9. Chromatic polynomials and related topics (Shanghai, 1994). MR1467982
[47] G.-C. Rota, *On the foundations of combinatorial theory. I. Theory of Möbius functions*, Z. Wahrscheinlichkeitstheorie und Verw. Gebiete **2** (1964), 340–368 (1964), DOI 10.1007/BF00531932. MR174487
[48] A. Satyanarayana, L. Schoppmann, and C. L. Suffel, *A reliability-improving graph transformation with applications to network reliability*, Networks **22** (1992), no. 2, 209–216, DOI 10.1002/net.3230220206. MR1148020
[49] R. Sazdanovic and M. Yip, *A categorification of the chromatic symmetric function*, J. Combin. Theory Ser. A **154** (2018), 218–246, DOI 10.1016/j.jcta.2017.08.014. MR3718066
[50] G. Singh, F. Mémoli, and G. Carlsson, *Topological methods for the analysis of high dimensional data sets and 3D object recognition*, SPBG **91** (2007), 100.
[51] R. P. Stanley, *A symmetric function generalization of the chromatic polynomial of a graph*, Adv. Math. **111** (1995), no. 1, 166–194, DOI 10.1006/aima.1995.1020. MR1317387
[52] W. A. Stein et al., *Sage Mathematics Software (Version 9.1)*, The Sage Development Team, 2020, `http://www.sagemath.org`.
[53] P. Turán, *Eine Extremalaufgabe aus der Graphentheorie* (Hungarian, with German summary), Mat. Fiz. Lapok **48** (1941), 436–452. MR18405
[54] W. T. Tutte, *A contribution to the theory of chromatic polynomials*, Canad. J. Math. **6** (1954), 80–91, DOI 10.4153/cjm-1954-010-9. MR61366
[55] L. Collatz and U. Sinogowitz, *Spektren endlicher Grafen* (German), Abh. Math. Sem. Univ. Hamburg **21** (1957), 63–77, DOI 10.1007/BF02941924. MR87952
[56] H. Whitney, *A logical expansion in mathematics*, Bull. Amer. Math. Soc. **38** (1932), no. 8, 572–579, DOI 10.1090/S0002-9904-1932-05460-X. MR1562461

Department of Mathematics, North Carolina State University, Raleigh, North Carolina 27695

Email address: `rsazdanovic@math.ncsu.edu`

Department of Mathematics, Francis Marion University, Florence, South Carolina 29506

Email address: `daniel.scofield@fmarion.edu`

Contemporary Mathematics
Volume **824**, 2025
https://doi.org/10.1090/conm/824/16456

Adventures with positivity: From computer science to skein algebras

Robert Owczarek

Dedicated to the memory of my dear friend Professor Zbigniew Oziewicz

ABSTRACT. A few examples of somewhat surprisingly interconnected appearances in mathemeatics, mathematical physics, and beyond of the notion of positivity are shortly discussed.

1. Informal introduction

I have been interested for many years in knot theory and its role in physics, especially in explaining phase transition in superfluid helium [**1**], [**2**], [**3**], [**4**], [**5**], [**6**], [**7**], [**8**], [**9**], [**10**], [**11**], [**12**]. In particular, in recent years I have been interested in understanding the skein algebras and possible expansion of the notion so that new knot invariants may be defined, and the role played in the process by Chebyshev polynomials, their generalizations, and topological quantum field theories in 2D that are identified with certain Frobenius algebras that may play a role in achieving this goal [**13**]. In the meantime, sort of aside, there are some interesting related issues I would like to talk about today, issues that touch also some other areas of mathematics, physics, and even computer science.

2. Some encounters with positivity

For the last couple of years, I have been meeting the notion(s) of positivity in different mathematical (and other) contexts Similarly, I was meeting the names of Frobenius and Perron in the different "positivity" contexts very frequently (and other contexts as well). Namely, a few years ago (probably in 2017), I heard a seminar at my former institute in Warsaw (Institute of Fundamental Technological Research of Polish Academy of Sciences) by my friend, Professor Tomasz Lipniacki about certain chains of biochemical reactions described by systems of ordinary differential equations. Another friend from my former Institute, Professor Janusz Szczepański asked a question if the system of equations shows chaotic behavior The answer was "no, probably because the coefficients of the system are positive", (the reason for that was the coefficients were concentrations, positive by their nature). Then at the AMS meeting at Portland, WA, in 2018, I heard an interesting lecture

2020 *Mathematics Subject Classification.* Primary 57K31; Secondary 68W05.
Key words and phrases. Frobenius–Perron dimension, Chebyshev polynomials, skein algebras.

about models of economy by yet another friend of mine and frequent participant of our sessions, Professor Simina Brânzei, where the apparently complicated dynamics, described by certain discrete dynamical systems was resulting in qualitatively surprisingly simple behavior (see her papers [**14**], [**15**]). To my question about the simplicity her answer was that probably it is due to positivity of the matrices involved. There are also interesting positivity related issues in statistical physics and computer science. Then I encountered positivity in the area I am more intensely involved recently, knot theory. Namely, there are several results in the literature about positive bases for surface skein algebras, while skein algebras appear in knot theory defined on 3-manifolds that are cylindrical, that means of the form $\Sigma \times I$, where Σ is a $2D$ surface. These results are particularly interesting to me because the positive bases typically involve Chebyshev polynomials, which recently were subject of my interest. Moreover, Perron–Frobenius theory was instrumental in classifying $2D$ surface automorphisms by William Thurston [**16**]. I think these appearances of positivity might be intrinsically connected though it is quite a complex issue (or rather a set if issues) I would like to understand.

To be more specific, consider a discrete dynamical system of the form

$$x_{n+1} = Ax_n, \quad n = 0, 1, 2, \ldots, \text{ with } x_i \in \mathbb{R}^N,\ A \in \mathbb{R}^{N\times N} \tag{2.1}$$

with all A elements positive. Then the following theorem, first proved by Perron [**17**] and Frobenius [**18**] (there are weaker versions of the theorem for weaker positivity assumptions on the matrix A) helps to understand the dynamics.

THEOREM 2.1. *The eigenvalue of largest absolute value of a positive (square) matrix A is both simple and positive and belongs to a positive eigenvector. All other eigenvalues are smaller in absolute value.*

The above dynamical system leads to the solution

$$x_n = A^n x_0 \tag{2.2}$$

and the criterion, following from the Perron theorem, reads "the iterative process will converge exactly when the (positive) dominant eigenvalue $\lambda = \rho$ is at most 1; the process converges to the value 0 if $\rho < 1$, and when $\rho = 1$, to the component of the initial condition in the (reducing) eigenspace spanned by the associated eigenvector $v > 0$". Thus, positivity does not guarantee stability, there is more needed for that. The famous example of Arnold's cat map [**19**] provides unstable map of the torus into itself. In this case torus is identified as $\mathbb{R}^2/\mathbb{Z}^2$. Let

$$A = \begin{bmatrix} 2 & 1 \\ 1 & 1 \end{bmatrix}. \tag{2.3}$$

The matrix A acts on vectors from $\mathbb{R}^2$ where the action is mod 1, so that it effectively is the action on the torus. The larger of the eigenvalues of the matrix is

$$\Lambda = \frac{3+\sqrt{5}}{2} > 1, \tag{2.4}$$

so the map is unstable, accordingly to the above criterion. It is known that this action on the torus provides an example of an Anosov diffeomorphism of the torus, and Anosov and their generalizations to higher genus surfaces (pseudo-Anosov diffeomorphisms) will be mentioned at the last portion of this paper.

3. Application in computer science: The PageRank algorithm

Larry Page and Sergey Brin developed PageRank algorithm at Stanford University in 1996 as part of a research project about a new kind of search engine [**20**]. This algorithm helped in starting the best in the market web search company, Google. The PageRank values are the entries of the dominant right eigenvector of the modified adjacency matrix rescaled so that each column adds up to one (stochastic matrix).

4. Positivity in statistical physics

The thermodynamics of the systems in statistical physics is governed by the function called partition function [**21**]. Often the partition function can be calculated as a trace of a power of a matrix, called the transition matrix, that acts in certain configuration space,

$$Z_n = TrP^n, \tag{4.1}$$

where n is the number of variables in the system (number of particles, spins, or other physical objects). It may happen for physical reasons that the matrix P is positive, and then it has dominant eigenvalue accordingly to the general theorem, say Λ. The thermodynamics of the system is usually considered in the thermodynamic limit, when $n \to \infty$. The quantities of interest in thermodynamics are contained in the limit

$$\lim_{n\to\infty} \frac{1}{n} lnZ_n, \tag{4.2}$$

which in case of positive matrix P will be equal to $ln\Lambda$.

5. Positivity in category theory/algebra: Frobenius–Perron dimension

The following is based on [**22**].

Definition 5.1. A unital $\mathbb{N}$-ring is a ring such that

- the underlying abelian group is a free abelian group;
- there exists a finite $\mathbb{N}$-basis: a finite set I of elements $X_i \in R$, $i \in I$, such that

$$X_iX_j = \sum_{k\in I} c_{ij}^k X_k \quad \text{for } c_{ij}^k \in \mathbb{N}; \tag{5.1}$$

- the ring unit 1 is among these basis elements.

Definition 5.2. Let R be a unital $\mathbb{N}$-ring with finite $\mathbb{N}$-basis I, $|I| = n \in \mathbb{N}$. For every $X \in R$ let $(N_X)_{ij} \in \mathbb{N}$ be its matrix of left multiplication, defined by $XX_j = \sum_{i\in I}(N_X)_{ij}X_i$. By the Frobenius–Perron theorem these matrices N_X have nonnegative eigenvalues. This implies that for each of them there is a maximal eigenvalue. This maximal eigenvalue is called the Frobenius–Perron dimension of X.

6. "Frobenius–Perron dimension" for Chebyshev polynomials of the first kind

The standard Chebyshev polynomials of the first kind satisfy the following product-to-sum rule (see, e.g., [**23**])

$$T_m(x)T_n(x) = T_{m+n}(x) + T_{|m-n|}(x), \quad \text{with } m, n = 0, 1, 2, \ldots. \tag{6.1}$$

The matrices of structure constants of this algebra are infinite dimensional in contrast to the definition of Frobenius–Perron dimension. However, the matrices are very sparse and have mostly ones as nonzero elements (the important exception is when one of the factors is $T_0(x)$, because then

$$T_0(x)T_n(x) = T_{0+n}(x) + T_{|0-n|}(x) = 2T_n(x), \tag{6.2}$$

which means the eigenvalue problem may be still well-defined. It turns out that the matrices have 2 as an eigenvalue, and it is likely the largest eigenvalue (it is not as easy to prove as in finite dimensional matrix case). This largest eigenvalue may be considered as the generalized Frobenius–Perron dimension in this particular case. However, it seems too early for a formal definition. In order to understand the stated above and expand, let us first remind what are Chebyshev polynomials of the first kind. Namely, the standard Chebyshev polynomials of the first kind appear as polynomials $T_n(x)$ satisfying

$$Tr(M^n) = T_n(TrM) \tag{6.3}$$

for $M \in SL(2, \mathbb{C})$.

There is an interesting generalization of the Chebyshev polynomials to two variables, $T_n(x, y)$, satisfying the generalized condition [**24**]

$$Tr(M^n) = T_n(TrM, DetM) \tag{6.4}$$

for $M \in Mat(2, \mathbb{C})$.

The generalized Chebyshev polynomials of the first kind satisfy the following product-to-sum relation, I worked out, which can be proved by straightforward though careful induction, which I skip here for the sake of space

$$T_m(x, y)T_n(x, y) = T_{m+n}(x, y) + y^{min(m,n)}T_{|m-n|}(x, y). \tag{6.5}$$

It is then a positive set over $\mathbb{Z}(y)$. There might exist a categorification of the generalized Chebyshev polynomials of the first kind.

It has been suggested by Thurston [**23**] that the product-to-sum relation for Chebyshev polynomials of the first kind, due to integers as "structure constants", may be a sign of possible categorification of the notion, which means there is a category with sum and tensor product such that lifted objects satisfy

$$\hat{T}_m \otimes \hat{T}_n = \hat{T}_{m+n} \oplus \hat{T}_{|m-n|}. \tag{6.6}$$

It turned out to be true thanks to a categorification of Chebyshev polynomials of the first kind.

7. Categorification of Chebyshev polynomials of the first kind

A categorification of Chebyshev polynomials of the first kind has been defined in the paper by Queffelec and Wedrich [**25**] (see also [**26**], [**27**], [**28**]) The construction is rather involved, so it is impossible to present it even in sketchy way. It uses the so-called essential affine Temperley–Lieb category, which is a relative of standard Temperley–Lieb category. In standard Temperley–Lieb category, the objects are natural numbers, morphisms from $n \in \mathbb{N}$ to $m \in \mathbb{N}$ are defined with the help of rectangular diagrams with n points at the bottom edge, m at the top edge, diagrams include arcs between the points, which are either through strings, or cups, or caps, nonintersecting each other. Multiplying the diagrams consists of putting one on top of the other and rearrangements, tensor product means putting diagrams side-by-side. There are rules for removing circles appearing in the process, like in skein relations in knot theory, discussed later. There is a functor from Temperley–Lieb category to the category of $sl(2,\mathbb{C})$ representations (which are basically the same as $SL(2,\mathbb{C})$ representations).

REMARK 7.1. In fact, the so-called Karoubi envelope of the Temperley–Lieb category is equivalent to the category of $sl(2,\mathbb{C})$ representations, while Temperley–Lieb category is a subcategory of the category of representations.

REMARK 7.2. The standard Temperley–Lieb algebra and the correspondence with $sl(2,\mathbb{C})$ representations is used in categorification of the Chebyshev polynomials of the second kind. In categorification of the Chebyshev polynomials of the first kind, the affine Temperley–Lieb category is used, in which the rectangular diagrams are replaced by annular ones, with multiplication and tensor product defined analogously to the rectangular case, but with some extra relations. In the case of the standard Temperley–Lieb category, there are Jones–Wenzl idempotents, which categorize Chebyshev polynomials of the second kind, and their counterparts for affine essential Temperley–Lieb category that categorify the Chebyshev polynomials of the first kind. In particular, the Jones–Wenzl idempotents for affine Temperley–Lieb category satisfy the relation

$$\hat{T}_m \otimes \hat{T}_n = \hat{T}_{m+n} \oplus \hat{T}_{|m-n|} \tag{7.1}$$

which is consistent with the categorification requirements.

8. Chebyshev polynomials and skein algebras

Chebyshev polynomials appear in the description of skein algebras. Let's remind then skein algebras, in particular for the case of the Kauffman bracket.

8.1. Kauffman bracket in $\mathbb{R}^3$. We project a knot or link on a plane. We resolve crossings using skein rules, see Figure 1. The necessity for defining skein modules can be best illustrated by looking at Figure 2 and Figure 3.

General definition of a skein module has been given first by Przytycki [**29**] and it reads:

Let M be an oriented 3-manifold and R a commutative ring with 1. For $r_0, r_1, \ldots, r_{k-1} \in R$ we define the kth skein module $S_k(M; R)(r_0, r_1, \ldots, r_{k-1})$ as follows:

Let $\mathcal{L}(M)$ be the set of all ambient isotopy classes of oriented links in M. Let $\mathcal{M}(\mathcal{L}, M)$ be a free R-module generated by $\mathcal{L}(M)$ and $S_{\mathcal{L}(M)}(r_0L_0 + r_1L_1 + \cdots + r_{k-1}L_{k-1})$ the submodule generated by linear skein expressions $r_0L_0 + r_1L_1 + \cdots + r_{k-1}L_{k-1}$, where $L_0, L_1, \ldots, L_{k-1}$ are classes of links identical except the parts shown in Figure 4.

DEFINITION 8.1. The R-module

$$S_k(M, R)(r_0, r_1, \ldots, r_{k-1}) = \mathcal{M}(\mathcal{L}, R)/S_{\mathcal{L}(M)}(r_0, r_1, \ldots, r_{k-1}) \tag{8.1}$$

is called the kth skein module of M.

When $M = \Sigma \times I$, there is an algebra structure on the skein module defined on links by stacking the first of the links on top of the second, which is best illustrated by an example. In Figure 5, $\Sigma = \mathbb{T}^2$, a $2D$ torus, and the skein rules are the Kauffman bracket ones.

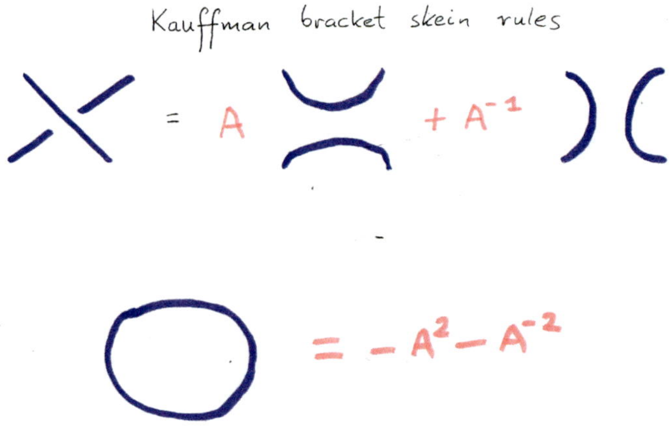

FIGURE 1. The skein relations for Kauffman bracket

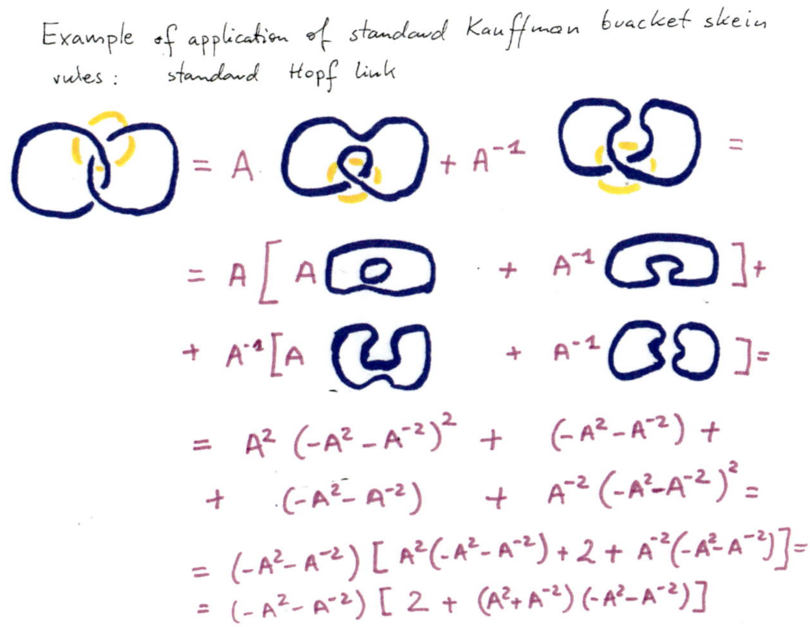

FIGURE 2. Resolving the standard Hopf link

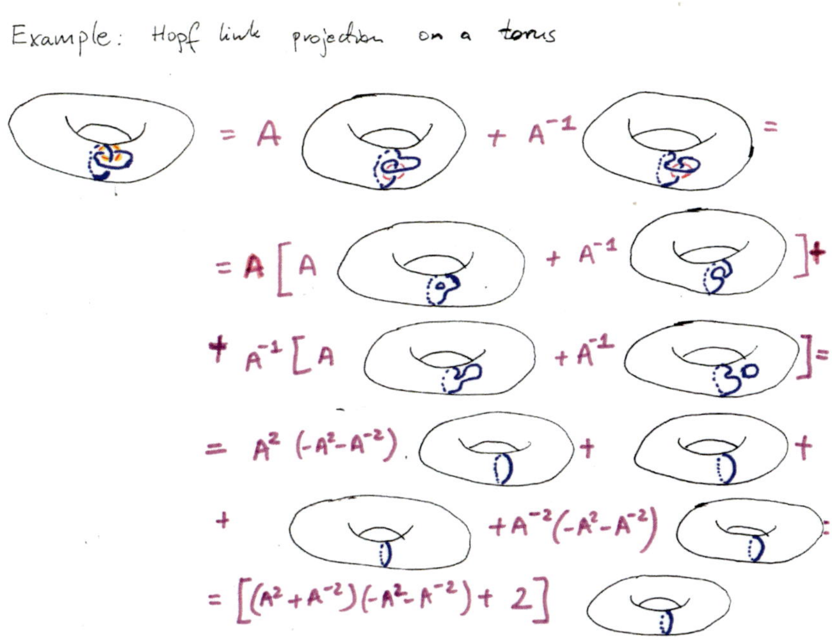

FIGURE 3. Resolving a Hopf link on the torus

FIGURE 4. General tangle

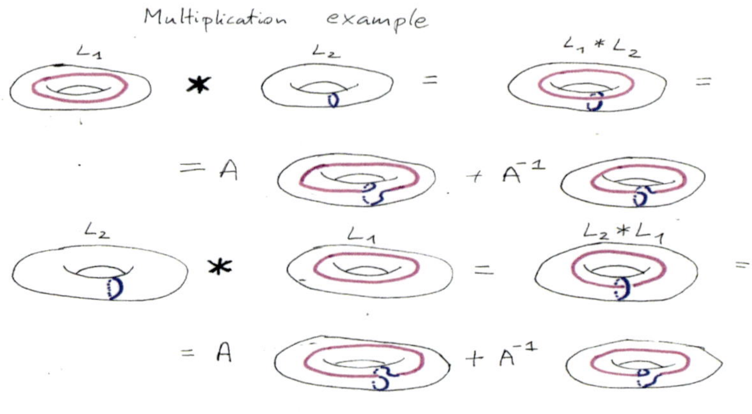

FIGURE 5. Product of links on a torus

9. Chebyshev polynomials and skein algebra for a torus

Chebyshev polynomials play a big role in the description of skein algebras of surfaces (perhaps also in general 3-manifolds, but little if anything is known about that). I am not sure who first paid attention to this role, but one important step has been made by Frohman and Gelca [**49**]. As I mentioned before, the skein module of the torus is defined by curves (or rather multicurves) which certain number of times go around the torus in the (positive or negative) direction of the longitude or meridian, and so by two integers, say p and q. The definition of basis elements in the skein module is then as follows:

$$(p,q)_T = T_{gcd(p,q)}\left(\left(\frac{p}{gcd(p,q)}, \frac{q}{gcd(p,q)}\right)\right), \tag{9.1}$$

where the argument of the Chebyshev polynomial is replaced by the curve and the raising to power means cabling the curve with identical ones in the number equal to the power. Frohman and Gelca proved a beautiful formula for the product in the skein algebra when applied to those basis elements, namely

$$(p,q)_T * (r,s)_T = A^{\begin{vmatrix} p & r \\ q & s \end{vmatrix}}(p+r, q+s)_T + A^{-\begin{vmatrix} p & r \\ q & s \end{vmatrix}}(p-r, q-s)_T. \tag{9.2}$$

The formula is connected to the formula for the product of Chebyshev polynomials. An interesting discussion of the formula can be found in [**50**]. What is also mentioned in this paper by Frohman and Gelca are connections of Chebyshev polynomials with Jones–Wenzl idempotents, which are important for categorification of Chebyshev polynomials, which was mentioned above.

There is an interesting connection of this formula with the formula for algebraic intersection number which is presented, e.g., in [**51**]. Namely the definition for the algebraic intersection number of two simple, closed, oriented, transverse curves in the surface Σ reads:

DEFINITION 9.1. Let α and β be a pair of transverse, oriented, simple closed curves in Σ. Recall that the **algebraic intersection number** $\hat{\imath}(\alpha,\beta)$ is defined as the sum of the indices of the intersection points of α and β, where an intersection point is of index $+1$ when the orientation of the intersection agrees with the orientation of Σ and is -1 otherwise.

A simple fact concerning simple closed curves on a torus is For two homotopy classes of simple, closed, oriented curves (p,q) and (p',q') on the torus we have

$$\hat{\imath}((p,q),(p',q')) = pq' - p'q, \tag{9.3}$$

which is obviously

$$\hat{\imath}((p,q),(p',q')) = \begin{vmatrix} p & p' \\ q & q' \end{vmatrix} \tag{9.4}$$

so that the Frohman–Gelca formula can be rewritten as

$$(p,q)_T * (r,s)_T = A^{\hat{\imath}((p,q),(r,s))}(p+r,q+s)_T + A^{-\hat{\imath}((p,q),(r,s))}(p-r,q-s)_T \tag{9.5}$$

showing how deformation of the algebra of Chebyshev polynomials depends on the intersection number of the simple curves on the torus.

10. Chebyshev polynomials and positive bases of surface skein algebras

It seems the first who realized the importance of Chebyshev polynomials as giving positive bases in the skein algebras was Dylan Paul Thurston [**23**]. What are positive bases? Accordingly to Thurston's Definition 1.2 from [**23**]:

DEFINITION 10.1. For an algebra A over $\mathbb{Z}$ (free as a $\mathbb{Z}$-module), a basis $\{x_i\}$ is positive if $x_i \cdot x_j = \sum_k m_{ij}^k x_k$, where $m_{ij}^k \geq 0$.

Thurston also proved in [**23**] Theorem 1, which reads:

THEOREM 10.2. *The bracelets basis is a natural positive basis for* $Sk(\Sigma)$.

Here $Sk(\Sigma)$ is the skein algebra at $A = 1$ accordingly to our notation, and bracelets are defined just as in Frohman–Gelca paper as Chebyshev polynomials of the first kind applied to curves. This result has been extended by Lê [**52**] in a number of ways, here it is difficult to go into details. Also, Bonahon and Wong showed many interesting relations of Chebyshev polynomials to Kauffman bracket skein algebras of surfaces in a series of papers [**30**], [**31**], [**32**].

DEFINITION 10.3. A **domain** is a nonzero ring in which $ab = 0$ implies $a = 0$ or $b = 0$.

A set of remarkable theorems have been proved by Lê, Thurston, Yu, in [**33**]. They show how positivity constraint almost fixes the basis of the skein algebra to consist of Chebyshev polynomials.

THEOREM 10.4 ([**33**]). *Suppose R is a commutative domain with a distinguished invertible element q and a positive part R_+. Assume either $g \geq 1$ or $p \geq 4$. Let (P_n) be a normalized sequence of polynomials that is positive on $\Sigma_{g,p}$ over R.*

(1) *One has $(P_n) \geq (\hat{T}_n)$ and $P_1(x) = x$.*
(2) *If $P_n = \hat{T}_n$ for $n \leq 3$ then $(P_n) = (\hat{T}_n)$.*

THEOREM 10.5 ([**33**]). *Suppose R is a commutative domain with a distinguished invertible element q and a positive part R_+. A normalized sequence of polynomials (P_n) is positive on the torus $\Sigma_{1,0}$ over R if and only if $(P_n) = (\hat{T}_n)$.*

THEOREM 10.6 ([**33**]). *Suppose $R = \mathbb{Z}[q^{\pm 1}]$ and (P_n) are normalized sequences of polynomials with integer coefficients. Assume either $g \geq 1$ or $p \geq 4$. If (P_n) is positive on $\Sigma_{g,p}$ over R, then $(P_n) \leq (S_n)$.*

In this theorem S_n denote Chebyshev polynomials of the second kind, that may be defined by

$$Tr(\rho_n(M)) = S_n(TrM) \quad \text{for } M \in SL(2, \mathbb{C}) \tag{10.1}$$

or by

$$S_0(x) = 1, \tag{10.2}$$

$$S_1(x) = x, \tag{10.3}$$

$$S_{n+1}(x) = x \cdot S_n(x) - S_{n-1}(x). \tag{10.4}$$

11. Positivity and automorphisms of 2D surfaces

Automorphisms (homeomorphisms) of $2D$ surfaces have been classified by William Thurston [**16**] There is, up to isotopy, only one automorphism (preserving orientation) of the $2D$ sphere, which is the identity. For the $2D$ torus there are three types of automorphisms: periodic (some composition power is isotopic to identity), reducible (preserves a $1D$ closed submanifold, thus reduces to actions on the remainder), and Anosov automorphisms. The latter preserve a pair of transverse foliations, one called stable, the other called unstable. The action on leaves of the stable foliation is stretching by a factor, the action on unstable foliation is shrinking by the inverse of the same factor. Despite the apparent complexity, the group of all orientation preserving automorphisms of the torus up to isotopy is $SL(2, \mathbb{Z})$ (remember the Arnold's cat map with the matrix $\begin{pmatrix} 2 & 1 \\ 1 & 1 \end{pmatrix}$ that acts on the universal cover $\mathbb{R}^2$ before appropriate quotient is applied?) In the higher genus case, the classification of automorphisms is into periodic, reducible, or pseudo-Anosov. The latter automorphisms leave invariant a pair of transverse geodesic laminations, called stable and unstable laminations (a geodesic lamination can be defined as a closed subset that is a union of disjoint geodesics). Again, the action of the pseudo-Anosov automorphism on the geodesics stretches and shrinks the corresponding geodesics of the laminations by certain factors (dilatations) that are inverse to each other.

The amazing fact, proved by Ackermann in his PhD thesis [**34**] is that the spectrum of the dilatations is discrete, and can be described by the eigenvalues of certain positive matrices that preserve a symplectic structure on the surface, for which the existence of the simple maximal eigenvalue is assured.

References

[1] R. Owczarek, *Knotted vortices and superfluid phase transition*, Modern Physics Letters B **7** (1993), 1523–1526.

[2] R. Owczarek, *Topological defects in superfluid helium*, Internat. J. Theoret. Phys. **30** (1991), no. 12, 1605–1612, DOI 10.1007/BF00673638. MR1138341

[3] R. Owczarek, *Knotted vortices and fermionic excitations in bulk superfluid helium*, Modern Physics Letters B **7** (1993), 1383–1386.

[4] R. Owczarek, *Frames and fermionic excitations of vortices in superfluid helium*, Journal of Physics: Condensed Matter **5**(1993), 8793.

[5] R. Owczarek, *Non-zero helicity generalization of the Nielsen-Olesen string solution*, Rep. Math. Phys. **34** (1994), no. 3, 305–310, DOI 10.1016/0034-4877(94)90003-5. MR1339467

[6] R. Owczarek, *Notes on superconductivity by quantum size effect and non-equivalent spin structures*, Journal of Technical Physics **36** (1995), 181–185.

[7] R. Owczarek, *On geometric methods in the description of quantum fluids*, Quantization, coherent states, and complex structures (Białowieża, 1994), Plenum, New York, 1995, pp. 265–269. MR1403169

[8] R. Owczarek, *Quantization of ideal fluids: Perturbative or non-perturbative approach*, Acta Phys. Polon. B **27** (1996), no. 10, 2387–2391. New ideas in the theory of fundamental interactions (Zakopane, 1995). MR1457303

[9] R. Owczarek, *On the role of helicity in stability and quantization problems of fluid flows*, Journal of technical physics **37** (1996), 539–542.

[10] R. Owczarek, *Spinor structures on flag manifolds of compact simple Lie groups*, J. Phys. A **32** (1999), no. 21, 3979–3991, DOI 10.1088/0305-4470/32/21/310. MR1696821

[11] G. A. Goldin, R. Owczarek, and D. H. Sharp, *Quantum kinematics of bosonic vortex loops*, J. Math. Phys. **46** (2005), no. 4, 042102, 14, DOI 10.1063/1.1853504. MR2131218

[12] R. Owczarek, *Statistical mechanics of superfluid helium near critical point*, Proceedings of IMBIC, Kolkata, Volume (ed. A. Adhikari), 2012.

[13] R. Owczarek, *Vortices, knots, Chebyshev polynomials, Frobenius algebras, topological quantum field theories*, Scientific legacy of Professor Zbigniew Oziewicz, Ser. Knots Everything, vol. 75, World Sci. Publ., Hackensack, NJ, [2024] ©2024, pp. 555–579. MR4678977

[14] S. Brânzei, R. Mehta, and N. Nisan, *Universal Growth in Production Economies*, Proceedings of the 32nd Conference on Neural Information Processing Systems (NeurIPS 2018), 2018.

[15] S. Brânzei, *Tit-for-tat dynamics and market volatility*, Nonlinearity **37** (2024), no. 3, Paper No. 035006, 22. MR4699112

[16] W. P. Thurston, *On the geometry and dynamics of diffeomorphisms of surfaces*, Bull. Amer. Math. Soc. (N.S.) **19** (1988), no. 2, 417–431, DOI 10.1090/S0273-0979-1988-15685-6. MR956596

[17] O. Perron, *Zur Theorie der Matrices* (German), Math. Ann. **64** (1907), no. 2, 248–263, DOI 10.1007/BF01449896. MR1511438

[18] G. Frobenius, *Ueber Matrizen aus nicht negativen Elementen*, Sitzungsberichte der Königlich Preussischen Akademie der Wissenschaften (1912), 456–477

[19] V. I. Arnold and A. Avez, *Problèmes ergodiques de la mécanique classique* (French), Monographies Internationales de Mathématiques Modernes, vol. 9, Gauthier-Villars, Éditeur, Paris, 1967. MR209436
[20] L. Page, PageRank: Bringing Order to the Web. Archived from the original on May 6, 2002. Retrieved 2016-09-11., Stanford Digital Library Project, talk. August 18, 1997 (archived 2002)
[21] K. Huang, *Statistical mechanics*, 2nd ed., John Wiley & Sons, Inc., New York, 1987. MR1042093
[22] D. Calaque and P. Etingof, *Lectures on tensor categories*, Quantum groups, IRMA Lect. Math. Theor. Phys., vol. 12, Eur. Math. Soc., Zürich, 2008, pp. 1–38, DOI 10.4171/047-1/1. MR2432988
[23] D. P. Thurston, *Positive basis for surface skein algebras*, Proc. Natl. Acad. Sci. USA **111** (2014), no. 27, 9725–9732, DOI 10.1073/pnas.1313070111. MR3263305
[24] R. Owczarek, *Remarks on Chebyshev polynomials, Fibonacci polynomials and Kauffman bracket skein modules*, J. Knot Theory Ramifications **27** (2018), no. 7, 1841007, 47, DOI 10.1142/S0218216518410079. MR3835749
[25] H. Queffelec and P. Wedrich, *Extremal weight projectors*, Math. Res. Lett. **25** (2018), no. 6, 1911–1936, DOI 10.4310/MRL.2018.v25.n6.a11. MR3934851
[26] M. Khovanov and R. Sazdanovic, *A categorification of one-variable polynomials* (English, with English and French summaries), Proceedings of FPSAC 2015, Discrete Math. Theor. Comput. Sci. Proc., Assoc. Discrete Math. Theor. Comput. Sci., Nancy, 2015, pp. 937–948. MR3470926
[27] M. Khovanov and R. Sazdanovic, *Categorifications of the polynomial ring*, Fund. Math. **230** (2015), no. 3, 251–280, DOI 10.4064/fm230-3-3. MR3351473
[28] M. Khovanov and R. Sazdanovic, *Diagrammatic categorification of the Chebyshev polynomials of the second kind*, J. Pure Appl. Algebra **225** (2021), no. 6, Paper No. 106592, 23, DOI 10.1016/j.jpaa.2020.106592. MR4171760
[29] J. H. Przytycki, *Skein modules of* 3*-manifolds*, Bull. Polish Acad. Sci. Math. **39** (1991), no. 1-2, 91–100. MR1194712
[30] F. Bonahon and H. Wong, *Representations of the Kauffman bracket skein algebra I: invariants and miraculous cancellations*, Invent. Math. **204** (2016), no. 1, 195–243, DOI 10.1007/s00222-015-0611-y. MR3480556
[31] F. Bonahon and H. Wong, *Representations of the Kauffman bracket skein algebra II: Punctured surfaces*, Algebr. Geom. Topol. **17** (2017), no. 6, 3399–3434, DOI 10.2140/agt.2017.17.3399. MR3709650
[32] F. Bonahon and H. Wong, *Representations of the Kauffman bracket skein algebra III: closed surfaces and naturality*, Quantum Topol. **10** (2019), no. 2, 325–398, DOI 10.4171/QT/125. MR3950651
[33] T. T. Q. Lê, D. P. Thurston, and T. Yu, *Lower and upper bounds for positive bases of skein algebras*, Int. Math. Res. Not. IMRN **4** (2021), 3186–3202, DOI 10.1093/imrn/rnaa082. MR4218350
[34] R. Ackermann, *On pseudo-Anosov maps, symplectic, Perron-Frobenius matrices, and compression bodies*, ProQuest LLC, Ann Arbor, MI, 2014. Thesis (Ph.D.)–University of California, Santa Barbara. MR3337606
[35] E. Witten, *Quantum field theory and the Jones polynomial*, Comm. Math. Phys. **121** (1989), no. 3, 351–399. MR990772
[36] G. A. Goldin, R. Menikoff, and D. H. Sharp, *Diffeomorphism groups and quantized vortex filaments*, Phys. Rev. Lett. **58** (1987), no. 21, 2162–2164, DOI 10.1103/PhysRevLett.58.2162. MR888717
[37] G. A. Goldin, R. Menikoff, and D. H. Sharp, *Quantum vortex configurations in three dimensions*, Phys. Rev. Lett. **67** (1991), no. 25, 3499–3502, DOI 10.1103/PhysRevLett.67.3499. MR1140179
[38] G. A. Goldin, *Quantum vortex configurations*, Acta Phys. Polon. B **27** (1996), no. 10, 2341–2355. New ideas in the theory of fundamental interactions (Zakopane, 1995). MR1457301
[39] H. Makaruk, Dirac description of energy levels of polarons in polyaniline, *Modern Physics Letters B*, **9**, (1995), 543-551.
[40] H. Weyl, *Invariants*, Duke Math. J. **5** (1939), 489–502.
[41] J. Kock, *Frobenius algebras and 2D topological quantum field theories*, London Mathematical Society Student Texts, vol. 59, Cambridge University Press, Cambridge, 2004. MR2037238

[42] L. Abrams, *Two-dimensional topological quantum field theories and Frobenius algebras*, J. Knot Theory Ramifications **5** (1996), no. 5, 569–587, DOI 10.1142/S0218216596000333. MR1414088

[43] A. A. Voronov, *Topological field theories, string backgrounds and homotopy algebras*, Adv. Appl. Clifford Algebras **4** (1994), no. Suppl. 1, 167–178. Differential geometric methods in theoretical physics (Ixtapa-Zihuatanejo, 1993). MR1337702

[44] L. H. Kauffman, *State models and the Jones polynomial*, Topology **26** (1987), no. 3, 395–407, DOI 10.1016/0040-9383(87)90009-7. MR899057

[45] V. G. Turaev, *The Conway and Kauffman modules of a solid torus* (Russian, with English summary), Zap. Nauchn. Sem. Leningrad. Otdel. Mat. Inst. Steklov. (LOMI) **167** (1988), no. Issled. Topol. 6, 79–89, 190, DOI 10.1007/BF01099241; English transl., J. Soviet Math. **52** (1990), no. 1, 2799–2805. MR964255

[46] J. H. Przytycki and A. S. Sikora, *On skein algebras and* $\mathrm{Sl}_2(\mathbf{C})$*-character varieties*, Topology **39** (2000), no. 1, 115–148, DOI 10.1016/S0040-9383(98)00062-7. MR1710996

[47] W. M. Goldman, *Geometric structures on manifolds*, Graduate Studies in Mathematics, vol. 227, American Mathematical Society, Providence, RI, [2022] ©2022, DOI 10.1090/gsm/227. MR4500072

[48] W. M. Goldman, *Geometric structures on manifolds and varieties of representations*, Geometry of group representations (Boulder, CO, 1987), Contemp. Math., vol. 74, Amer. Math. Soc., Providence, RI, 1988, pp. 169–198, DOI 10.1090/conm/074/957518. MR957518

[49] C. Frohman and R. Gelca, *Skein modules and the noncommutative torus*, Trans. Amer. Math. Soc. **352** (2000), no. 10, 4877–4888, DOI 10.1090/S0002-9947-00-02512-5. MR1675190

[50] H. Queffelec and H. M. Russell, *Chebyshev polynomials and the Frohman-Gelca formula*, J. Knot Theory Ramifications **24** (2015), no. 4, 1550023, 17, DOI 10.1142/S0218216515500236. MR3346923

[51] B. Farb and D. Margalit, *A primer on mapping class groups*, Princeton Mathematical Series, vol. 49, Princeton University Press, Princeton, NJ, 2012. MR2850125

[52] T. T. Q. Lê, *On positivity of Kauffman bracket skein algebras of surfaces*, Int. Math. Res. Not. IMRN **5** (2018), 1314–1328, DOI 10.1093/imrn/rnw280. MR3801463

[53] J. Kocik, *Lucas, Fibonacci, and Chebyshev polynomials from matrices*, `arXiv:2105.13989v1` [math.GM], 2021.

[54] M. Atiyah, *Topological quantum field theories*, Inst. Hautes Études Sci. Publ. Math. **68** (1988), 175–186 (1989). MR1001453

[55] M. Khovanov, *A categorification of the Jones polynomial*, Duke Math. J. **101** (2000), no. 3, 359–426, DOI 10.1215/S0012-7094-00-10131-7. MR1740682

[56] P. Turner, *Five lectures on Khovanov homology*, J. Knot Theory Ramifications **26** (2017), no. 3, 1741009, 41, DOI 10.1142/S0218216517410097. MR3627709

[57] L. H. Kauffman, *Khovanov homology*, Introductory lectures on knot theory, Ser. Knots Everything, vol. 46, World Sci. Publ., Hackensack, NJ, 2012, pp. 248–280, DOI 10.1142/9789814313001_0010. MR2885237

[58] E. S. Lee, *An endomorphism of the Khovanov invariant*, Adv. Math. **197** (2005), no. 2, 554–586, DOI 10.1016/j.aim.2004.10.015. MR2173845

Department of Mathematics, University of New Mexico, Los Alamos, New Mexico 87544; and Department of Mechanical Engineering, University of New Mexico, Albuquerque, New Mexico 87131

Email address: `rowczare@unm.edu`

Contemporary Mathematics
Volume **824**, 2025
https://doi.org/10.1090/conm/824/16450

Effective ultrapowers of graphs and other structures

Valentina Harizanov and Keshav Srinivasan

Abstract. The effective ultraproduct construction for structures is a computability-theoretic analog of the classical ultraproduct construction, where the structures are uniformly computable and the role of an ultrafilter is played by a cohesive set. A cohesive set is an infinite set of natural numbers, which cannot be split into two infinite parts by any computably enumerable set. While the elements of the classical ultraproduct are equivalence classes of arbitrary sequences of elements of structures, in the effective case, these sequences are partial computable, and in some cases computable. Hence the effective ultraproduct construction allows us to build countable nonstandard models with interesting properties. We investigate the isomorphism types and other model-theoretic properties of cohesive powers of computable structures, focusing on certain graphs and equivalence structures. We show how computable structures can have effective ultrapowers with new properties, while preserving some old ones, as expressed in a formal language. For graphs we show that every graph can be embedded into the cohesive power of a certain kind of special graph.

1. Introduction: Classical ultraproducts of structures

The ultraproduct construction is a major mathematical construction that is used in algebra, mathematical logic, and other areas. It allows us to construct new models of theories from old ones. The ultraproduct is a quotient of the direct product of structures modulo an equivalence relation determined by an ultrafilter. The ultraproduct construction has some striking applications, such as a very elegant proof of the compactness theorem in first-order logic, and the construction of a nonstandard model of analysis. We will recall some of the formal notions of model theory, including structures, languages, ultrafilters, the construction itself, and how it relates to the language of the structures.

A *structure* consists of a nonempty set of elements, called the domain or universe of the structure, with certain relations, functions and constants on the domain, named by the corresponding symbols. Their symbols form the language of the structure. First-order formulas allow existential and universal quantification

2020 *Mathematics Subject Classification.* Primary 03C57, Secondary 03D45.
Key words and phrases. Computable structure, cohesive power, undirected graph, equivalence structure.
The authors were partially supported by FRG NSF grant DMS-2152095. The second author was also partially supported by NSF grant DMS-1600625.

only over variables for elements of the domain. A *sentence* is a formula where all variables are quantified, so it can be assigned a truth value (true or false) in a structure for the same language. The theory of a structure $\mathcal{A}$ is the set of all first-order sentences true in $\mathcal{A}$. In general, a theory is a consistent set of sentences. We say that a first-order formula in the prenex form is Σ_n^0 if it has n alternating blocks of like quantifiers beginning with a block of existential ($\exists$) quantifiers, followed by a quantifier-free formula. Dually, a formula is said to be a Π_n^0 formula if it has n alternating blocks of like quantifiers beginning with a block of universal ($\forall$) quantifiers, followed by a quantifier-free formula. In other words, Σ_1^0 formulas are existential formulas and Π_1^0 formulas are universal formulas, Π_2^0 formulas are $\forall\exists$ formulas, Σ_2^0 formulas are $\exists\forall$ formulas, Π_3^0 formulas are $\forall\exists\forall$ formulas, etc. The quantifier-free formulas are also called Σ_0^0 and Π_0^0 formulas.

Two structures $\mathcal{A}$ and $\mathcal{B}$ for the same language L, with domains A and B respectively, are *isomorphic* if there is a bijection h from A onto B such that for each m-ary relation symbol R in L, each m-ary function symbol F in L, and each constant symbol c in L, we have for all $a_1, \ldots, a_m \in A$:

$$R^{\mathcal{A}}(a_1, \ldots, a_m) \Leftrightarrow R^{\mathcal{B}}(h(a_1), \ldots, h(a_m)),$$
$$F^{\mathcal{A}}(a_1, \ldots, a_m) = b \Leftrightarrow F^{\mathcal{B}}(h(a_1), \ldots, h(a_m)) = h(b),$$
$$h(c^{\mathcal{A}}) = c^{\mathcal{B}}.$$

Moreover, it can be shown that then for every (first-order) formula $\theta(x_1, \ldots, x_n)$ and $a_1, \ldots, a_n \in A$ we have:

$$\mathcal{A} \models \theta(a_1, \ldots, a_n) \quad \text{if and only if} \quad \mathcal{B} \models \theta(h(a_1), \ldots, h(a_n)).$$

Hence isomorphic structures satisfy the same sentences. We denote by $\mathcal{A} \cong \mathcal{B}$ that $\mathcal{A}$ and $\mathcal{B}$ are isomorophic.

Certain sets of sets called ultrafilters are used to determine truth in the ultraproduct of structures. First, consider a nonempty set I, called the index set, which will be used to index structures. An *ultrafilter* over I is a nonempty set $U$of subsets of I such that U is closed under the intersection of two sets and under the superset relation, and for every subset X of I, U contains either X or its complement. An ultrafilter is *nonprincipal* if it does not contain any singletons. Hence it must contain co-finite subsets of I.

The notion of the ultraproduct of structures for the same language L in general has been introduced by Łoś in [**16**]. Let I be an infinite set, and let $\mathcal{A}_i$ for $i \in I$ be a family of structures for language L with domains A_i. Let U be an ultrafilter over I. Then the *ultraproduct* of $\mathcal{A}_i$ over U, denoted by $\mathcal{B} = \prod_U \mathcal{A}_i$, is a structure for L defined as follows.

(1) The domain of the ultraproduct consists of the equivalence classes of functions f such that $f(i) \in A_i$ for every $i \in I$, modulo the equivalence relation $=_U$ defined below. We can write f as a sequence $(f(i))_{i \in I}$ choosing for each i an element of A_i. The *equality modulo U* relation $=_U$ is defined as follows:

$$f =_U g \text{ iff } \{i : f(i) = g(i)\} \in U.$$

It is easy to show that $=_U$ is an equivalence relation. The equivalence class of a function f is denoted by $[f]$. The equivalence classes are the elements of the domain $\prod_U A_i$ of the ultraproduct.

(2) Let R be an m-ary relation symbol in L. We define its interpretation in $\mathcal{B}$:

$$R^{\mathcal{B}}([f_1],\ldots,[f_m]) \quad \text{holds iff}$$
$$\{i \in I : R^{\mathcal{A}_i}(f_1(i),\ldots,f_m(i))\} \in U.$$

We can show that this relation is well-defined.

(3) Let F be an m-ary function symbol in L. We define its interpretation in $\mathcal{B}$:

$$F^{\mathcal{B}}([f_1],\ldots,[f_m]) = [g], \quad \text{where for every } i \in I \text{ we have}$$
$$g(i) = F^{\mathcal{A}_i}(f_1(i),\ldots,f_m(i)).$$

We can show that this function is well-defined.

(4) Let c be a constant symbol in L. We define its interpretation in $\mathcal{B}$:

$$c^{\mathcal{B}} = [f] \quad \text{where for every } i \in I,$$
$$f(i) = c^{\mathcal{A}_i}.$$

The following theorem, Łoś's Theorem, sometimes called the fundamental theorem of ultraproducts, establishes the satisfiability in the ultraproduct. Its proof uses the properties of an ultrafilter to run induction on the complexity of formulas (see [**1**]).

THEOREM 1 (Łoś [**16**]). *If $\alpha(x_1,\ldots,x_n)$ is a formula in language L, then:*

$$\mathcal{B} \models \alpha([f_1],\ldots,[f_n]) \quad \textit{iff} \quad \{i : \mathcal{A}_i \models \alpha(f_1(i),\ldots,f_n(i))\} \in U.$$

Hence if σ is a sentence, then:

$$\mathcal{B} \models \sigma \quad \textit{iff} \quad \{i : \mathcal{A}_i \models \sigma\} \in U.$$

If all structures $\mathcal{A}_i$ are the same structure $\mathcal{A}$, then the ultraproduct is called an *ultrapower* and is denoted by $\prod_U \mathcal{A}$. We say that two structures in the same language are *elementarily equivalent* if they satisfy the same first-order sentences. Clearly, isomorphic structures are elementarily equivalent. If finite structures are elementarily equivalent, then they are isomorphic. On the other hand, there are elementarily equivalent (infinite) structures that are not isomorphic. We denote by $\mathcal{A} \equiv \mathcal{B}$ that $\mathcal{A}$ and $\mathcal{B}$ are elementarily equivalent.

An important consequence of Łoś's Theorem is the following statement.

If $\mathcal{B} = \prod_U \mathcal{A}$ is an ultrapower, then $\mathcal{A}$ and $\mathcal{B}$ are elementarily equivalent.

An *embedding* of a structure $\mathcal{A}$ into a structure $\mathcal{C}$ is an isomorphism between $\mathcal{A}$ and a substructure of $\mathcal{C}$. Equivalently, there is a (one-to-one) function $h : A \to C$ such that for all quantifier-free formulas $\delta(x_1,\ldots,x_n)$ and $a_1,\ldots,a_n \in A$ we have:

$$\mathcal{A} \models \delta(a_1,\ldots,a_n) \quad \text{if and only if} \quad \mathcal{C} \models \delta(h(a_1),\ldots,h(a_n)).$$

There is a natural embedding of $\mathcal{A}$ into the ultrapower $\prod_U \mathcal{A}$.

If U is a principal ultrafilter such that $\{i_0\} \in U$, then $\prod_U \mathcal{A}_i$ is isomorphic to $\mathcal{A}_{i_0}$. The construction of nontrivial ultraproducts requires non-principal ultrafilters. The existence of non-principal ultrafilters, however, cannot be proven from the Zermelo-Fraenkel axioms of set theory (ZF), see [**2**]. However, the axiom of choice, which states that every nonempty family of nonempty sets has a choice function, proves not just the existence of non-principal ultrafilters, but a stronger statement called the ultrafilter lemma, which says that every proper filter is contained in an ultrafilter [**15**]. The ultrafilter lemma is strictly weaker than the full axiom of choice [**14**], while being strictly stronger than the axiom of choice for finite sets, another consequence of the axiom of choice, which states that every nonempty family of nonempty finite sets has a choice function [**15**]. Effective ultraproducts, also called cohesive products, are based on a more constructive approach.

In Section 2, we present the effective ultraproduct construction. In Section 3, we give the fundamental theorem for effective ultraproducts and present some earlier results. In Section 4, we focus on graphs and show that every graph can be embedded into the cohesive power of a certain kind of special graph. In Section 5, we focus on equivalence structures and the isomorphism types of their cohesive powers.

2. Effective ultraproducts of structures

Dimitrov [**6**] introduced the notion of an effective ultrapower (also called cohesive power) of computable structures. It is an effective ultrapower construction, which starting with computable structures produces countable structures. The role of an ultrafilter is played by a cohesive set. Recall the following notions from computability theory. The complement of a set $X \subseteq N$ is denoted by $\overline{X}$.

All Turing machine programs can be algorithmically listed by systematically coding them. This listing gives us an algorithmic enumeration of all unary *partial computable* functions on natural numbers, which are computed by the corresponding Turing machine programs:

$$\varphi_0, \varphi_1, \varphi_2, \ldots .$$

A partial function φ_e may not be defined on some inputs x; that is, the corresponding Turing machine does not halt on such inputs, sometimes written as $\varphi_e(x)\uparrow$. Otherwise, x belongs to the domain of φ_e, which we also write as $\varphi_e(x)\downarrow$, and its value is the output $y = \varphi_e(x)$. Some of these partial computable functions are total (defined on all inputs) and thus *computable.* We can use a diagonalization argument to show that there is no algorithmic enumeration of all computable functions. A set of natural numbers is *computable* if its characteristic function is computable. A nonempty set of natural numbers is *computably enumerable* if there is a computable function that enumerates it. As usual, we abbreviate computably enumerable by c.e. We can show that a set is c.e. if and only if it is the domain of a partial computable function. Hence we can computably enumerate all c.e. sets as:

$$W_0, W_1, W_2, \ldots .$$

It can be shown that a set X is computable if and only if both X and $\overline{X}$ are c.e. There are c.e. sets that are not computable, such as the *halting set*, defined as:

$$K = \{e : e \in W_e\} = \{e : \varphi_e(e)\downarrow\}.$$

In the effective ultraproduct construction, the role of an ultrafilter is played by a cohesive set. Cohesive sets are infinite sets of natural numbers, which are indecomposable by c.e. sets.

DEFINITION 2. An infinite set $C \subseteq N$ is *cohesive* if for every c.e. set W, either $W \cap C$ or $\overline{W} \cap C$ is finite.

An infinite subset of a cohesive set is cohesive. There are continuum many cohesive sets. A cohesive set can have a c.e. complement, as established by Friedberg (see [**17**]). These complements are called maximal sets.

DEFINITION 3. A set $M \subseteq N$ is *maximal* if M is c.e. and $\overline{M}$ is cohesive.

We say that a countable structure $\mathcal{A}$ for a computable language L is *computable* if its domain A is a computable set and its relations and functions are uniformly computable. Equivalently, there is a decision procedure for determining the truth of the quantifier-free sentences with constant symbols added as names for all elements of the domain. Examples of computable structures include the ordered set of natural numbers $(N, <)$, the group of integers $(Z, +)$, and the field of rational numbers $(Q, +, \cdot)$. Two computable structures are *computably isomorphic* if there is a computable isomorphism between them.

Cohesive powers of fields were used in [**5**] to characterize principal filters of maximal and, more generally, quasimaximal vector spaces. These fields, which turned out to be cohesive powers of the computable field for a vector space motivated the definition of cohesive powers of computable structures (see [**6**]). We will now give a more general definition of cohesive products of computable structures introduced in [**7**]. For two partial functions ψ_1 and ψ_2, by $\psi_1(x)\downarrow = \psi_2(x)\downarrow$ we denote that on input x, both ψ_1 and ψ_2 are defined and have the same value. We say that ψ_1 and ψ_2 are equal, in symbols $\psi_1 \simeq \psi_2$, if for every x, either both $\psi_1(x)$ and $\psi_2(x)$ are undefined or $\psi_1(x)\downarrow = \psi_2(x)\downarrow$.

DEFINITION 4. Let $\mathcal{A}_i$ for $i \in N$ be a uniformly computable sequence of computable structures in a computable language L and let $C \subseteq N$ be a cohesive set. The *cohesive product* $\mathcal{B}$ *of* $\mathcal{A}_i$ *over* C, in symbols $\mathcal{B} = \prod_C \mathcal{A}_i$, is a structure defined as follows.

(1) Let
$D = \{\psi \mid \psi : N \to \bigcup_{i \in N} A_i,\ \psi$ is a partial computable function, $C \subseteq^* dom(\psi)$, and if $\psi(i)\downarrow$ then $\psi(i) \in A_i\}$.
For $\psi_1, \psi_2 \in D$, let

$$\psi_1 =_C \psi_2 \quad \text{iff} \quad C \subseteq^* \{i : \psi_1(i)\downarrow = \psi_2(i)\downarrow\}.$$

The domain of $\prod_C \mathcal{A}_i$ is the quotient set $D/_{=_C}$ and is denoted here by B.

(2) If $f \in L$ is an m-ary function symbol, then $f^{\mathcal{B}}$ is an m-ary function on B such that for every $[\psi_1], \dots, [\psi_m] \in B$, we have

$$f^{\mathcal{B}}([\psi_1], \dots, [\psi_m]) = [\psi] \quad \text{iff} \quad (\forall i \in N)(\psi(i) \simeq f^{\mathcal{A}_i}(\psi_1(i), \dots, \psi_m(i))).$$

(3) If $R \in L$ is an m-ary relation symbol, then $R^{\mathcal{B}}$ is an m-ary relation on B such that for every $[\psi_1], \dots, [\psi_m] \in B$,

$$R^{\mathcal{B}}([\psi_1], \dots, [\psi_m]) \quad \text{iff} \quad C \subseteq^* \{i \in N \mid R^{\mathcal{A}_i}(\psi_1(i), \dots, \psi_m(i))\}.$$

(4) If $c \in L$ is a constant symbol, then $c^{\mathcal{B}}$ is the equivalence class (with respect to $=_C$) of the computable function $g : N \to \bigcup_{i \in N} A_i$ such that $g(i) = c^{\mathcal{A}_i}$ for every $i \in N$.

We can show that $f^{\mathcal{B}}$ and $R^{\mathcal{B}}$ are well-defined.

If $\mathcal{A}_i = \mathcal{A}$ for $i \in N$, then $\prod_C \mathcal{A}_i$ is called the *cohesive power of* $\mathcal{A}$ *over* C and is denoted by $\prod_C \mathcal{A}$. We will present a restricted version of Łoś's theorem for cohesive products in the next section.

3. Fundamental properties of effective ultraproducts

The following theorem is a version of Łoś's theorem for cohesive products of uniformly computable structures.

THEOREM 5 ([**7**]). *Let $\mathcal{A}_i$ for $i \in N$ be a uniformly computable sequence of computable structures in a computable language L and let $C \subseteq N$ be a cohesive set. Let $\mathcal{B}$ be the cohesive product.*

(1) *If $\theta(x_1, \ldots, x_n)$ is a formula in L, which is a Boolean combination of Σ_1^0 formulas, and $[\psi_1], \ldots, [\psi_n] \in B$, then*

$$\mathcal{B} \models \theta([\psi_1], \ldots, [\psi_n]) \text{ iff } C \subseteq^* \{i : \mathcal{A}_i \models \theta(\psi_1(i), \ldots, \psi_n(i))\}.$$

(2) *If σ is a Π_2^0 sentence in L, then $C \subseteq^* \{i : \mathcal{A}_i \models \sigma\}$ implies $\mathcal{B} \models \sigma$.*

Let $\mathcal{B} = \prod_C \mathcal{A}$ be a cohesive power of $\mathcal{A}$. For $a \in A$ let $[c_a] \in B$ be the equivalence class of the total function c_a such that $c_a(i) = a$ for every $i \in N$. The function $d : A \to B$ such that $d(a) = [c_a]$ is an embedding of $\mathcal{A}$ into $\mathcal{B}$. It is called the *canonical embedding.*

The following theorem, due to Dimitrov, is the fundamental theorem for cohesive powers.

THEOREM 6 (Dimitrov [**6**]). *Let $\mathcal{B} = \prod_C \mathcal{A}$ be a cohesive power of $\mathcal{A}$.*

(1) *If σ is a Π_3^0 sentence in L, then $\mathcal{B} \models \sigma$ implies $\mathcal{A} \models \sigma$.*
(2) *By contrapositive of* (1), *if σ is a Σ_3^0 sentence in L, then $\mathcal{A} \models \sigma$ implies $\mathcal{B} \models \sigma$.*
(3) *If σ is a Π_2^0 (or Σ_2^0) sentence in L, then $\mathcal{B} \models \sigma$ iff $\mathcal{A} \models \sigma$.*
(4) *Let $\theta(x_1, \ldots, x_n)$ be a Π_2^0 or a Σ_2^0 formula in L, and let $a_1, \ldots, a_n \in A$. Then*

$$\mathcal{A} \models \theta(a_1, \ldots, a_n) \quad \text{iff} \quad \mathcal{B} \models \theta([c_{a_1}], \ldots, [c_{a_n}]).$$

(5) *Let $\theta(x_1, \ldots, x_n)$ be a Σ_3^0 formula in L, and let $a_1, \ldots, a_n \in A$. Then $\mathcal{A} \models \theta(a_1, \ldots, a_n)$ implies $\mathcal{B} \models \theta([c_{a_1}], \ldots, [c_{a_n}])$.*

Feferman, Scott, and Tennenbaum [**12**] found a Π_3^0 sentence that is true in the structure $\mathcal{N} = (N, +, \cdot, 0, 1)$ but not in its cohesive power $\prod_C \mathcal{N}$. More recently, simpler examples of such sentences have been produced for linear orders in [**10**].

An equivalence structure is a set with an equivalence relation on it. If $\mathcal{A}$ is a computable equivalence structure, then a cohesive power $\prod_C \mathcal{A}$ is an equivalence structure. This is true since the theory of equivalence structures is Π_1^0-axiomatizable, so we can apply Theorem 6. Similarly, a cohesive power of a computable linear order is a linear order, a cohesive power of a computable field is a field, and a cohesive power of a computable graph is a graph.

A countable structure $\mathcal{A}$ for a computable language L is called *k-decidable* if its domain A is a computable set and there is a decision procedure for determining the truth of the first-order Σ_k^0 sentences with constant symbols added as names for all elements of the domain. The next theorem is a version of Łoś's theorem for cohesive powers of k-decidable structures.

THEOREM 7 ([**10**]). *Let $\mathcal{A}$ be a k-decidable structure in a computable language L, and let C be a cohesive set. Let $\mathcal{B} = \prod_C \mathcal{A}$.*

(1) *If $\theta(x_1, \ldots, x_n)$ is a formula in L, which is a Boolean combination of Σ^0_{k+1} formulas, and $[\psi_1], \ldots, [\psi_n] \in B$, then*

$$\prod_C \mathcal{A} \models \theta([\psi_1], \ldots, [\psi_n]) \quad \text{iff} \quad C \subseteq^* \{i : \mathcal{A} \models \theta(\psi_1(i), \ldots, \psi_n(i))\}.$$

(2) *If σ is a Σ^0_{k+2} sentence or a Π^0_{k+2} sentence, then $\prod_C \mathcal{A} \models \sigma$ iff $\mathcal{A} \models \sigma$.*
(3) *Let σ be a Σ^0_{k+3} sentence. If $\mathcal{A} \models \sigma$ then $\prod_C \mathcal{A} \models \sigma$.*

PROPOSITION 8 ([**6**]). *Let $\mathcal{A}$ be a finite structure, and let $\mathcal{B} = \prod_C \mathcal{A}$. Then $\mathcal{B} \cong \mathcal{A}$.*

PROOF. Let $[\varphi] \in B$ be arbitrary. For any $a \in A$, let $I_a = \{i : \varphi(i) = a\}$. Note that the set I_a is c.e. Since $dom(\varphi) = \bigcup_{a \in A} I_a$ and A is finite, for some $a_0 \in A$, the set $I_{a_0} \cap C$ is infinite. Since C is cohesive, we have that $C \subseteq^* I_{a_0}$, and so $[\varphi] = [\varphi_{a_0}]$. Thus, the canonical embedding of $\mathcal{A}$ into $\mathcal{B}$ is onto, and so $\mathcal{B} \cong \mathcal{A}$. □

PROPOSITION 9. *If there is a computable permutation p of N such that $p(C_1) =^* C_2$ then $\prod_{C_1} \mathcal{A} \cong \prod_{C_2} \mathcal{A}$.*

PROOF. Let the function $\Phi : \prod_{C_2} \mathcal{A} \to \prod_{C_1} \mathcal{A}$ be such that $\Phi([\psi]) = [\psi \circ p]$. The proof that Φ is an isomorphism is straightforward. □

PROPOSITION 10 ([**7**]). *Let C be the complement of a maximal set. Then for every $[\psi] \in \prod_C \mathcal{A}$ there is a computable function f such that $[f] = [\psi]$.*

PROOF. By assumption, C is a co-c.e. cohesive set. Let $[\psi] \in \prod_C \mathcal{A}$, and let $a_0 \in A$ be an arbitrary fixed element. Define a (total) computable function f:

$$f(i) = \begin{cases} \psi(i) & \text{if } \psi(i) \downarrow \text{first}, \\ a_0 & \text{if } i \text{ is enumerated into } \overline{C} \text{ first}, \end{cases}$$

where $i \in N$. □

PROPOSITION 11. *Let C be a cohesive set. If computable structures $\mathcal{A}_1$ and $\mathcal{A}_2$ are computably isomorphic, then $\prod_C \mathcal{A}_1 \cong \prod_C \mathcal{A}_2$.*

In [**9**], the authors obtained important results about the isomorphism types of cohesive powers of the field of rational numbers over co-maximal sets, as well as about the automorphisms of these cohesive powers. They further applied these results in [**8**] to study the lattice of c.e. vector spaces. In [**10, 11**], the authors investigated cohesive powers of computable linear orders with special emphasis on computable copies of the ordered set of natural numbers. They showed how computable properties of structures might affect the properties of their cohesive powers.

4. Effective ultrapowers of graphs

As usual, a *graph* (or undirected graph) $G = (V, E)$ is a nonempty set V of vertices (nodes) with a symmetric binary relation E called the *edge relation*. Thus, it can be axiomatized by the following universal sentence:

$$\forall x \forall y [E(x, y) \Rightarrow E(y, x)].$$

Hence, by Theorem 6, a cohesive power of a computable graph is a graph.

If $(a, b) \in E$, then vertices a and b are adjacent to each other. The *degree of a vertex* is the number of vertices it is adjacent to. A graph G is called *locally finite* if the degree of each vertex in G is finite. A *connected component* of a graph is a subgraph in which each pair of vertices is connected with each other via a path, and no other vertices outside of the subgraph are adjacent to any vertex in the subgraph.

DEFINITION 12. A graph G is called *strongly locally finite* if all connected components of G are finite.

Here is an example of a locally finite graph which is not strongly locally finite. Let us define a relation R on the set of natural numbers as: xRy if and only if either $x = y + 2$ or $y = x + 2$. Then (N, R) is a graph with two connected components that are both infinite, the set of even numbers and the set of odd numbers. Hence it is not strongly locally finite, yet every vertex has degree either 1 or 2, so it is locally finite. This example can be easily modified to have any finite number of connected components, and, using the pairing function, can be modified to have infinitely many connected components.

DEFINITION 13. The *disjoint union* of graphs (V_1, E_1) and (V_2, E_2) where $V_1 \cap V_2 = \varnothing$ is a graph $G = (V_1 \cup V_2, E_1 \cup E_2)$.

Hence there are no edges between V_1 and V_2 in the disjoint union. We write $(V_1, E_1) \amalg (V_2, E_2)$ for G and also view it as a decomposition of G into the disjoint union.

If $V_1 \cap V_2 = \varnothing$, then a union of graphs $G_1 = (V_1, E_1)$ and $G_2 = (V_2, E_2)$ is a graph $G = (V_1 \cup V_2, E)$ such that $E_1 \cup E_2 \subseteq E$ and $E \restriction (V_i \times V_i) = E_i$ for $i = 1, 2$. We simply write $G = G_1 \cup G_2$.

The following result demonstrates the universal feature with respect to embeddability of cohesive powers of certain computable graphs.

THEOREM 14. *Let G be a computable graph. Let C be a cohesive set. Then there is a computable strongly locally finite graph $\mathcal{A}$ such that $\Pi_C \mathcal{A}$ is isomorphic to $G \cup H$ for some graph H or $H = \varnothing$.*

PROOF. If G is finite, then G is strongly locally finite and $\Pi_C G \cong G$.

Now, assume that $G = (V, R)$ is infinite, and fix a computable enumeration f of its vertices, $V = \{f(n) : n \geq 1\}$. Using this enumeration we will build a graph $\mathcal{A}$ with the domain $N - \{0\}$ and an edge set E. Consider vertices $a, b \in N - \{0\}$. If a, b can be written as $a = \frac{k(k+1)}{2} + m$ and $b = \frac{k(k+1)}{2} + n$ for some k, m, n such that $1 \leq m, n \leq k + 1$, then let

$$(a, b) \in E \Leftrightarrow (f(m), f(n)) \in R.$$

Otherwise, let $(a, b) \notin E$.

The graph $\mathcal{A}$ is computable since its domain is a computable set, and we can effectively determine whether any two vertices a and b are connected by an E-edge. That is, we can write each positive number as the sum of the first k numbers plus a positive number that is $\leq k+1$. If a and b do not have the same k, then they are not connected by an E-edge. If a and b have the same k, then they are connected by an E-edge exactly when their f-images are connected by an R-edge.

For $n \geq 1$, let a computable function $\psi_n : N \to A$ be defined as:

$$\psi_n(i) = \frac{i(i+1)}{2} + n.$$

Then the equivalence class $[\psi_n]$ is an element of the cohesive power $\Pi_C \mathcal{A}$. Let $S = \{[\psi_n] : n \geq 1\}$. Then $\mathcal{S} = (S, E^{\mathcal{A}} \upharpoonright S)$ is a subgraph of $\Pi_C \mathcal{A}$.

Define $\rho : \{[\psi_n] : n \geq 1\} \to G$ by

$$\rho([\psi_n]) = f(n).$$

We will now show that ρ is a graph isomorphism between $\mathcal{S}$ and G. Clearly, ρ is a bijection. Let us show that ρ is a homomorphism. Let $m, n \geq 1$.

Assume that $(\rho([\psi_m]), \rho([\psi_n])) \in R$. Then $(f(m), f(n)) \in R$ by the definition of ρ. By the definition of E, it follows that for every k such that $1 \leq m, n \leq k+1$ we have $(\frac{k(k+1)}{2} + m, \frac{k(k+1)}{2} + n) \in E$. Hence for $I =_{def} \{i \in N : (\frac{i(i+1)}{2} + m, \frac{i(i+1)}{2} + n) \in E\}$, we have $I \supseteq \{i \in N : i+1 \geq \max\{m,n\}\}$, so I is co-infinite, hence $C \subseteq^* I$. Since $I = \{i \in N : (\psi_m(i), \psi_n(i)) \in E\}$, we have that $([\psi_m], [\psi_n]) \in E^{\mathcal{A}}$.

Now, assume that $(\rho([\psi_m]), \rho([\psi_n])) \notin R$. Then $(f(m), f(n)) \notin R$. By the definition of E, for every k with $1 \leq m, n \leq k+1$ we have $(\frac{k(k+1)}{2} + m, \frac{k(k+1)}{2} + n) \notin E$. Hence for $I =_{def} \{i \in N : (\frac{i(i+1)}{2} + m, \frac{i(i+1)}{2} + n) \in E\}$, we have $I \subseteq \{i \in N : i+1 < \max\{m,n\}\}$, so I is finite. Hence $([\psi_m], [\psi_n]) \notin E^{\mathcal{A}}$.

Thus, $\Pi_C \mathcal{A}$ is isomorphic to $G \cup H$ where the domain of H is $dom(\Pi_C \mathcal{A}) - \{[\psi_n] : n \geq 1\}$. □

If a computable graph G is such that each of its vertices has a finite degree, then we have a stronger embedding result.

THEOREM 15. *Let G be an infinite computable graph that is locally finite. Let C be a cohesive set. Then there is a computable strongly locally finite graph $\mathcal{A}$ such that $\Pi_C \mathcal{A}$ is isomorphic to the disjoint union $G \sqcup H$ for some graph H.*

PROOF. Let $G = (V, R)$. Let $f, E, \mathcal{A}, \psi_n, \rho$ be defined as in the proof of Theorem 14. Let φ be a partial computable function such that $[\varphi] \in dom(\Pi_C \mathcal{A})$ and $E^{\Pi_C \mathcal{A}}([\varphi], [\psi_m])$ for some $m \geq 1$. Then $C \subseteq^* \{i \in N : \varphi(i)\downarrow \wedge (\varphi(i), \psi_m(i)) \in E\}$. Since G is locally finite, we have that the set on the right-hand side of $\subseteq^*$ is the following finite disjoint union:

$$\coprod_{n : f(m) R f(n)} \{i \in N : \varphi(i) = \frac{i(i+1)}{2} + n, \quad \text{where } 1 \leq m, n \leq i+1\}.$$

For some n_0 such that $1 \leq n_0 \leq i+1$, we have

$$C \subseteq^* \{i : \varphi(i)\downarrow = \frac{i(i+1)}{2} + n_0\} \subseteq \{i : \varphi(i)\downarrow = \psi_{n_0}(i)\}.$$

Such n_0 exists since C is cohesive. Hence $[\varphi] = [\psi_{n_0}]$, so $[\varphi] \in \{[\psi_n] : n \geq 1\}$. Clearly, $\Pi_C \mathcal{A}$ is a disjoint union, where

$$dom(\Pi_C \mathcal{A}) = \{[\psi_n] : n \geq 1\} \cup (dom(\Pi_C \mathcal{A}) - \{[\psi_n] : n \geq 1\}).$$

Hence $\Pi_C \mathcal{A}$ is isomorphic to the disjoint union of G and another graph, so G embeds into the cohesive power $\Pi_C \mathcal{A}$ in a strong way in this sense. □

5. Effective ultrapowers of equivalence structures

An equivalence structure $\mathcal{A} = (A, E^{\mathcal{A}})$ consists of a nonempty set A with a binary relation $E^{\mathcal{A}}$ that is reflexive, symmetric, and transitive. An equivalence structure $\mathcal{A}$ is *computable* if A is a computable set and $E^{\mathcal{A}}$ is a computable relation. If A is an infinite set, we may assume, without loss of generality, that $A = N$. The equivalence class of $a \in A$ is

$$eqv^{\mathcal{A}}(a) = \{b \in A : bE^{\mathcal{A}}a\}.$$

We often omit the superscript $^{\mathcal{A}}$ when it can be inferred from the context.

Let $\langle z_1, z_2 \rangle$ be a computable bijection from N^2 onto N, which is strictly increasing with respect to each coordinate and such that $z_1, z_2 \leq \langle z_1, z_2 \rangle$.

DEFINITION 16.

(i) Let $\mathcal{A}$ be an equivalence structure. The *character* of $\mathcal{A}$ is the set

$$\chi(\mathcal{A}) = \{\langle k, n \rangle : \ n, k > 0 \text{ and } \mathcal{A} \text{ has at least } n \text{ equivalence classes of size } k\}.$$

(ii) We say that $\mathcal{A}$ has a *bounded character* if there is some finite k such that all finite equivalence classes of $\mathcal{A}$ have size at most k.

Let

$$Inf^{\mathcal{A}} = \{a : eqv^{\mathcal{A}}(a) \text{ is infinite}\} \quad \text{and} \quad Fin^{\mathcal{A}} = \{a : eqv^{\mathcal{A}}(a) \text{ is finite}\}.$$

The following proposition from [4] gives us some important complexities. By $|X|$ we denote the size (cardinality) of a set X.

PROPOSITION 17. *For any computable equivalence structure $\mathcal{A}$ and $k \in N - \{0\}$, we have:*

(a) $\{\langle k, a \rangle : |eqv^{\mathcal{A}}(a)| \leq k\}$ *is a Π^0_1 set, and* $\{\langle k, a \rangle : |eqv^{\mathcal{A}}(a)| \geq k\}$ *is a Σ^0_1 set;*
(b) $Inf^{\mathcal{A}}$ *is a Π^0_2 set, and $Fin^{\mathcal{A}}$ is a Σ^0_2 set;*
(c) $\chi(\mathcal{A})$ *is a Σ^0_2 set.*

We say that a subset K of N is a *character* if there is some equivalence structure with character K. This is the same as saying that $K \subseteq \{\langle k, n \rangle : k, n \geq 1\}$ and for all $n > 0$ and k:

$$\langle k, n+1 \rangle \in K \Rightarrow \langle k, n \rangle \in K.$$

THEOREM 18 ([4]). *For any Σ^0_2 character K, there is a computable equivalence structure $\mathcal{A}$ with character K, which has infinitely many infinite equivalence classes. Furthermore, $Fin^{\mathcal{A}}$ is a Π^0_1 set.*

A computable structure $\mathcal{A}$ is called *computably categorical* if every computable structure isomorphic to $\mathcal{A}$ is computably isomorphic to $\mathcal{A}$.

Let C be a cohesive set. It follows from Proposition 11 that if $\mathcal{A}_1$ is a computably categorical structure and $\mathcal{A}_2$ is a computable structure such that $\mathcal{A}_1 \cong \mathcal{A}_2$, then $\Pi_C \mathcal{A}_1 \cong \Pi_C \mathcal{A}_2$.

THEOREM 19 ([**4**]). *Let $\mathcal{A}$ be a computable equivalence structure. The structure $\mathcal{A}$ is computably categorical if and only if one of the following holds:*

(1) *$\mathcal{A}$ has only finitely many finite equivalence classes;*
(2) *$\mathcal{A}$ has finitely many infinite classes, bounded character, and at most one finite k such that there are infinitely many classes of size k.*

Clearly, by Theorem 8, if an equivalence structure $\mathcal{A}$ has finitely many finite equivalence classes and no infinite equivalence classes, then $\Pi_C \mathcal{A} \cong \mathcal{A}$.

The following proposition establishes some more basic properties of cohesive powers of equivalence structures.

PROPOSITION 20. *Let $\mathcal{A}$ be a computable equivalence structure. Let C be a cohesive set, and $\mathcal{B} = \Pi_C \mathcal{A}$.*

(a) *Then $\chi(\mathcal{B}) = \chi(\mathcal{A})$.*
(b) *If $\mathcal{A}$ has infinitely many infinite equivalence classes, then $\mathcal{B}$ has infinitely many infinite equivalence classes.*
If $\mathcal{A}$ has exactly n infinite equivalence classes, where $n \in N$, then $\mathcal{B}$ has at least n infinite equivalence classes.
(c) *If $\mathcal{A}$ has infinitely many infinite equivalence classes, then $\Pi_C \mathcal{A} \cong \mathcal{A}$.*

PROOF. (a) The character of a computable equivalence structure is definable by a Σ^0_2 sentence; see [**4**].

(b) This is true since there is an embedding of $\mathcal{A}$ into $\mathcal{B}$ (e.g., the canonical embedding).

(c) This follows from (a) and (b) . □

THEOREM 21. *Let $\mathcal{A} = (A, E^{\mathcal{A}})$ be a computable equivalence structure. Let C be a cohesive set. If $\mathcal{A}$ has a bounded character, then $\Pi_C \mathcal{A} \cong \mathcal{A}$.*

PROOF. Since the character of $\mathcal{A}$ is bounded, if it is nonempty, let $k \in N$ be the largest size of a finite equivalence class of $\mathcal{A}$. If the character of $\mathcal{A}$ is empty, let $k = 1$.

By Proposition 20 (a), $\mathcal{A}$ and $\Pi_C \mathcal{A}$ have the same character. If $\mathcal{A}$ has infinitely many infinite equivalence classes, then $\Pi_C \mathcal{A}$ also has infinitely many infinite equivalence classes, and since $\mathcal{A}$ and $\Pi_C \mathcal{A}$ have the same character, they are isomorphic. Thus, assume that $\mathcal{A}$ has at most finitely many infinite equivalence classes.

If $\mathcal{A}$ has no infinite equivalence classes, then $\mathcal{A}$ satisfies the following Π^0_1 sentence, saying that there are no $k+1$ equivalent elements:

$$(\forall x_1) \cdots (\forall x_{k+1}) \left[\bigwedge_{1 \le i < j \le k+1} x_i \neq x_j \Rightarrow \bigvee_{1 \le i < j \le k+1} \neg(x_i E x_j) \right]$$

Since $\Pi_C \mathcal{A}$ satisfies the same sentence, it has no infinite equivalence classes, so $\mathcal{A}$ and $\Pi_C \mathcal{A}$ are isomorphic.

Thus, assume that $\mathcal{A}$ has m infinite equivalence classes, where $m \in N$ and $m > 0$. Hence $\mathcal{A}$ satisfies the following Σ_2^0 sentence, saying that there are exactly m equivalence classes with at least $k+1$ elements, hence infinite. We will use symbols x_i^l to denote variables.

$$(\exists x_1^1)\cdots(\exists x_{k+1}^1)\cdots(\exists x_1^m)\cdots(\exists x_{k+1}^m)(\forall y_1)\cdots(\forall y_{k+1})$$

$$\times\left[\bigwedge_{\substack{1\le l\le m\\ 1\le i<j\le k+1}} (x_i^l \neq x_j^l \wedge x_i^l E x_j^l) \wedge \left(\bigwedge_{1\le i<j\le k+1} (y_i \neq y_j \wedge y_i E y_j) \Rightarrow \bigvee_{l=1,\ldots,m} y_1 E x_1^l\right)\right]$$

Hence $\Pi_C\mathcal{A}$ satisfies the same sentence. Thus, $\Pi_C\mathcal{A}$ has exactly m infinite equivalence classes, so it is isomorphic to $\mathcal{A}$. □

THEOREM 22. *Let $\mathcal{A} = (A, E^{\mathcal{A}})$ be a computable equivalence structure. Let C be a cohesive set.*

If $\mathcal{A}$ has an unbounded character, then $\Pi_C\mathcal{A}$ has infinitely many infinite equivalence classes.

Hence if $\mathcal{A}$ has an unbounded character and finitely many (possibly zero) infinite equivalence classes, then $\Pi_C\mathcal{A} \not\cong \mathcal{A}$.

PROOF. Without loss of generality, we will assume that $A = N$. Then we can order the elements in A by the usual ordering of the natural numbers. Let $\mathcal{B} = \Pi_C\mathcal{A}$, and $\mathcal{B} = (B, E^{\mathcal{B}})$. Choose a computable sequence of elements in A:

$$a(0), a(1), a(2), a(3), \ldots$$

such that they all belong to distinct equivalence classes, and for every k, the equivalence class of $a(k)$ has $> k$ elements. Since $E^{\mathcal{A}}$ is computable and the character of $\mathcal{A}$ is unbounded, such a sequence can be obtained by enumerating A and checking the conditions. We can think of $a(\langle m, i\rangle)$ as a representative of the equivalence class in $\mathcal{A}$ where a partial function $\psi_{m,*}$ might take its ith coordinate value (for $i \geq 0$). That is, define partial computable functions $\psi_{m,n}(i)$, for $m, i \geq 0$ and $n \geq 1$, as follows:

$$\psi_{m,n}(i) = \begin{cases} c_n & c_n \text{ is the } n^{th} \text{ element in } eqv(a\langle m,i\rangle) \text{ if it exists;} \\ \uparrow & \text{otherwise.} \end{cases}$$

Every $\psi_{m,n}$ is defined for all except possibly finitely many initial values, so $C \subseteq^* dom(\psi_{m,n})$. Hence $[\psi_{m,n}] \in B$.

Fix m. Let $n_1 \neq n_2$. Then we have that $[\psi_{m,n_1}] \neq [\psi_{m,n_2}]$ since $|eqv(a(\langle m,i\rangle))| > \langle m,i\rangle \geq i$, so starting with some i, $eqv(a\langle m,i\rangle)$ will have the n_1^{th} and the n_2^{th} elements and they will be distinct. Also, $[\psi_{m,n_1}]E^{\mathcal{B}}[\psi_{m,n_2}]$ since the values of ψ_{m,n_1} and ψ_{m,n_2}, when defined, are from the same equivalence class in $\mathcal{A}$. Hence the equivalence class of $[\psi_{m,1}]$ is infinite.

Now, let $m_1 \neq m_2$. Then $\neg([\psi_{m_1,1}]E^{\mathcal{B}}[\psi_{m_2,1}])$ since the values of $\psi_{m_1,1}$ and $\psi_{m_2,1}$, when defined, are from different equivalence classes in $\mathcal{A}$. □

COROLLARY 23. *Let* $\mathcal{A} = (A, E^{\mathcal{A}})$ *be a computable equivalence structure. Let* C *be a cohesive set. Then* $\Pi_C\mathcal{A} \cong \mathcal{A}$ *if and only if* $\mathcal{A}$ *has a bounded character or infinitely many infinite equivalence classes.*

It was shown in [**3**] that the following model-theoretic result holds for the equivalence structures. Let the formulas $\gamma_k(x)$ state that the equivalence class of x has at least k elements, where $k \in N$ and $k \geq 1$. Then the language of equivalence relations $\{E\}$ expanded with unary predicates $\{\gamma_k : k \geq 1\}$ has quantifier elimination; i.e., every first-order formula in the original language is logically equivalent to a quantifier-free formula in the new language. Since the formulas $\gamma_k(x)$ are Σ_1^0 formulas, it follows that a computable equivalence structure $\mathcal{A}$ and its cohesive powers satisfy the same first-order sentences. However, in some cases, the distinction can be made by using computable infinitary sentences. For more on computable infinitary language see [**13**].

Computable Σ_0 and Π_0 formulas are just the finitary quantifier-free formulas. Let $n > 1$.

A *computable* Π_n *formula* is a c.e. conjunction of formulas

$$\forall \overline{u}\, \phi(\overline{x}, \overline{u}),$$

where ϕ is a computable Σ_m formula for some $m < n$.

A *computable* Σ_n *formula* is a c.e. disjunction of formulas

$$\exists \overline{v}\, \theta(\overline{y}, \overline{v}),$$

where θ is a computable Π_m formula for some $m < n$.

COROLLARY 24. *Let* $\mathcal{A} = (A, E^{\mathcal{A}})$ *be a computable equivalence structure with unbounded character and no infinite equivalence classes. Let* C *be a cohesive set. Then there is a computable* Σ_3 *sentence* α *such that* $\Pi_C\mathcal{A} \models \alpha$ *and* $\mathcal{A} \models \neg\alpha$.

PROOF. We have that $\Pi_C\mathcal{A}$ has an infinite equivalence class, while $\mathcal{A}$ does not. Let α be the sentence saying that there is an infinite equivalence class:

$$\exists x \bigwedge_{n \in N} [\exists y_1 \cdots \exists y_n \bigwedge_{1 \leq i < j \leq n} (y_i \neq y_j \wedge xEy_i)].$$

Hence α is a computable Σ_3 sentence true in $\Pi_C\mathcal{A}$ but false in $\mathcal{A}$. □

References

[1] J. L. Bell and A. B. Slomson, *Models and ultraproducts: An introduction*, North-Holland Publishing Co., Amsterdam-London, 1969. MR269486

[2] A. Blass, *A model without ultrafilters* (English, with Russian summary), Bull. Acad. Polon. Sci. Sér. Sci. Math. Astronom. Phys. **25** (1977), no. 4, 329–331. MR476510

[3] W. Calvert, D. Cenzer, and V. Harizanov, *Densely computable structures*, J. Logic Comput. **32** (2022), no. 3, 581–607, DOI 10.1093/logcom/exab057. MR4406803

[4] W. Calvert, D. Cenzer, V. Harizanov, and A. Morozov, *Effective categoricity of equivalence structures*, Ann. Pure Appl. Logic **141** (2006), no. 1-2, 61–78, DOI 10.1016/j.apal.2005.10.002. MR2229930

[5] R. Dimitrov, *A class of* Σ_3^0 *modular lattices embeddable as principal filters in* $\mathcal{L}^*(V_\infty)$, Arch. Math. Logic **47** (2008), no. 2, 111–132, DOI 10.1007/s00153-008-0078-2. MR2410809

[6] R. Dimitrov, *Cohesive powers of computable structures*, Annuaire Univ. Sofia Fac. Math. Inform. **99** (2009), 193–201. MR2528900

[7] R. D. Dimitrov and V. Harizanov, *Effective ultrapowers and applications*, Aspects of computation and automata theory with applications, Lect. Notes Ser. Inst. Math. Sci. Natl. Univ. Singap., vol. 42, World Sci. Publ., Hackensack, NJ, [2024] ©2024, pp. 201–221. MR4679144

[8] R. D. Dimitrov and V. Harizanov, *Orbits of maximal vector spaces* (Russian, with English and Russian summaries), Algebra Logika **54** (2015), no. 6, 680–732, 783–784, DOI 10.1007/s10469-016-9366-9; English transl., Algebra Logic **54** (2016), no. 6, 440–477. MR3497816

[9] R. Dimitrov, V. Harizanov, R. Miller, and K. J. Mourad, *Isomorphisms on non-standard fields and Ash's conjecture*, Language, life, limits, Lecture Notes in Comput. Sci., vol. 8493, Springer, Cham, 2014, pp. 143–152, DOI 10.1007/978-3-319-08019-2_15. MR3219029

[10] R. Dimitrov, V. Harizanov, A. Morozov, P. Shafer, A. A. Soskova, and S. V. Vatev, *On cohesive powers of linear orders*, J. Symb. Log. **88** (2023), no. 3, 947–1004, DOI 10.1017/jsl.2023.14. MR4636622

[11] R. Dimitrov, V. Harizanov, A. Morozov, P. Shafer, A. Soskova, and S. Vatev, *Cohesive powers of linear orders*, Computing with foresight and industry, Lecture Notes in Comput. Sci., vol. 11558, Springer, Cham, 2019, pp. 168–180, DOI 10.1007/978-3-030-22996-2_15. MR3981880

[12] S. Feferman, D. S. Scott, and S. Tennenbaum, *Models of arithmetic through function rings*, Notices of the American Mathematical Society **6** (1959), no. 2, 173–174, Abstract #556-31.

[13] E. B. Fokina, V. Harizanov, and A. Melnikov, *Computable model theory*, Turing's legacy: developments from Turing's ideas in logic, Lect. Notes Log., vol. 42, Assoc. Symbol. Logic, La Jolla, CA, 2014, pp. 124–194. MR3497661

[14] J. D. Halpern and A. Lévy, *The Boolean prime ideal theorem does not imply the axiom of choice*, Axiomatic Set Theory (Proc. Sympos. Pure Math., Vol. XIII, Part I, Univ. California, Los Angeles, Calif., 1967), Proc. Sympos. Pure Math., XIII, Part I, Amer. Math. Soc., Providence, RI, 1971, pp. 83–134. MR284328

[15] T. J. Jech, *The axiom of choice*, Studies in Logic and the Foundations of Mathematics, Vol. 75, North-Holland Publishing Co., Amsterdam-London; American Elsevier Publishing Co., Inc., New York, 1973. MR396271

[16] J. Łoś, *Quelques remarques, théorèmes et problèmes sur les classes définissables d'algèbres* (French), Mathematical interpretation of formal systems, North-Holland, Amsterdam, 1955, pp. 98–113. MR75156

[17] R. I. Soare, *Recursively enumerable sets and degrees: A study of computable functions and computably generated sets*, Perspectives in Mathematical Logic, Springer-Verlag, Berlin, 1987, DOI 10.1007/978-3-662-02460-7. MR882921

Department of Mathematics, The George Washington University, Washington, DC 20052

Email address: `harizanv@gwu.edu`

Department of Mathematics, The George Washington University, Washington, DC 20052

Email address: `ksrinivasan@gwu.edu`

Contemporary Mathematics
Volume **824**, 2025
https://doi.org/10.1090/conm/824/16448

Complexity of properties of computable magmas

Jennifer Chubb, Valentina Harizanov, and Dario Verta

ABSTRACT. A magma is an algebraic structure with a single binary operation. Magmas include semigroups, groups, quandles, and other structures of importance in algebra, topology, and physics. Quandles are certain right self-distributive magmas, which are not necessarily associative. A magma is computable if its domain is a computable set and its operation is computable. We investigate the algorithmic complexity of natural properties on computable magmas, which are Π_1^0 or Π_2^0 in Kleene-Mostowski arithmetical hierarchy. A property is Π_1^0 if it can be stated using a universal quantifier followed by a decidable predicate. A property is Σ_2^0 if it can be stated using an existential quantifier followed by a Π_1^0 property. Inspired by the study of Markov properties of groups, we formulate general conditions for determining Π_1^0-hardness and Π_2^0-hardness of certain problems on computable magmas. We obtain hardness by using computable reductions of known hard problems in computability theory. Hardness together with definability of properties allows us to calibrate exact complexity of various properties.

1. Introduction and preliminaries

For a given natural class of algebraic structure closed under isomorphism, we are interested in the exact complexity of its members with certain properties. In other words, in computable algebra we would like to find out from the algorithmic description of a structure how hard it is to determine whether the structure exhibits a certain important algebraic property. When such a property cannot be decided by an algorithm, i.e., the problem under consideration is undecidable, we use computability-theoretic methods to measure its complexity. A similar investigation dates back to the study of famous problems in combinatorial group theory involving interplay of algebra and decidability, such as the word problem, the conjugacy problem, and the isomorphism problem for finitely presented groups. A group is *finitely presented* if it is given by finitely many generators and finitely many relators. The *word problem* is to determine whether any two words in the generators are equal. (For more on finitely presented groups see [**13**].)

Computability theory was developed as the rigorous mathematical theory of algorithms in the 1930s by Turing, Church, Gödel, Kleene, Markov and others. A problem is decidable or computable if it can be solved by an algorithm. In the 1950s,

2020 *Mathematics Subject Classification.* Primary 03C57; Secondary 03D45.
Key words and phrases. Computable structure, magma, arithmetical hierarchy.
The second author was partially supported by the FRG NSF grant DMS-2152095. The third author was supported by the NSF Graduate Research Fellowship.

the word problem was studied by Novikov [**16**] and Boone [**5**] who independently proved that there exist finitely presented groups with undecidable word problem. Adyan [**1**] further investigated undecidability of various group-theoretic problems. For example, he showed that it is impossible to find an algorithm that will answer the question for an arbitrary given group whether the word problem is decidable. Adyan and Rabin showed that there is a collection of properties of groups called Markov properties that are undecidable. A property of a finitely presented group is abstract if it is invariant under different group presentations. We say that an abstract property P of finitely presented groups is a *Markov property* if there are two finitely presented groups, G^+, G^-, such that G^+ has property P, and if G^- is embedded in a finitely presented group H then H does not have property P. We call G^+ a positive witness and G^- a negative witness for the Markov property P.

This investigation of problems regarding finitely presented groups has been extended to include the class of recursively presented groups, and the class of computable groups and other computable structures. A *recursively presented* group is given by a computable set of generators and by a set of relators that can be enumerated by an algorithm. While a recursively presented group is given in terms of generators and relators, a *computable* group is given by a computable domain and a computable group operation. In the 1960s, Rabin [**17**] and Mal'cev [**14**] studied computable groups and computable algebraic structures in general.

Undecidable properties of structures can be further described using Kleene-Mostowski arithmetical hierarchy. This description can be made more precise by using the computability-theoretic notion of completeness. Examples of results on the exact complexity of certain properties for finitely presented groups include results by Boone and Rogers [**4**], a number of results in C. F. Miller [**15**], and a result by Lempp [**12**]. Recently, Bilanovic, Chubb, and Roven [**2, 3**] investigated the exact arithmetical complexity of various properties for recursively presented groups and for computable groups.

In [**7**], Carson et al. investigated the exact arithmetical complexity of properties of certain free groups in the class of computable groups, as well as in the class of all computable free groups. For other related results of complexity of algebraic and model-theoretic properties in classes of computable structures see [**6, 9, 10**].

A *magma* $(M, *)$ is a non-empty set M, its domain, and a binary operation,

$$* : M \times M \to M.$$

A well-known example of a magma is the additive monoid of natural numbers $(N, +)$. Magmas with specific properties are of interest in various areas of mathematics and physics. The class of magmas includes not only semigroups and groups, but also nonassociative structures. Important examples of magmas where the operation is self-distributive come from knot theory and are called quandles.

A magma $(Q, *)$ is a *quandle* if the following three axioms are satisfied, where the symbol "$\exists!$" stands for "there is unique."

(1) $(\forall a)$ $[a * a = a]$ (That is, $*$ is idempotent.)
(2) $(\forall b, c)(\exists! a)$ $[a * b = c]$ (That is, for every $b \in Q$, the function $*_b : Q \to Q$ defined by $*_b(a) = a * b$ is a bijection.)
(3) $(\forall a, b, c)$ $[(a * b) * c = (a * c) * (b * c)]$ (That is, $*$ is right self-distributive.)

Quandles were first studied in 1982 by Joyce and Matveev. An example of a quandle is a *trivial quandle* $(Q, *)$, which satisfies

$$(\forall a, b)\ [a * b = a].$$

For more on quandles and other related magmas see [**8**].

We will next review some basic notions of computability theory, which we will later use. For $k \geq 1$, a k-ary function f on natural numbers, $f : N^k \to N$, is computable if there is a Turing machine (program) T that computes it. That is, when we run T on any input $(x_1, \dots, x_k) \in N^k$, T halts in finitely many steps and outputs the value $y = f(x_1, \dots, x_k)$. Moreover, by algorithmically enumerating all distinct Turing machines

$$T_0, T_1, \dots, T_e, \dots,$$

we also obtain an algorithmic enumeration of all k-ary *partial computable functions* on natural numbers:

$$\varphi_0^{(k)}, \varphi_1^{(k)}, \dots, \varphi_e^{(k)}, \dots.$$

It is not hard to see that every k-ary partial computable function appears infinitely often in this list. *Computable* functions are exactly the total functions in the above list. We omit the superscript for $k = 1$, i.e., in the enumeration of all unary partial computable functions.

On an input x, a Turing machine T_e may halt and output the value $\varphi_e(x)$, or it may compute forever. By $\varphi_e(x) \downarrow$ we denote that T_e halts on input x, and by $\varphi_e(x) \uparrow$ we denote that T_e does not halt on input x. It is an undecidable problem to determine for any e and input x, whether T_e halts on x. This problem is known as the *halting problem*. However, for $s \geq 1$, by $\varphi_{e,s}(x) \downarrow$ we denote that $e, x < s$, and T_e halts on input x after $< s$ steps and outputs the value that is $< s$. Otherwise, we write $\varphi_{e,s}(x) \uparrow$.

A nonempty set is *computably enumerable* if it is the range of a computable function. We can think of this function as an algorithmic mechanism for enumerating the set. Hence there are only countably many computably enumerabe sets. A set A is *computable* or *decidable* if its characteristic function χ_A is computable, where

$$\chi_A(a) = \begin{cases} 1, & \text{if } a \in A; \\ 0, & \text{if } a \notin A. \end{cases}$$

Clearly, a computable set is computably enumerable, and the complement of a computable set is computable. Moreover, we can show that a set is computable if both the set and its complement are computably enumerable. Namely, from the algorithmic enumerations of the set and its complement, we can devise a decision procedure for recognizing the elements of the set.

It can be shown that computably enumerable sets are exactly the domains of partial computable functions. Let $W_e = dom(\varphi_e)$. Hence we have an algorithmic enumeration of all computably enumerable sets,

$$W_0, W_1, \dots, W_e, \dots.$$

Let $W_{e,s} = dom(\varphi_{e,s})$. Clearly, $W_{e,s} \subseteq W_{e,s+1}$ and $W = \bigcup_{s \geq 0} W_{e,s}$.

The (diagonal) *halting set*, denoted by K, is defined as

$$K = \{e : \varphi_e(e)\downarrow\} = \{e : e \in W_e\}.$$

The halting set K is also called the first Turing jump of the empty set. It is a computably enumerable set. We can show that K is not computable by showing that its complement $\overline{K}$ is not computably enumerable.

For $n \geq 1$, we say that a set is Σ^0_n, or Π^0_n, if it can be expressed using n blocks of like quantifiers, beginning with $\exists$, or $\forall$, respectively, followed by a computable relation. More precisely, a set of natural numbers A is Σ^0_n if there is a computable $(n+1)$-ary relation R such that for every $a \in N$,

$$a \in A \Leftrightarrow (\exists x_1)(\forall x_2)\cdots(Qx_n)R(a, x_1, x_2, \dots, x_n),$$

where Q is $\exists$ if n is an odd number, and Q is $\forall$ if n is an even number.

We similarly define Π^0_n sets starting with the universal quantifier. Clearly, the complement of a Σ^0_n set is a Π^0_n set and vice versa. It follows that Σ^0_1 sets are the computably enumerable sets, and Π^0_1 sets are the complements of the computably enumerable sets. A set is called *arithmetical* if it is Π^0_m for some m (or, equivalently, Σ^0_k for some k). For example, the halting set is Σ^0_1 because it can be expressed as

$$e \in K \Leftrightarrow \exists s\ (\varphi_{e,s}(e)\downarrow)$$

and $\{(e, s) : \varphi_{e,s}(e)\downarrow\}$ is a computable relation.

It is an undecidable problem to determine whether W_e is finite or infinite. More formally, let $FIN = \{e : W_e \text{ is finite}\}$ and $INF = \{e : W_e \text{ is infinite}\}$. We have that FIN is a Σ^0_2 set since it can be expressed as

$$e \in FIN \Leftrightarrow (\exists s)(\forall t)\ (t \leq s \vee W_{e,t} = W_{e,s}).$$

Hence INF is a Π^0_2 set. For more on computability theory and arithmetical hierarchy see [**18**].

We also classify first-order formulas in the prenex normal form as Σ^0_n formulas, or Π^0_n formulas, if they have n blocks of like quantifiers beginning with a block of $\exists$, or $\forall$, quantifiers. Hence Σ^0_1 formulas are existential formulas, Π^0_1 formulas are universal formulas, Π^0_2 formulas are $\forall\exists$ formulas, etc.

A countable structure for a finite language is computable if its domain is a computable set and its operations and relations are computable. Hence a countable magma $\mathcal{M} = (M, *)$ is *computable* if its domain M is a computable set and its operation $*$ is computable. Equivalently, a structure is computable if its atomic diagram is computable. The *atomic diagram* of a structure is the set of all atomic sentences and negations of atomic sentences in the language of the structure augmented with new constant symbols for all elements of the domain, such that these sentences are true in the structure with added constants. Hence a magma $\mathcal{M} = (M, *)$ is computable if its atomic diagram $D(\mathcal{M})$ is computable. For an infinite M, simple sentences in $D(\mathcal{M})$ include the sentences of the form

$$m_i * m_j = m_k \quad \text{and} \quad m_i * m_j \neq m_l,$$
$$\text{for } i, j, k, l \in N \text{ and } m_0, m_1, \ldots \in M.$$

Without loss of generality, for any infinite computable magma we may assume, by using algorithmic coding by natural numbers, that the domain is the set N. For more on computable structures see [**9**].

In Section 2, we will formulate a general condition for computable magmas with properties that are hard at the Π_1^0 level of the arithmetical hierarchy. In Section 3, we will formulate a general condition and give examples of magmas with properties that are hard at the Π_2^0 level of the arithmetical hierarchy. These general conditions are inspired by Markov properties for groups. Together with definability, these results will yield completeness results for some important properties of computable magmas.

2. Π_1^0-complete properties

Regarding the classification problem for countable magmas with specific properties, we can consider the index set approach. Assume that $\widehat{\mathbb{K}}$ is a set of countable structures closed under isomorphism, and $\mathbb{K}$ is the set of its computable members. For a computable structure $\mathcal{A}$, an *index* is a number e such that $\varphi_e = \chi_{D(\mathcal{A})}$, where $D(\mathcal{A})$ is the atomic diagram of $\mathcal{A}$ and $\chi_{D(\mathcal{A})}$ is its characteristic function. Let P be a property. The *index set* for P, denoted by $I(P)$, is the set of all indices for structures in $\mathbb{K}$ that have property P. The *index set* of $\mathbb{K}$, denoted by $I(\mathbb{K})$, is the set of all indices for members of $\mathbb{K}$.

We can often use first-order sentences to describe structures with a property of interest. To show that our description is optimal we can consider the notion of m-completeness. For a complexity class $\mathfrak{C}$, we define $I(P)$ to be $\mathfrak{C}$ *within* $\mathbb{K}$ (or $\widehat{\mathbb{K}}$) if

$$I(P) = C \cap I(\mathbb{K}) \quad \text{for some } C \in \mathfrak{C}.$$

Furthermore, such $I(P)$ is *m-complete* $\mathfrak{C}$ *within* $\mathbb{K}$ if it is also $\mathfrak{C}$-hard within $\mathbb{K}$. That is, for every $X \in \mathfrak{C}$, there is a computable function $f : N \to I(\mathbb{K})$ such that:

$$n \in X \Leftrightarrow f(n) \in I(P).$$

Here, the reduction function f does not have to be one-one $(1-1)$; it can be many-one, so we use the term m-reducibility and m-completeness, or simply completeness. For example, the halting set K is a Σ_1^0-complete set, hence $\overline{K}$ is a Π_1^0-complete set. The set INF is Π_2^0-complete or, equivalently, the set FIN is Σ_2^0-complete. (See [**18**].)

We can identify a magma $\mathcal{M} = (M, *)$ with the graph of its operation $*$ on M and denote if by $G_{\mathcal{M}}$. Then a magma on M is a function $* : M^2 \to M$, or a special relation on $M^3 = M^2 \times M$ (total on M^2 and single-valued in M). Thus, we can write $(m_i, m_j, m_k) \in G_{\mathcal{M}}$ as $m_i * m_j = m_k$, and view it as a sentence in the atomic diagram of $\mathcal{M}$, which is also an entry in the "multiplication table" for $*$.

A *partial magma* on M is any single-valued (non necessarily total) relation on $M^2 \times M$. A partial magma on M can be extended to a magma on M.

Let $(M_1, *_1), (M_2, *_2)$ be magmas. The *direct product* of these magmas, which we write as $(M_1, *_1) \times (M_2, *_2)$, is defined as usual. The underlying set is the Cartesian product $M_1 \times M_2$, and the binary operation $*$ is defined component-wise:

$$(a, b) * (c, d) = (a *_1 c, b *_2 d).$$

An *embedding* of a magma $(M_1, *_1)$ into a magma $(M_2, *_2)$ is an injective homomorphism, that is, a $1-1$ function $h : M_1 \to M_2$ such that for all $a, c \in M_1$:

$$h(a *_1 c) = h(a) *_2 h(c).$$

An embedding that is surjective (onto) is an *isomorphism*.

We often refer to a magma $\mathcal{M} = (M, *)$ briefly as just magma M, and similarly for partial magmas.

THEOREM 1. *Let* $\mathbb{K}$ *be the class of all computable magmas, and consider a property* P *which may be exhibited by a magma and is preserved under isomorphism. Suppose that for* $e \in N$ *there is a uniformly computable sequence* $(M_s^e)_{s \in N}$ *of finite partial magmas* M_s^e *with* $M_0^e = \varnothing$*, and* $M^e \in \mathbb{K}$ *where* $M^e = \bigcup_{s \geq 0} M_s^e$*. Assume further that for every* s *we can compute finite partial magmas* $A_{s+1}^{e,+}$, $A_{s+1}^{e,-}$*, each of which extends* M_s^e*, and one of which equals* M_{s+1}^e*. If* M^e *has property* P *if and only if for every* s, $M_{s+1}^e = A_{s+1}^{e,+}$*, then detecting property* P *in the class* $\mathbb{K}$ *is* Π_1^0*-hard.*

PROOF. Recall that the complement $\overline{K}$ of the halting set K is a Π_1^0-complete set. We will describe an m-reduction of $\overline{K}$ to the set of indices of (computable) magmas in $\mathbb{K}$ with property P.

For each e, we will build a corresponding magma M^e, stage by stage, by extending finite partial magmas. We can assume that the domain of each M^e is the set of all natural numbers.

Construction.

Stage 0: Let M_0^e be $\varnothing$ for every e.

Stage $s+1$: Consider each $e \leq s$. Assume inductively that we have a finite partial magma M_s^e such that $M_s^e \in \{A_s^{e,+}, A_s^{e,-}\}$. If $\varphi_{e,s+1}(e)\uparrow$, choose M_{s+1}^e to be the positive partial magma $A_{s+1}^{e,+}$. If $\varphi_{e,s+1}(e)\downarrow$, choose M_{s+1}^e to be the negative partial magma $A_{s+1}^{e,-}$.

End of the construction.

Let $M^e =_{def} \bigcup_{s \geq 0} M_s^e$.

Then

$$e \in \overline{K} \Leftrightarrow (M^e \text{ has property } P).$$

□

In Section 3, we will generalize Theorem 1. However, we can also formulate a version of Markov property for magmas similar to the one for groups.

DEFINITION 2. Let $\mathbb{K}$ be a set of computable magmas. A property P is called a *Markov property* with respect to $\mathbb{K}$ if there is an infinite magma $A^+ \in \mathbb{K}$ such that A^+ has property P, and there is $A^- \in \mathbb{K}$ such that for any magma M if there is an embedding of A^- into M, then M does not have property P.

In the previous definition, we can assume that A^- is infinite since otherwise, we can take the direct product of A^- and some infinite magma, say $A^- \times \mathbb{N}$ where $\mathbb{N} = (N, +)$ is the additive monoid of natural numbers.

We can show that Markov properties for magmas are Π_1^0-hard, which is a theorem established by Ha [**11**]. We include its proof and emphasize how it relates to Theorem 1.

THEOREM 3 ([**11**]). *Let* P *be a Markov property for computable magmas. Then detecting* P *in the class of computable magmas is* Π_1^0*-hard.*

PROOF. Let infinite A^+ and A^- be as in the definition of Markov property. Further, let $A^{-,id}$ be the expansion of A^- into $A^- \cup \{id\}$ where id is a new element that acts as an identity element of $A^{-,id}$. If A^- already has an identity element, we do not need to expand A^- and will fix such an element id.

Consider an algorithmic enumeration $\sigma_0, \sigma_1, \sigma_2, \ldots$ of all distinct sentences of the atomic diagram of $A^+ \times \{id\}$ which form its multiplication table. Similarly, consider an algorithmic enumeration of all distinct sentences $\tau_0, \tau_1, \tau_2, \ldots$ of the atomic diagram of $A^+ \times A^{-,id}$ which form its multiplication table.

As in Theorem 1, given a finite partial magma M^e_s, define $A^{e,+}_{s+1}$ to be the extension of M^e_s obtained by adding the first new sentence σ_i (i.e., one not already included in M^e_s), and define $A^{e,-}_{s+1}$ to be the extension of M^e_s obtained by adding the first new sentence τ_j. If $\varphi_{e,s}(e) \uparrow$ choose M^e_{s+1} to be $A^{e,+}_{s+1}$, and if $\varphi_{e,s}(e) \downarrow$ choose M^e_{s+1} to be $A^{e,-}_{s+1}$.

Thus, as long as $\varphi_{e,s}(e)\uparrow$, we continue to build M^e as a copy of $A^+ \times \{id\}$. If $\varphi_{e,s_0}(e) \downarrow$ for the first time at some stage s_0, then we switch to building M^e as a copy of $A^+ \times A^{-,id}$ and continue this way at all subsequent stages. In both cases, $M^e = \bigcup_{s\geq 0} M^e_s$ is a computable magma.

If we choose for every stage s a partial magma $M^{e,+}_s$ then M^e is isomorphic to $A^+ \times \{id\}$, so isomorphic to A^+ and has property P. Otherwise, M^e is isomorphic to $A^+ \times A^{-,id}$, and hence contains a copy of A^-, so it does not have property P. $\square$

We can apply Theorem 1 or Theorem 3 to show that particular properties of magmas are Π^0_1-hard or Π^0_1-complete. As an application of Theorem 3, Ha [**11**] showed that the following properties are Π^0_1-complete in the class of computable magmas:

(1) being a semigroup (associative magma);
(2) being a commutative magma;
(3) being an idempotent magma;
(4) being a right self-distributive magma (also called a right shelf).

We further add more examples of important Π^0_1-complete properties of magmas.

COROLLARY 4. *The following properties are Π^0_1-complete in the class of all computable magmas:*

(1) *being a trivial quandle;*
(2) *being a Kei;*
(3) *being a medial magma;*
(4) *being a right alternative magma.*

PROOF.
(1) We have that a magma $(M, *)$ is a trivial quandle if

$$\forall x \forall y \quad [x * y = x],$$

so it is characterized by a Π^0_1 sentence.

To apply Theorem 3, we can take as M^+ the trivial quandle on the set N of natural numbers, and as M^- any magma $(\{x, y\}, *)$ where $x \neq y$ and $x * x = y$.

(2) A magma $(M, *)$ is a *Kei* if

$$\forall x \ [x * x = x],$$
$$\forall x \forall y \ [(x * y) * y = x], \quad \text{and}$$
$$\forall x \forall y \forall z \ [(x * y) * z = (x * z) * (y * z)].$$

The conjunction of finitely many Π^0_1 sentences is a Π^0_1 sentence, so being a Kei is Π^0_1.

To apply Theorem 3, we can take as M^+ the trivial quandle, and as M^- any magma $(\{x, y\}, *)$ where $x \neq y$ and $y * y = x$.

(3) A magma $(M, *)$ is a *medial* magma if

$$\forall x \forall y \forall u \forall v \; [(x * y) * (u * v) = (x * u) * (y * v)],$$

so it is characterized by a Π^0_1 sentence.

We can take M^+ to be the medial magma obtained from an infinite commutative group $(G, +)$ and two distinct fixed integers m, n with magma operation defined by $x * y = mx + ny$. We can take M^- to be the magma $(\{x, y\}, *)$ where $x \neq y$ and

$$x * x = y, \quad y * y = x, \quad \text{and}$$
$$x * y = y * x = y,$$

so that

$$(x * y) * (x * y) \neq (x * x) * (y * y).$$

(4) A magma $(M, *)$ is a *right alternative* magma if

$$\forall x \forall y [x * (y * y) = (x * y) * y],$$

so this property is characterized by a Π^0_1 sentence.

We can take M^+ to be an infinite associative magma. We can take M^- to be the magma $(\{x, y\}, *)$ where $x \neq y$ and

$$x * x = y, \quad \text{and}$$
$$x * y = y * x = y * y = x,$$

so that

$$x * (y * y) \neq (x * y) * y.$$

□

3. Π^0_2-complete problems

We will now generalize Theorem 1 to obtain a general condition for Π^0_2-hard problems on computable magmas.

THEOREM 5. *Let $\mathbb{K}$ be the class of all computable magmas, and consider a property P which may be exhibited by a magma and is preserved under isomorphism. Suppose that for $e \in N$ there is a uniformly computable sequence $(M^e_s)_{s \in N}$ of finite partial magmas M^e_s with $M^e_0 = \varnothing$, and $M^e \in \mathbb{K}$ where $M^e = \bigcup_{s \geq 0} M^e_s$. Assume further that for every s we can compute finite partial magmas $A^{e,+}_{s+1}, A^{e,-}_{s+1}$, each of which extends M^e_s and one of which equals M^e_{s+1}. If M^e has property P if and only if for infinitely many s, $M^e_{s+1} = A^{e,+}_{s+1}$, then detecting property P in the class $\mathbb{K}$ is Π^0_2-hard.*

PROOF. Recall that the set $INF = \{e : W_e \text{ is infinite}\}$ is Π_2^0-complete.

For each W_e, we will build a corresponding magma M^e, stage by stage, by extending finite magmas. We may assume, without loss of generality, that the domain of each M^e is the set of all natural numbers.

Construction.

Stage 0: Let M_0^e be $\varnothing$.

Stage $s+1$: Consider each $e \leq s$. Assume inductively that we have a finite magma M_s^e. If a new element is enumerated in W_e at stage $s+1$, choose M_{s+1}^e to be the positive magma $A_{s+1}^{e,+}$. If no new element is enumerated in W_e at stage $s+1$, choose M_{s+1}^e to be the negative magma $A_{s+1}^{e,-}$.

End of the construction.

Let $M^e =_{def} \bigcup_{s \geq 0} M_s^e$.

Then

$$e \in INF \Leftrightarrow \ M^e \text{ has property } P.$$

□

We can apply Theorem 5 to obtain the following completeness results.

COROLLARY 6.

(1) *The index set of magmas with left inverses (right inverses) is Π_2^0-complete in the class of all computable magmas.*

(2) *The index set of magmas with unique left inverses (unique right inverses) is Π_2^0-complete in the class of all computable magmas.*

(3) *The index set of magmas without left units (right units) is Π_2^0-complete in the class of all computable magmas.*

(4) *The index set of uniquely 2-divisible magmas is Π_2^0-complete in the class of all computable magmas.*

PROOF. Let $m_1, m_2, \ldots$ be a computable $1-1$ enumeration of all elements of magmas to be built, which form a fixed computable domain. We will label the rows and columns of the multiplication table of every magma M^e by these elements.

(1) The property that each element has a left inverse can be expressed by the following Π_2^0 sentence:

$$\forall y \forall z \exists x \ (x * y = z).$$

We will show, using Theorem 5, that detecting this property is Π_2^0-hard.

To assure the existence of left inverses, we have to make sure that all elements of magma appear in each column of its multiplication table. To ruin the left inverse property, it is enough to have at least one column with at least one element missing. We will choose the first column. Consider a computable $1-1$ enumeration $p_1, p_2, \ldots$ of the pairs (m_i, m_j) for $i \geq 1$ and $j > 1$.

Let $M_0^e = \varnothing$.

At stage $s+1$, given a finite partial magma M_s^e (i.e., its multiplication table finitely filled in), let (m_i, m_j) be the s-th pair p_s in the enumeration of pairs of elements in the above list.

We define $A_{s+1}^{e,+}$ by adding the following entries to the multiplication table of M_s^e. First, add $m_i * m_j = m_k$ where m_k is the first element in the above enumeration of the domain which has not been used in the column of m_j. Then, for the first column, put in the first unfilled entry of the column, the first element in the enumeration of the domain not used so far in the column.

We define $A^{e,-}_{s+1}$ by adding the following entries to the multiplication table of M^e_s. First, add $m_i * m_j = m_k$ where m_k is the first element not used in the column of m_j. Then, for the first column, put in the first unfilled entry of the column, the second element in the enumeration of the domain not used so far in the column, so skipping one element of the domain.

In any case, there is no repetition of elements in the columns. If we choose $A^{e,+}_t$ infinitely often (i.e., for infinitely many t), every column is filled with all elements of the domain, while if we choose $A^{e,-}_t$ after some stage t_0 (i.e., at every $t \geq t_0$), we will miss exactly one element m in the first column, so there will be no element n for which $n * m_1 = m$.

(2) The property that each element has a *unique* left inverse can be expressed by the following Π^0_2 sentence:

$$\forall y \forall z \forall u \exists x\ [(x * y = z \wedge u * y \neq z) \vee (x * y = z \wedge x = u)].$$

Π^0_2-hardness follows from the construction in (1) since no element is repeated in the same column.

(3) Having a left unit element can be expressed by the following Σ^0_2 sentence:

$$\exists u \forall y\ (u * y = y).$$

Hence not having a left unit element can be expressed by a Π^0_2 sentence. We will now show that detecting this property is Π^0_2-hard.

Fix a computable $1-1$ enumeration $p_1, p_2, \ldots$ of the pairs (m_i, m_j) for $i, j \geq 1$. Again, we will label the rows and columns of the multiplication table of $M^e = \bigcup_{s\geq 0} M^e_s$ by elements $m_1, m_2, \ldots$. To assure the non-existence of a left unit element, we have to make sure that there is no row of the multiplication table of M^e in which the elements appear exactly as $m_1, m_2, \ldots$. To assure the existence of a left unit element, it is enough to have at least one row with the elements $m_1, m_2, \ldots$.

Let $M^e_0 = \varnothing$.

At stage $s+1$, given a finite magma M^e_s (i.e., its multiplication table finitely filled in), let (m_i, m_j) be the s-th pair of elements.

We define $A^{e,+}_{s+1}$ by adding the following entry to the multiplication table of M^e_s. First, if $m_i * m_j$ is not already defined, add $m_i * m_j = m_j$ to M^e_s. Then let i_0 be the first row in M^e_s which is consistent with m_{i_0} being a left unit element. Let $m_{i_0} * m_{j_0}$ be the first undefined entry. Choose the first element m_{k_0} such that $m_{k_0} \neq m_{j_0}$ and add $m_{i_0} * m_{j_0} = m_{k_0}$ to M^e_s.

We define $A^{e,-}_{s+1}$ by adding $m_i * m_j = m_j$ to M^e_s, if $m_i * m_j$ is not already defined; otherwise, let $A^{e,-}_{s+1}$ be M^e_s.

If we choose $A^{e,+}_t$ infinitely often, all rows are filled in a way that prevents $m_1, m_2, \ldots$ from being left unit elements, while if we choose $A^{e,-}_t$ after some stage on, we will assure that every newly filled row is labelled by a left unit element.

(4) A magma is uniquely 2-divisible if it satisfies the sentence:

$$\forall x \exists! y\ (x = y * y).$$

This is equivalent to

$$\forall x \forall z \exists y [(x = y * y \wedge x \neq z * z) \vee (x = y * y \wedge z = y)].$$

Hence being a uniquely 2-divisible can be expressed by a Π^0_2 sentence. We will now show that detecting this property is Π^0_2-hard.

Fix a computable $1-1$ enumeration $p_1, p_2, \ldots$ of the pairs (m_i, m_j) for $i, j \geq 1$ and $i \neq j$. Again, we will label the rows and columns of the multiplication table of M^e by elements $m_1, m_2, \ldots$. To assure the property of being a uniquely 2-divisible, we have to make sure that every element appears exactly once on the main diagonal of the multiplication table. If some element is missing on the main diagonal or appears at least twice, then the unique 2-divisibility does not hold.

Let $M_0^e = \varnothing$.

At stage $s+1$, given a finite partial magma M_s^e (i.e., its multiplication table finitely filled in), let (m_i, m_j) be the s-th pair p_s in the enumeration of pairs of elements in the above list.

We define $A_{s+1}^{e,+}$ by adding the following entries to the multiplication table of M_s^e. First, add $m_i * m_j = m_k$ where m_k is any element. Then, for the main diagonal, put in the first unfilled entry of the main diagonal the first element of the domain not used so far in the diagonal.

We define $A_{s+1}^{e,-}$ by adding the following entries to the multiplication table of M_s^e. First, add $m_i * m_j = m_k$ where m_k is any element. Then, for the main diagonal, put in the first unfilled entry of the main diagonal the second element of the domain not used so far in the column.

If we choose $A_t^{e,+}$ infinitely often, then the main diagonal is filled with all elements without repetition, while if we choose $A_t^{e,-}$ from some stage on, we will miss exactly one element m in the main diagonal, so there will be no n for which $n * n = m$. □

The previous corollary provides a sample of properties of magmas to which Theorem 5 applies. Parts (1)–(3) of Corollary 6 have also been obtained by Ha in [**11**] by direct separate constructions. It would be worthwhile to formulate general results about Π_n^0-hardness of properties of magmas for $n > 2$.

References

[1] S. I. Adyan, *Algorithmic unsolvability of problems of recognition of certain properties of groups* (Russian), Dokl. Akad. Nauk SSSR (N.S.) **103** (1955), 533–535. MR81851

[2] I. I. Bilanovic, *Detecting Properties of Algorithmically Presented Algebraic and Relational Structures*, ProQuest LLC, Ann Arbor, MI, 2020. Thesis (Ph.D.)–The George Washington University. MR4158002

[3] I. Bilanovic, J. Chubb, and S. Roven, *Detecting properties from descriptions of groups*, Arch. Math. Logic **59** (2020), no. 3-4, 293–312, DOI 10.1007/s00153-019-00690-x. MR4081061

[4] W. W. Boone and H. Rogers Jr., *On a problem of J. H. C. Whitehead and a problem of Alonzo Church*, Math. Scand. **19** (1966), 185–192, DOI 10.7146/math.scand.a-10808. MR210578

[5] W. W. Boone, *The word problem*, Proc. Nat. Acad. Sci. U.S.A. **44** (1958), 1061–1065, DOI 10.1073/pnas.44.10.1061. MR101267

[6] W. Calvert, V. S. Harizanov, J. F. Knight, and S. Miller, *Index sets of computable models* (Russian, with Russian summary), Algebra Logika **45** (2006), no. 5, 538–574, 631–632, DOI 10.1007/s10469-006-0029-0; English transl., Algebra Logic **45** (2006), no. 5, 306–325. MR2307694

[7] J. Carson, V. Harizanov, J. Knight, K. Lange, C. McCoy, A. Morozov, S. Quinn, C. Safranski, and J. Wallbaum, *Describing free groups*, Trans. Amer. Math. Soc. **364** (2012), no. 11, 5715–5728, DOI 10.1090/S0002-9947-2012-05456-0. MR2946928

[8] M. Elhamdadi and S. Nelson, *Quandles—an introduction to the algebra of knots*, Student Mathematical Library, vol. 74, American Mathematical Society, Providence, RI, 2015, DOI 10.1090/stml/074. MR3379534

[9] E. B. Fokina, V. Harizanov, and A. Melnikov, *Computable model theory*, Turing's legacy: developments from Turing's ideas in logic, Lect. Notes Log., vol. 42, Assoc. Symbol. Logic, La Jolla, CA, 2014, pp. 124–194. MR3497661

[10] S. S. Goncharov and J. F. Knight, *Computable structure and antistructure theorems* (Russian, with Russian summary), Algebra Logika **41** (2002), no. 6, 639–681, 757, DOI 10.1023/A:1021758312697; English transl., Algebra Logic **41** (2002), no. 6, 351–373. MR1967769

[11] T. Ha, *On Algorithmic Properties of Computable Magmas*, ProQuest LLC, Ann Arbor, MI, 2018. Thesis (Ph.D.)–The George Washington University. MR3858241

[12] S. Lempp, *The computational complexity of torsion-freeness of finitely presented groups*, Bull. Austral. Math. Soc. **56** (1997), no. 2, 273–277, DOI 10.1017/S0004972700031014. MR1470080

[13] R. C. Lyndon and P. E. Schupp, *Combinatorial group theory*, Ergebnisse der Mathematik und ihrer Grenzgebiete [Results in Mathematics and Related Areas], Band 89, Springer-Verlag, Berlin-New York, 1977. MR577064

[14] A. I. Mal'cev, *On recursive Abelian groups* (English translation), Soviet Mathematics Doklady **3** (1962), 1431–1434.

[15] C. F. Miller III, *Decision problems for groups—survey and reflections*, Algorithms and classification in combinatorial group theory (Berkeley, CA, 1989), Math. Sci. Res. Inst. Publ., vol. 23, Springer, New York, 1992, pp. 1–59, DOI 10.1007/978-1-4613-9730-4_1. MR1230627

[16] P. S. Novikov, *Ob algoritmičeskoĭ nerazrešimosti problemy toždestva slov v teorii grupp* (Russian), Izdat. Akad. Nauk SSSR, Moscow, 1955. Trudy Mat. Inst. Steklov. no. 44. MR75197

[17] M. O. Rabin, *Recursive unsolvability of group theoretic problems*, Ann. of Math. (2) **67** (1958), 172–194, DOI 10.2307/1969933. MR110743

[18] R. I. Soare, *Turing computability: Theory and applications*, Theory and Applications of Computability, Springer-Verlag, Berlin, 2016, DOI 10.1007/978-3-642-31933-4. MR3496974

Department of Mathematics, The Pennsylvania State University, University Park, PA 16802

Email address: jcc31@psu.edu

Department of Mathematics, The George Washington University, Washington, DC 20052

Email address: harizanv@gwu.edu

Department of Mathematics, The George Washington University, Washington, DC 20052

Email address: dario.verta@hotmail.com

Selected Published Titles in This Series

824 **Hanna E. Makaruk and Robert M. Owczarek, Editors,** Inverse Problems: In Memory of Professor Zbigniew Oziewicz, 2025

819 **Howard S. Cohl, Roberto S. Costas-Santos, and Robert S. Maier, Editors,** Applications and q-Extensions of Hypergeometric Functions, 2025

818 **Howard S. Cohl, Roberto S. Costas-Santos, and Robert S. Maier, Editors,** Classical Hypergeometric Functions and Generalizations, 2025

817 **Mikhail Khovanov, Joshua Sussan, and Anton Zeitlin, Editors,** From Representation Theory to Mathematical Physics and Back, 2025

816 **Luca F. Di Cerbo and Laurenţiu G. Maxim, Editors,** Geometry and Topology of Aspherical Manifolds, 2025

815 **Daniel Orr and Joshua Jeishing Wen, Editors,** Macdonald Theory and Beyond, 2025

814 **Susan Montgomery and Siu-Hung Ng, Editors,** Quantum Groups, Hopf Algebras, and Applications, 2025

813 **Thomas Creutzig, Julia Plavnik, and David Ridout, Editors,** Quantum Symmetries, 2025

812 **Ross Geoghegan, Craig R. Guilbault, and Kim Ruane, Editors,** Topology at Infinity of Discrete Groups, 2025

811 **Krishnendu Gongopadhyay and Shashikant A. Katre, Editors,** Geometry, Groups and Mathematical Philosophy, 2025

810 **Robert J. Berman and Yanir A. Rubinstein, Editors,** Convex and Complex: Perspectives on Positivity in Geometry, 2025

809 **Fabrice Baudoin and Luca Rizzi, Editors,** New Trends in Sub-Riemannian Geometry, 2025

808 **Loring W. Tu, Editor,** Group Actions and Equivariant Cohomology, 2024

807 **Galina Filipuk, Editor,** Recent Progress in Special Functions, 2024

806 **Maximilian Engel, Hildeberto Jardón-Kojakhmetov, and Cinzia Soresina, Editors,** Topics in Multiple Time Scale Dynamics, 2024

805 **Anthony Iarrobino, Pedro Macias Marques, Maria Evelina Rossi, and Jean Vallès, Editors,** Deformation of Artinian Algebras and Jordan Type, 2024

804 **Mahir Bilen Can and Jörg Feldvoss, Editors,** A Glimpse into Geometric Representation Theory, 2024

803 **Peter Gothen, Margarida Melo, and Montserrat Teixidor i Bigas, Editors,** Moduli Spaces and Vector Bundles—New Trends, 2024

802 **Ralph M. Kaufmann, Martin Markl, and Alexander A. Voronov, Editors,** Higher Structures in Topology, Geometry, and Physics, 2024

801 **K. A. Brown, T. J. Hodges, M. Vancliff, and J. J. Zhang, Editors,** Recent Advances in Noncommutative Algebra and Geometry, 2024

800 **Avinoam Mann, Louis H. Rowen, David J. Saltman, Aner Shalev, Lance W. Small, and Uzi Vishne, Editors,** Amitsur Centennial Symposium, 2024

799 **Alberto A. Condori, Elodie Pozzi, William T. Ross, and Alan A. Sola, Editors,** Recent Progress in Function Theory and Operator Theory, 2024

798 **Marat V. Markin, Igor V. Nikolaev, and Carsten Trunk, Editors,** Advances in Functional Analysis and Operator Theory, 2024

797 **Sangita Jha, Mrinal Kanti Roychowdhury, and Saurabh Verma, Editors,** Recent Developments in Fractal Geometry and Dynamical Systems, 2024

796 **John Cremona, John Jones, Jennifer Paulhus, Andrew V. Sutherland, and John Voight, Editors,** LuCaNT: LMFDB, Computation, and Number Theory, 2024

For a complete list of titles in this series, visit the
AMS Bookstore at **www.ams.org/bookstore/conmseries/**.

FSC
www.fsc.org
MIX
Papier | Fördert gute Waldnutzung
FSC® C083411

Kolibri 360 GmbH
Ferdinand-Jühlke-Straße 7
99095 Erfurt, Deutschland
produktsicherheit@kolibri360.de